Christianity,
Wilderness,
and Wildlife

Christianity, Wilderness, and Wildlife

The Original Desert Solitaire

Susan Power Bratton

UNIVERSITY OF SCRANTON PRESS
Scranton and London

.

Library of Congress Cataloging-in-Publication Data

Bratton, Susan.
Christianity, wilderness, and wildlife : the original desert
solitaire / Susan Power Bratton.
    p. cm.
Includes bibliographical references and index.
ISBN 0-940866-14-5 Hardcover, ISBN 978-1-58966-177-6 paperback
  1. Wilderness (Theology)—History of doctrines. 2. Spirituality.
I. Title.
BT695.5.B73 1993
231.7—dc20   91-67107
                                                         CIP

UNIVERSITY OF SCRANTON PRESS
Chicago Distribution Center
11030 S. Langley
Chicago IL 60628

PRINTED IN THE UNITED STATES OF AMERICA

For my mother,
Julie Hallahan Bratton
who left anthologies of Irish literature lying around the house,
and taught me to care for plants and animals

Contents

Abbreviations
for Translations
of the Scriptures

Abbreviation	Source	Chapters
None	Herbert G. May and Bruce M. Metzger, eds., *The New Oxford Annotated Bible, The Holy Bible, Revised Standard Version* (New York: Oxford University Press, 1973); and Bruce M. Metzger, ed., *The Apocrypha of the Old Testament, Revised Standard Version* (New York: Oxford University Press, 1977).	2–9, 16
CHL	Brevard C. Childs, *The Book of Exodus, A Critical Theological Commentary* (Philadelphia: Westminster Press, 1974).	3
DeV	Simon J. DeVries, *Word Biblical Commentary No. 12, 1 Kings* (Waco, TX: Word Publishers, 1985).	6
JB	Alexander Jones, ed., *The Jerusalem Bible* (Garden City, NY: Doubleday & Co., 1968).	5
NIV	*New International Version, The Holy Bible* (Grand Rapids, MI: Zondervan Bible Publishers, 1978).	5, 7, 16
SG	J. Alberto Soggins, *Judges* (Philadelphia: Westminster Press, 1981).	6

WS   Claus Westermann, *Genesis 1–11,* trans-   2, 16
                    lated by John J. Scullion (Minneapolis,
                    MN: Augsburg Publishing House, 1984),
                    and *Genesis 12–36* (Minneapolis,
                    MN: Augsburg Publishing House,
                    1985).

WSR   Artur Weiser, *The Psalms* (Philadel   5, 16
                    phia: Westminster Press, 1962).

Acknowledgments

I would like to thank all the people who encouraged me to work on wilderness ethics, expanded my view of environmental history, reviewed portions of the manuscript, offered useful suggestions for improvements, or provided social support during the project, including James Bradley, Walter Brueggemann, Rita Cantu, Eugene Hargrove, John Hart, J. Donald Hughes, Robert Hurt, Alice Ingerson, Teri Mobley, Marcy Nejat, Oliver Rackham, and Loren Wilkinson. I would also like to express my appreciation to Bob Barr, Calvin DeWitt, and all the other staff of AuSable Institute for their continuing interest in both the topics covered in this volume and in Christian environmental ethics in general. The graduate students and office staff of the Institute of Ecology, University of Georgia, including Bart, Liu, Aditi, Toni, Bill, Albert, Sandy, Neil, Thelma, John Paul, Judy, Martha, Winnona, Wendy, and Janice, kept me company, provided entertainment, corrected errors, and helped to preserve my sanity while I was working holidays, nights, and weekends at the computer with improved drafts of this manuscript.

* * *

Scripture quotations from the Revised Standard Version of the Bible© 1946, 1952, 1971, and 1973 by the Division of Christian Education of the National Council of Churches of Christ in the U.S.A. and are used by permission.

Scripture quotations marked *JB* are from *The Jerusalem Bible,* 1966, 1967, and 1968 by Darton, Longman & Todd, Ltd. and Doubleday and Co., Inc. Used by permission of the publishers.

Scripture quotations marked *WS* from *Genesis 1–11* © 1974 by Neukrichen Verlag; English translation © 1983 by Augsburg Publishing House and SPCK. Used by permission of Augsburg Fortress, Minneapolis, Minnesota.

Scripture quotations marked *WS* from *Genesis 12–36* © 1981 by Neukirchen Verlag; English translation © 1985 by Augsburg Publishing House and SPCK. Used by permission of Augsburg Fortress, Minneapolis, Minnesota.

Scripture quotations marked *NIV* are from *Holy Bible, New International Version,* © 1973, 1978, and 1984 by the International Bible Society. Used by permission of Zondervan Bible Publishers.

Scripture quotations marked *deV* are from *Word Bible Commentary 12, 1 Kings* by Simon DeVries, © 1985 by Word Books. Used by permission of Word Books, Dallas, Texas.

Scripture quotations marked *CHL* are from *The Book of Exodus* by Brevard Childs, © 1974 to The Westminster Press. Scripture quotations marked *SG* are from J. Alberto Soggins. *Judges,* © 1981 to The Westminster Press. Scripture quotations marked WSR are from Artur Weiser, *The Psalms,* © 1962 to The Westminster Press. All used by permission of John Knox/Westminster Press, Louisville, Kentucky.

The translation of the poem beginning "Delightful to me on an island hill . . ." is by Myles Dillion and is found in *Early Irish Literature,* © 1948 by The University of Chicago Press. Used by permission of the publishers.

The translation of the poem "Guiare and Marban" is from *Early Irish Verse* by Ruth Lehman, © 1982 by the University of Texas Press, Austin. Used by permission of the publishers.

The translations from *Bethada naem nErenn: Lives of Irish Saints* by Carolus Plummer, © 1968, are used with the permission of Oxford University Press, Oxford, England.

The translations from "The Life of Paul the Hermit" and "Monks of Egypt" are from *The Desert Fathers* by Helen Wadell © 1957 by the University of Michigan Press. Used by permission of the publishers.

Passages from *Little Flowers of Saint Francis,* © 1985 by Professor E. M. Blaiklock and Professor A. C. Keys, are published by Servant Publications, Box 8617, Ann Arbor, Michigan. Used with permission of the publishers.

The translation of "The Canticle of Brother Sun" by St. Francis is by Regis J. Armstrong, OFM, Cap. and Ignatius C. Brady, OFM, and is found in *Francis and Clare: The Complete Works,* © 1982 by The Missionary Society of St. Paul the Apostle in the State of New York. Used by permission of Paulist Press (world English rights, except for the United Kingdom), and SPCK, Holy Trinity Church, Marylebone Rd., London, England (United Kingdom rights).

Christianity, Wilderness, and Wildlife

1
Introduction

On Top of Mount LeConte

Sitting on top of Mount LeConte on a summer Sunday afternoon, I slowly read a passage in Isaiah. Storm clouds build below me, threatening rain at my feet. I go through the text once and enjoy it so much that I read it again. No need to hurry. This is as high as I can climb, and I don't have to be home until dark. Isaiah makes better sense at the upper elevations, I think to myself. Thunder rolls somewhere across the ridge in North Carolina while I rest comfortably in Tennessee.

The tip of a cloud creeps up over the rock ledge and chases away the hikers who have come up behind me. The mist touches the fir trees and starts to weave through them. A raven circles on an updraft. I stop reading for a moment and watch his casual flight as he disappears over a spruce covered ridge.

Slowly, the cloud engulfs the peak. If I didn't know I was sitting at the edge of a cliff, I would be tempted to step out onto what looks like fog covered ground. A light sprinkle of rain wets Isaiah 54 and 55. I close my Bible and start to pray. The cloud has discouraged other hikers, interested primarily in the view. I am alone with God at the top of the world.

* * *

In my younger Christian days, I had no idea there was any question about Christian relationship to wilderness or to wild nature. The mountain and I were great friends. I never thought about either dominion or idolatry—the mountain and I praised God together.

In academic circles, however, naivete never lasts long. Some of my associates informed me that Christianity was the worst thing that ever happened to wilderness or to nature in general, and I got around to reading "The Historical Roots of Our Ecological Crisis," by Lynn White, Jr.[1] Then I heard Roderick Nash, author of *Wilderness and the American Mind*,[2] present his views on historic Western attitudes towards wilderness at the National Park Service training center at Grand Canyon National Park. Nash blasted Christianity for placing nature in a hierarchy with humans at the top and the snake at the bottom. He presented historic evidence to document Christianity's continuing antagonism to wilderness and its association of wilderness with the demonic. My first reaction was anger, not so much because Nash was offending the faith, but because he was threatening the legitimacy of one of my favorite spiritual occupations—going out into the mountains to pray. In the aftermath of this encounter, I came to realize Christians do not know or understand their own wilderness heritage. Further, there are numerous serious theological questions concerning both Christian attitudes toward wild nature and Christian use of wilderness.

Most of the available literature on Christian environmental ethics emphasizes God's intent for material world and for our relationship with our fellow creatures. The discussions center on creation theology and human care for the earth, particularly in cultivated and "useful" landscapes. The purpose of this book is, in contrast, to analyze wilderness as a setting for spiritual events and to determine if isolated wilderness sojourns or contact with wild nature can be beneficial to Christian spirituality. The book also contains a look at the human relationship to wild or "nonuseful" nature and a discussion of possible ethical models for Christian interaction with the portions of the environment that remain undomesticated and free of human control.

The role of wilderness in the Bible and in Christian tradition is quite complex and has already generated numerous articles in biblical studies journals. The Exodus alone has several "wilderness wandering" motifs, including the continual murmurings in the desert and the series of spectacular theophanies (appearances of a divine being) at Mount Sinai. Moses' wilderness experiences have precursors in the narratives of the patriarchs and replicates in the lives of later prophets, including Christ.

With the exception of two books, *Wilderness and Paradise in Christian Thought,* by George Williams,[3] and *Christ in the Wilderness,* by Ulrich Mauser,[4] very little comparative study of wilderness in the different books of the Bible or in different eras of Christian thought has been undertaken. Most Biblical scholars, including Mauser, end with the New Testament and make no attempt to relate their findings on wilderness either to the history of the Christianity or to modern wilderness use. The one major recent theological discussion of western spirituality and wilderness protection, *Baptized into Wilderness: A Christian Perspective on John Muir,*[5] centers on a leading figure in the American wilderness movement (whom many scholars believe had rejected Christianity) and therefore does not review biblical sources. To add to the difficulties, commentators on Hebrew and Christian attitudes towards wilderness have neglected some key works, such as the Old Testament books of 1 and 2 Samuel and monastic religious poetry, in favor of better known sources, such as the Gospel accounts of Satan's temptation of Christ in the wilderness. The protection of the "wild" has not been an issue in most of the theological discussions published to date.

The dispersion of sources on wilderness spirituality and wilderness ethics has left Christianity without a coherent response to Roderick Nash and other critics who suggest the Judeo-Christian tradition has been influential in "arousing and nourishing antipathy"[6] towards wilderness and towards any part of nature not resting firmly in human "control." The assumption that Christianity dismisses wilderness as demonic has left wilderness users with the impression Christianity is incompatible with wilderness spirituality and with wilderness preservation. Ironically, the developers of the United States Wilderness Act and the contemporary wilderness movement have long suggested wilderness experience has "spiritual, aesthetic, and mystical dimensions" and can provide "mental and moral restoration."[7] According to the text *Wilderness Management,* published by the United States Forest Service: "In today's bustling world, [wilderness] offers a place where important human values can be rediscovered and where a simpler, less complicated life exists, at least momentarily; it offers a chance to be rehumanized."[8] Despite this concern for the spiritual, wilderness advocates have made little attempt to relate wilderness in any positive way to western religious traditions. The environmental community is willing to claim spiritual values for wilderness but, apparently, lacks the self-confidence and perhaps the vocabulary to define what these are.

The irony deepens when one considers the importance of Judaism and Christianity as world religions and then reviews the role of wilderness in the history of the Hebrew people and of the early church. Passages about wild nature appear in the first chapters of Genesis, and tales of wilderness journeys appear in the lives of the patriarchs. The Exodus was a formative period for the nation of Israel, and the long wilderness wanderings helped to produce a structured religion. David hid from Saul in the caves above the Dead Sea and defeated his own rebellious son battling on heavily forested slopes. The psalmists and prophets mention wilderness and wild nature numerous times and use the imagery to instruct the faithful in the graces of the coming kingdom. Both Christ and John the Baptist prepared for ministry in the wilderness of Judea, and the fathers and mothers of the early church cultivated ascetic life styles in the isolated deserts of Palestine and Egypt. We might consider it historic coincidence, but a great many events of importance to Christianity transpired in the wilderness. This volume will thus attempt to correct several of the deficiencies in the literature by reviewing a wide variety of biblical texts, relating the biblical materials to a selection of postbiblical sources, and discussing contemporary theological issues in light of historic Christian wilderness use.

Wilderness?

Since this book is a late twentieth-century American production, while the literature it investigates is the bounty of many centuries of Old World religious thought, we must first recognize that an environmentalist from a modern industrial nation and a biblical text (or its translator) may not mean the same thing when they refer to "wilderness." In the United States, wilderness is actually a legal designation for federal and state lands subject to little or no human economic use and implies an area where natural ecological processes operate with as little human interference as possible. Most of the first wilderness areas in the United States system were unlogged forests or relatively undisturbed mountain ranges, and, although sites disturbed by grazing, logging, or mining are now incorporated into the system, the wilderness designation limits future economic exploitation and encourages little human intrusion other than camping and non-motorized recreation. Aldo Leopold, one of the first advocates for designated wilderness, speaks for the contemporary perspective when he portrays wilderness as a healthy natural ecosystem, where wild creatures, such as wolves and grizzly bears, are able to survive. Leopold considers wilderness necessary for outdoor recreation, scientific study, and wildlife conservation. He also portrays techno-industrial humans as striving for "safety, prosperity, comfort, long life and dullness" and suggests wilderness may help to save us from our adventure-depauperate life paths.[9]

In contrast to Leopold, the biblical writer who mentions "the wilderness," or a specific region such as the Wilderness of Maon, is usually discussing a specific geographic locality. The biblical wilderness is arid, although it may be a grassland rather than an open, sandy desert. The Old Testament uses a number of words for "wilderness" or "desert," which may or may not have the same implications the term "wilderness" has in English. The most common New Testament term for wilderness, *eremia,* would be better translated "desert" or "isolated place."[10] The "wilderness" of the Bible includes a variety of landscapes and habitats—from the extremely dry valley of the Dead Sea to moister steppe grasslands—yet is very limited in comparison to what contemporary environmentalists might consider wilderness. The biblical wilderness is also a grazed region and subject to human economic use, even if that use is very dispersed.

To fit the objectives of this analysis to the available materials, in this book wilderness is not defined solely in terms of the modern designated "Wilderness Area," which is really a legal entity, nor solely in terms of biblical "wilderness," which is restricted to arid sites. To investigate wilderness as a setting for spiritual events, we must expand the investigation past texts where translators have chosen to use the term "wilderness," and add passages describing wild land-scapes or people left in isolation in wild nature. If we enlarge our search for wilderness settings to include all the types of environments acceptable for con-temporary wilderness status, the Bible discusses not only desert wilderness, but also mountains, forests, grasslands, caves, wadis (small canyons), cliffs, rivers, lakes, the open ocean, and the "isolated places" away from villages. In this book, I investigate all these wild landscapes, although one needs to recognize specific types of sites, such as mountains, may be associated with different types of spiri-tual events than, say, rivers. The book also includes a review of texts concerning wild animals and plants and human (or divine) interactions with them. The term "wild nature" will thus refer to all types of wild creatures and natural objects, in-cluding large predators, song birds, natural water courses and geologic features, such as caves.

To avoid confusion, the word "wilderness" used in a general sense of wild, uncultivated areas, will be distinguished from the biblical concept of specific desert regions, such as the Wilderness of Judea or the Wilderness of Sin. The lat-ter will be called "desert," "desert wilderness," or "biblical wilderness." In bibli-cal quotations, the word "wilderness" means "desert" or "arid lands." A compro-mise with one contemporary criterion for wilderness is also necessary—ancient wilderness was often the home of shepherds or nomads, while monastic wilder-ness dwellers usually kept small gardens. This investigation will therefore allow sheep, a few date palms, or an isolated human residence in wilderness.

Although this approach may superficially appear to force a modern concept of wilderness on biblical and other historic materials, it is quite appropriate from a literary point of view. It distinguishes sites with a strong human presence from those without and also distinguishes living creatures and natural objects under direct human control or management from those free of domestication. These are distinctions ancient writers recognize and purposefully employ. Biblical authors are, in fact, often very geographically specific. Further, as will be noticeable in the biblical texts cited, certain wilderness environments, such as mountains and deserts, may serve very similar functions in some literary contexts.

The Purpose of the Journey

The first task undertaken in this book is to investigate how the Bible relates wilderness to spirituality. In the secular wilderness literature, the spiritual, the aesthetic, and the emotional are often confused. For our purposes, wilderness spirituality will be defined as those portions of the wilderness experience having an impact on one's relationship to God. This could result from a direct encounter with the divine or the holy while in the wilderness but might also emerge from a formative event which influenced an individual's spiritual outlook. Wilderness spirituality could forward development of personal religious understanding and knowledge of God. In Christian thought, this would be expressed in a deeper understanding of the person of Christ or some knowledge of God's characteristics, such as God's grace, mercy, omnipotence, or creative power. A change in the individual's or the community's understanding of their relationship to God is also part of Christian spiritual experience. This might be a recognition of God's love for the individual or for all creation, or an understanding of one's inadequacy before God, accompanied by repentance of sin. Wilderness spirituality can also incorporate the reception of God's grace and providence. A spiritual transformation in the wilderness may inspire the individual or the community to act for God or to undertake a ministry. Traditionally, wilderness spirituality has promoted spiritual practices or disciplines, such as prayer, fasting, and worship. Monastic Christianity has, for centuries, sought the wilderness for contemplation.

As we search the various books of the Bible for models of wilderness use, we will ask: Why did these people go into the wilderness? What type of interaction did they have with God in the wilderness? What was God's intent for the wilderness sojourn? What did the people learn about God? What was the people's attitude toward the wilderness? How did God interact with wild nature? As we examine different portions of the Bible, we will encounter different patterns of wilderness use and different understandings of God's activity in the wilderness.

In pursuing these questions, I will not attempt to examine every text mentioning the wilderness or describing an element of wild nature, such as predatory animals. The texts selected for discussion belong to narratives that utilize the wilderness repeatedly for a setting or to portions of the Bible, including poetic or prophetic texts, that emphasize "wilderness" or lack of human presence or control. Isolated passages providing geographic reference points or mentioning a natural feature are not incorporated in this overview. In selecting biblical books and passages for analysis, the existing academic literature on wilderness provided general guidance, although I have added material on figures such as David and Jonathan that would, for example, rarely be compared to the wilderness sojourns in Genesis and Exodus. A primary purpose of this work is to trace the recurrence of key wilderness motifs through the Bible, as well as note the development of new themes and new variants. The general exegetical strategy was to select texts that concerned similar types of wilderness sojourns, wild natural features, or landscapes and to look for similarities and differences between them—particularly in human social responses to the wild settings or features and in human interaction with the divine. The repeated patterns extracted from these comparisons, summarized at the end of each chapter, are termed "models" because they usually consist of several associated themes or motifs.

Selection of passages, therefore, favors those whose wilderness imagery is repeated elsewhere or which are precursors for wilderness imagery in other biblical books. The selection of passages from Genesis, for example, emphasizes patterns which will appear, perhaps in modified form, in Exodus. The selection of passages concerning the prophets incorporates those related to New Testament attitudes towards wilderness. In terms of secondary sources, this book relies primarily on commentaries that consider the meaning of the texts in their original social and environmental context and on theological works that analyze Hebrew and early Christian relationship to the land. Claus Westermann's commentaries on Genesis, for example, provide extensive references to natural features and geographic locations, as well as detailed discussion of Hebrew creation theology. They are, therefore, repeatedly cited in Chapters 2 and 16.

In investigating biblical attitudes towards wilderness and wild nature, we have to recognize that the purpose of the biblical writers was to discuss God's relationship to humans and to the created world and that the information provided about the wild is often incidental to the main point of the text. The Bible is theocentric rather than anthropocentric or biocentric, and any exposition of its contents must assume the ultimate purpose of the Holy Scriptures is to discuss God and God's will for humankind and the creation. Wilderness is a repeated site, however, for events of importance to the biblical writers, and it is frequently associated with human interactions with the divine. A major thesis of this book is that wilderness does play an important role as a setting for specific types of spiritual events and that the Biblical writers, perhaps without realizing it would someday be an issue, tell us a great deal about the relationship between God and the untamed residents of the cosmos.

Biblical scholars, if they chance to read such a wide-ranging summary as this, will probably be disappointed by the relatively limited attention paid to the origins of the texts (whether they are from J or P, for example) and the lack of definite chronologies for the appearance of variations in wilderness motifs. Determining the exact relationship between the context of the passages and specific Hebrew or Greek vocabulary is beyond the scope of a review with this historic breadth."[11] The primary thrust of this work is to extract common patterns, rather than to identify divergent sources, and to relate the Scriptural materials to contemporary theological concerns. I assumed in my writing that some readers will be very familiar with environmental or wilderness issues but have little back ground in Christian writings. Thus, I have made an effort to both define "technical" theological terms and present Scriptural examples in as complete a context as possible.

Christian Foundations

As a second task, this book contains a review of Christian literature written from the end of the Apostolic Age to the Middle Ages and repeats the questions directed at the biblical texts. This analysis is somewhat modified, as these are not canonical writings (those officially included in the Holy Scriptures) and do not carry the same theological weight as the Bible. It is valuable, in fact, to compare the postbiblical writings to biblical models for wilderness use and to determine to what extent they vary from the biblical patterns.

I have not attempted to review all the historic Christian wilderness movements but have selected cases for study. Christians who have lived in, prayed in, or otherwise experienced wilderness tend to have different (and more positive) attitudes from those who have viewed wilderness from a distance. The book, therefore, dedicates little text to allegorical uses of "wilderness" or to the vocabulary of wild nature (other than from biblical sources) and concentrates on literature that discusses actual wilderness sojourns or experiences with the wild. The historic survey concludes with Saint Francis of Assisi, who in many ways represents the transition to the modern era. Romanticism and the struggles of American Christianity are left for another volume.

The book's third purpose is to evaluate human responsibilities towards wild nature and to translate some of the questions posed by philosophical environmental ethics into a theological framework. We will inquire how Hebrew and Christian traditions have historically valued wild nature and determine if these values have been positive or negative. We can also extract biblical and traditional Christian characterizations of the wild and investigate the possible theological importance of these. By careful comparison of relevant biblical texts, we may gain some insight into God's intent for the wild, and thereby our relationship to it.

The fourth task represented in this book is to relate the biblical and historic models of wilderness and wildness to contemporary Christianity. Today's Christians spend much of their time in church buildings and organized gatherings. Community life and group activities (if not middle-class values) become the center of spiritual concern. What place do wilderness and isolated spiritual practice have in this sort of a social framework? Does wild nature have any special role in spiritual experience? Does isolation from the mainstream of human life have any value in spiritual development? Can wilderness provide "moral and mental" (or spiritual) restoration? Do biblical and historic models for wilderness spirituality have any relevance today? From a Christian perspective, is there a legitimate role for a wilderness sojourn?

In the final chapters of this book I examine contemporary Christian concerns about wilderness and wild nature and discuss Christian options for wilderness use. An attempt is made to engage both issues arising in the academic wilderness literature—such as the role of sacred places—and traditional Christian concerns—such as the conflict between withdrawing to the wilderness for prayer and participating in active social ministry. Although some types of wilderness spiritual experience—such as the appearance of God at Sinai—are difficult to implement in a contemporary spiritual context, others—such as Christ's repeated withdrawal to isolated places for prayer—are easy to replicate.

Translations, Theophanies, and Theological Conventions

Despite its well-established status in Roman Catholic and Orthodox contemplative traditions, wilderness spiritual experience remains a confused and oft rejected concept in the denominations spawned by the reformers. All factions, however, accept the Bible as establishing precedents for contemporary spirituality. My own theological training has been primarily in Evangelical institutions, and I am more familiar with Protestant schools of thought concerning biblical interpretation than with those of other Christian traditions. I hope, however, that, by relying strongly on the actual biblical accounts of wilderness sojourns and divine interaction with wild nature, I can identify concepts and models of wilderness spiritual experience that can be discussed in a variety of denominational frameworks. Some denominational differences in interpretation of these materials are, of course, to be expected, particularly concerning the postbiblical writings of the early church.

An effort has been made, throughout the book, to cite the Holy Scriptures and historic Christian manuscripts. These quotes will provide ideas for "wilderness reading" as well as verify the conclusions presented in the text. Although the *Revised Standard Version* is the most frequently cited Bible translation, I frequently employ other versions, either to provide greater accuracy or more poetic interpretation. When repeatedly citing a specific commentary, I usually quote the translation employed by the commentary. The name used for God may vary from translation to translation as may other terms.

As always, working with some types of premodern texts poses problems for the contemporary reader. One difficulty is the use of inclusive language. When I am writing about the interactions of people with the natural environmental, I find the term "humankind" much preferable to "mankind," since we are all in this together. In citing historic sources and translations, however, I prefer not to deviate from the gender chosen by the commentator or translator, which is, one hopes, the gender employed in the original language of the biblical or historic text. Unfortunately, the commentaries and the original sources are hardly inclusive, and it is difficult to take a consistent approach to pronouns. One solution, in the case of the divine, is to use proper names for God as much as possible, which is a formal, but very respectful approach. This, of course does not resolve the question of what to do about biblical texts that discuss only Adam, rather than Adam and Eve. In the end, however, this book isn't about gender, and there isn't much evidence that wilderness has ever been a "gender" issue. One of the important patterns found in this book, in fact, is the universality of wilderness spiritual experience since it concerns figures ranging from Hagar the serving maid, to the boy David, to bearded Elijah, to Christ at the height of his ministry. I hope readers from a variety of social backgrounds will find examples of wilderness spiritual encounter cited here that speak to them as individuals.[12]

A last issue of interpretation concerns appearances of the divine. The standard theological term for this is "theophany," which means literally a visual appearance of a divine being. Another term sometimes used for this is "epiphany," which may also be defined as an appearance or manifestation of a divine being, although this word often implies a deep grasp of the presence or reality of God (and often in contemporary usage does not mean literally "seeing" God). The biblical examples of divine manifestations do not fit this selection of theological vocabulary as well as one might like. Some biblical encounters with the divine are primarily visual while some are auditory, and some have elements of both. Further, encounters with divine messengers (angels) and with Satan or the demonic have much in common with encounters with God, at least in terms of the way they occur. New Testament meetings with Christ are not usually counted as theophanies, yet encounters with the transfigured or risen Christ are certainly intersections between the mortal and the otherworldly. The risen Christ, in fact, behaves much like an Old Testament angelic messenger, who appears and disappears suddenly, can be distinguished from a mortal primarily from the content of his speech, makes physical provision for those with whom he is meeting, and utters prophetic statements. Conversely, there are some Old Testament passages where there is confusion over the exact status of a messenger. In some cases, these "angels" are called "men," but postencounter passages imply a meeting with God. Typological biblical interpretation also favors treating some of these "men" as manifestations of the divine, i.e., the man with whom Jacob wrestled, although not everyone agrees with this.

The word "theophany" is not derived from the Greek vocabulary of the New Testament and does not have any special biblical meaning relative to the divine. The Greek words, *epiphaneia, epiphanes,* and *epiphano,* do occur in the New Testament, and the former is used in several places in the Epistles to refer to the future manifestation of Christ. The root *phaino* (to appear or to shine) is used for appearances of the risen Christ, Elijah, and angels, but is also used to describe weeds, sinners, and tombs. Biblical vocabulary for manifestations of the divine is relatively unspecialized and favors words that mean simply to see, to hear, to shine, etc. What we can assume is that whatever is described, be it Yahweh in glory or a host of angels, is actually there. Visions and dreams are generally distinguished from face-to-face confrontations, although the voice or vision of God in a dream functions just as a direct encounter would. There is evolution of encounter with the divine through the Scriptures. Yahweh is much more of a physical presence in the Genesis and Exodus narratives than in the later history of the Hebrew people. In intertestamental times, prophecy and the ability to "see and hear" God seem to depart from the nation of Israel. Appearances of Satan and the demonic are primarily found in the Gospels, and much of the direct encounter with the divine in the New Testament centers on Christ and not on Yahweh God.

If an anthropologist were reviewing these materials, she would probably place all otherworldly appearances in one category. Christian theology, however, strongly distinguishes between the Trinity, the heavenly court, and the fallen demons. I try, in analyzing individual passages, to distinguish between encounters with one of the persons of God and meetings with other sorts of otherworldly beings. The term "theophany" is, however, used in summary sections to refer to "something manifesting or revealing deity,"[13] including angels. This isn't intended to imply the angels are *theos*; there isn't another suitable collective name for these phenomena. Further, it is assumed that a physical expression of Yahweh is actually a manifestation of God present at the site in question (and is, therefore, the same as God, although not, in any earthly epiphany, a complete expression of God's essence).

Beginning the Sojourn

Vocabulary problems aside, I hope this book enhances the reader's appreciation of wild nature as a work of God as creator and raises further questions about our treatment of our fellow creatures and our care of wilderness areas. I also hope, it helps to unravel some of the confusion over Christian spiritual practice in wilderness and provides some guidance to individual Christians about the use of natural settings in pursuit of a deeper Christian spirituality. The reader is encouraged to expand this exposition and to personally investigate Hebrew and Christian wilderness texts. We will never understand our spiritual heritage unless we begin to appreciate the trials, the songs, and the victories of those who walked the mountains and deserts before us.

2

She Went and Wandered in the Wilderness

To Goshen Prong

It was Christmas Eve, and I wasn't working, so I invited a friend to go hiking for a few hours before the evening church service in town. Snow had covered the peaks of the Great Smoky Mountains, making access to the higher ridges difficult. We decided to travel an easy, low elevation route to a creek named Goshen Prong.

The weather was cold, and the air was still; only the splashing of Little River broke the silence of the forest. We were completely alone on the trail—everyone else was home making last-minute preparations for the holidays. Having no specific goal for the hike and no real reason to get to Goshen Prong that afternoon, we stopped frequently to admire the leafless yellow poplars, rising straight-trunked above the carpet of dead leaves and snow patches.

We wandered for a while and then sat on the rocks at the river's edge and appreciated the ice sculptures. Cold water pouring over falls and cascades had produced sparkling columns, curtains, and arches. Drenched repeatedly with spray, each log and boulder had acquired its own jeweled necklace of translucent frozen beads. Returning to the path, we found the ice formations were forcing their way out of the forest floor also. Water in the soil surface had solidified, causing small ice crystals to burst from the ground and grow upright, until they looked like colonies of tiny stalagmites. Fern fronds, broken branches, and blades of dried grass, glazed by freezing rain, decorated the trail side.

Lost in the beauty of the wintry mountain valley, our conversation turned to personal and then spiritual matters. We talked about past holidays with old boy friends and the associated social disasters. We talked about God and the church service that night. My friend was planning to sing, but she wasn't certain she believed. The God of her childhood was behind her, pushed back by adult relationships and adult doubts.

13

Late in the afternoon, barely half way to Goshen, we decided to turn back. Even though it was still well before supper time, the low light was beginning to blend the gray forms of the trees into the outlines of the surrounding ridges. We were only two days from the solstice, and the mountains to the west would soon block the sun. Reaching a trail intersection, I suggested we climb up out of the valley for a few minutes and try to get a view over the river. We strolled up on a shoulder where we could see the crimson glow of the sunset reflecting off the higher peaks. The scene before us was an ordinary mountain landscape, but the fiery red haze drew my vision out beyond the tops of the farthest ridges. I leaned on my walking stick and let myself enjoy the color. My friend stopped behind me. Without the crunch of our boots on the frozen trail, the forest was perfectly still.

Suddenly, I felt something change. I wasn't just absorbing the intensity of the dying light—God was calling me. Somewhere out over the mountain, God was speaking, and my heart was responding. I forgot about my friend, turned my mind from the forest and the snow, and began to pray silently. The presence of the Divine filled the mountain. The sunset grew pale against the greater glory. I felt as if the Holy Spirit had completely surrounded me and moved through me. Then the grasp of God loosened . . . and I looked back at my friend. She was on her knees, crying.

"Susan, are you praying?" she asked.

She sang beautifully that night, filling the little church with "Angels we have heard on high," while I sat in the back convinced of the goodness of God. From small graces to great ones, from the snowy solitary paths to the candle-lit halls of community, it was indeed the time of Emmanuel—"God with us." Within the range of my spiritual experience, that day on the mountain was exceptionally intense and joyous. Yet many Christians will never have such a direct and memorable encounter with the divine, and many will doubt the legitimacy of such a "wilderness experience."

One suspects the isolation and the beauty of the forest made us more willing to deal with the spiritual. How important was the setting? Was it easier for me to "hear" God when I was out in the forest with my friend? She recognized we had moved beyond the aesthetic. Was it better for us to be alone?

This encounter with God was not in a church or an organized meeting. Is this kind of experience within the framework of biblical spirituality? Is there any precedent for suddenly meeting God in the wilderness? Was this just an individual event in individual circumstances, or was there something more here, something the very technologically and theologically sophisticated western Christianity has forgotten?

Among the Trees Planted by God

To develop a framework for interactions between God and humankind in wild nature, we must first return to the book of Genesis and the Garden of Eden. Adam and Eve lived in a privileged environment where the plants of the garden provided food for them, and they resided in peace with the entire animal kingdom. The existence of Adam and Eve was not a passive one; God had commanded them to both till and watch or preserve the garden. Yet the garden before the Fall was an unusual situation, not just in terms of human relationship to animals, but also in terms of human relationship to God.

This primeval story portrays God in direct interaction with Adam and Eve. God moved about in the garden "at the time of the day breeze" (Gen. 3:8, *WS),* and the two new human creations were able to perceive God's presence and react. Through the entire story of Eden, God spoke directly to humans and they answered their Creator. When God forbade taking the fruit of the tree of the knowledge of good and evil, He spoke directly to Adam. When Yahweh detected humankind's failure to abide by this command, God called to Adam and asked: "Where are you?" God also questioned Eve and spoke directly to the snake. Discovering something had gone wrong, God interrogated the guilty and pronounced the punishment directly. Only in the first few chapters of Genesis does God command or confront people in this manner. In all the Old Testament, only in the garden does humankind have such close and continual contact with God.[1]

In Eden, nothing could be distinguished as spiritual experience or closeness to God, because God was always completely available. Adam and Eve could see and hear the Holy One without special effort. Spiritual disciplines, such as prayer, were not necessary and humankind was free of the complexity and restrictions of religious law. God placed only one "thou shalt not," concerning the fruit of the tree, before Adam. Adam and Eve's primary "thou shalt" was to care for creation.

In Eden, there was also no distinction between the wild and the tame, or friendly and unfriendly nature. Adam named the mighty lion as well as the domestic goat. The snake spoke to Eve, and she answered him.

The fall from grace damaged this freedom of communication as human relationships with both God and creation were broken. Adam and Eve's final experience with the Divine at the garden was unpleasant. Cherubim and a flaming sword blocked them out—not only from Eden, but away from the tree of life and the presence of Yahweh. (Yahweh is an old Hebrew name for the Creator God.) In the beginning, God, wild nature, and humankind could all communicate with each other and coexist together. After the fall from grace, humankind could no longer clearly hear or see God.

And She Went and Wandered in the Wilderness

The key figures of the Bible are the men and women who can understand the Word of God despite broken spiritual relationships and the problems of sin and human failing. In the narratives of the patriarchs, we find the Lord choosing Abraham (called Abram before making a covenant with God) and his family line. The Lord spoke to Abram, and Abram responded, moving from Ur to the land of Canaan. Abram and his family were herders who migrated across the landscape. Genesis 13:1 mentions Abram and his brother's son Lot traveling with their flocks to Egypt and then moving back to Canaan through the Negeb desert but does not mention any difficulties with the arid environment. Abraham and the other patriarchs lived in tents, rather than in settled agricultural or urban settings, so distinguishing events in uncultivated areas in the life of Abraham and his sons is not, by itself, very meaningful.

Genesis does, however, report a series of incidents which took place in wild sites away from the family camps, including the two journeys of Hagar in the desert, Abraham's journey to Mount Moriah, Jacob's dream while sleeping in the open, and Jacob's struggle at wadi Jabbock. The first real wilderness sojourn in the Bible is, in fact, not that of a prophet, but of Hagar, the serving maid of Abraham's wife Sarah.

Sarah (called Sarai at the time) was advanced in years and had not borne Abram any children, although God had already promised Abram a son. Sarai sent her maid, Hagar, in to Abram as a substitute for herself, in hope that Hagar would conceive and Sarai could bear children through her. Hagar became pregnant and, realizing her superior social situation, "looked with contempt on her mistress" (Gen. 16:4). Sarai, feeling her position as Abraham's favored wife threatened, dealt harshly with Hagar and Hagar fled (Gen. 16:6).

According to Genesis 16:7: "The angel of the Lord found her by a spring of water in the wilderness, the spring on the way to Shur." The angel, or messenger, addressed Hagar by name, indicating he had special knowledge of the situation, and told her to return to Sarai. He also made a promise saying (Gen. 16:10–12): "I will so greatly multiply your descendants that they cannot be numbered for the multitude. . . . Behold, you are with child, and shall bear a son; and you shall call his name Ishmael; because the Lord has given heed to your affliction. He shall be a wild ass of a man, his hand against every man and every man's hand against him; and he shall dwell over against all his kinsmen."

When Sarai sent Hagar in to Abram, the faith of both Abram and Sarai had failed. The patriarch did not trust God for his heir through his wife. Prior to this point in the Genesis texts concerning the origins of the nation of Israel, God had spoken only to Abram about what was going to happen. When the outcast Hagar fled to the desert, the serving maid encountered a messenger of God, thus becoming party to communication with the divine. The promise she received was a promise to her, not to the patriarch.

The wilderness, in this case, was critical to what transpired. First, the setting proved Hagar's desperation. Second, it is unlikely a messenger of God would have spoken directly to Hagar in Abram's tent.[2] Hagar's isolation allowed her to become a participant in the holy promises. Third, the messenger indicated Ishmael, translated "God hears," would become a "wild ass of a man"—a desert dweller. The arid lands would be his home, and the wandering bedouins his heirs. God was willing to extend His blessings beyond His promise to Abram, and the blessing fell on a woman and her son wandering as outcasts in the wilderness.

This account of Hagar in the desert ends with an interesting passage which provides problems in translation. The *Revised Standard Version* of Genesis 16:13 reads: "So she [Hagar] called the name of the Lord who spoke to her, 'Thou art a God of seeing'; for she said, 'Have I really seen God and remained alive after seeing him?' " Westermann's translation reads: "And she called the name of Yahweh who was speaking to her: You are the God who sees me. She said indeed: Truly I have seen 'God' after he saw me." The sentences really refer to the naming of a spring, presumably a holy place in ancient times. Working from the English, one might assume the messenger was actually God, but this seems unlikely. A better interpretation might be Hagar, praising God, named the spring "the God who sees me lives," or that Hagar, "naming God" or describing who God is, said "For me God is the one who sees me, that is the one who helped me in my distress."[3]

In this wilderness sojourn, we find the hand of a god who both "hears" and "sees" extended to an unlikely outcast. Angelic messengers appear elsewhere in the story of Abraham and are hardly limited to wilderness settings. The meeting between Hagar and the messenger, however, preceded the three men, also messengers, who came to visit Abraham in his tent by the oaks of Mamre and the visit of two messengers to Sodom. Wilderness, in the case of Hagar, was the setting for exceptional communication with God and for exceptional providence or help from God. Wilderness was also the site for an encounter that probably could not have happened at Abraham's tent.

Hagar's troubles did not end with her return home and the birth of Ishmael. Sarah also bore a son—Isaac—and became jealous of the natural rights of Ishmael, the first born. Sarah requested Abraham cast out the "slave woman with her son" (Gen. 21:10), and, although Abraham was very displeased with the request, he complied after the Lord promised also to make a nation of Ishmael.

Given water and bread, Hagar and her son wandered in the wilderness of Beersheba.

> When the water in the skin was gone, she cast the child under one of the bushes. Then she went and sat down over against him a good way off, about the distance of a bowshot; for she said, "Let me not look upon the death of the child." And as she sat over against him, the child lifted up his voice and wept. And God heard the voice of the lad; and the angel of God called to Hagar from heaven, and said to her, "What troubles you, Hagar? Fear not, for God has heard the voice of the lad where he is. Arise, lift up the lad, and hold him fast with your hand; for I will make him a great nation." Then God opened her eyes, and she saw a well of water; and she went, and filled the skin with water, and gave the lad a drink. And God was with the lad, and he grew up; he lived in the wilderness, and he became an expert with the bow. He lived in the wilderness of Paran; and his mother took a wife for him from the land of Egypt.   (Gen. 21:15–21)

This narrative of the desert is again the story of an outcast who receives miraculous deliverance. The boy was only the son of the maid servant and would not produce Abraham's main lineage, but he was also the child of a promise and, therefore, not forgotten. God personally cared about the welfare of the boy.

Again, the passage incorporates the themes of seeing and hearing. A messenger spoke, this time from heaven. There is a transition from the man in the previous wilderness sojourn of Hagar, who could be recognized as the messenger by his knowledge of the situation, to one who clearly resided in a divine location. God not only heard Ishmael, Hagar heard the voice from above and then saw the well, which was not visible when she first arrived at the place, and laid Ishmael under the bush.

Hagar and Ishmael also made a transition in life circumstances. From nearly dying in the desert, Ishmael was suddenly able to cope with the harsh conditions and provide food for himself (and presumably for his mother). He became a bowman and a skilled hand at wilderness survival. The text clearly indicates this was because "God was with him." We find Ishmael—without his natural father or family group—growing up in the most strenuous of environments. With God's help, the wilderness provided first water and then a livelihood. The wilderness proved the depths of God's love.

Blood on Mount Moriah

Ishmael's rescue by God was followed by Isaac's rescue from sacrifice at the hands of his father. The setting was a mountain that was selected by God and stood three days travel from Abraham's camp. God commanded Abraham to go and offer Isaac as a sacrifice, and the obedient Abraham set about the task. Leaving his two servants behind with the donkey, he climbed the mountain with his son, built an altar, bound the boy, and laid him on the firewood. Just as Abraham stretched out his hand with a knife to kill the boy:

> the angel of Yahweh cried to him from heaven and said: Abraham! Abraham! And he answered: Here I am! And he said: Do not raise your hand against the boy, and do him no harm! For now I know you fear God, because you have not refused me your son, your only son. And Abraham raised his eyes and looked— there was a ram entangled in the bushes by the horns. And Abraham went and took the ram and offered it in sacrifice in place of his son. And Abraham gave this place the name "Yahweh sees," of which one says today: On the mountain Yahweh made himself seen.
>
> (Gen. 22:11–14, *WS*)

The angel of the Lord then calls to Abraham a second time and blesses not only Abraham and his "seed," but Gen. 22:18 (WS) also promises that "all the nations of the earth shall find blessing" through the descendants of Abraham.

Here we find many similarities with the story of Hagar. The situation was desperate, and the death of a son would result. The site was isolated, away from both the family camp and other groups of people. Abraham heard the angel speak from heaven, as Hagar did. The mountain became the place where "Yahweh sees" (and has mercy) and where Yahweh (or the work of Yahweh) was seen. Just as Hagar did not see the spring when she thought Ishmael was going to die, Abraham did not see the ram until after he spoke to the angelic messenger. Ishmael needed water to live, while Isaac needed a substitute for the sacrifice. The modern reader may miss the implication here of a divine act. It is unlikely Abraham would not have noticed a ram on the site, much less that Hagar, thirsty and desperate, would have overlooked a spring. In each case, they "see" God first and then "see" salvation. In each story, God made a promise concerning the son and God blessed each boy with a future.

A primary difference between the two narratives is the reason the children were in the wild site alone with a parent. Abraham cast out Ishmael and Hagar. They were in extreme danger in the difficult desert environment. Abraham willfully took Isaac to the mountain in order to worship God, thus the only danger lay in God's requirements. God chose the mountain as the site and directed Abraham to it. Again, the patriarch probably had to be alone—away from the family camp—for this interaction with God. Sarah and perhaps others might have tried to stop a sacrifice. The oracle and the promise were meant for Abraham.

How Awesome Is This Place

A third figure from Genesis who spends time alone in the out-of-doors is Jacob. Jacob was fleeing his brother Esau and traveling to the camp of a relative, where he hoped to obtain a wife. He stopped on the way and slept with his head on a stone. The site was certainly not desert, but was probably on rich agricultural land in

Canaan. The text does not indicate, however, the presence of any camp or village there at the time. Jacob did not see a messenger of God standing before him, but he dreamt he saw a stairway to heaven with the angels of God going up and down. In the dream (Gen. 28:13, *WS*): "Yahweh stood before him and said: I am Yahweh the God of Abraham your father, and the God of Isaac. The land on which you are lying I give to you and your descendants." Yahweh also repeated the promises made to Abraham of many descendants, and that "all the families of the earth shall bless themselves in you (and your descendants)" (Gen. 28:14, *WS*).

Although this was a dream, Jacob saw not only angels but also God. As was the case with Abraham and Hagar, a promise was made, this time from the mouth of Yahweh. The promise, like the others, was of a future nation and of a God who will be faithful—no matter what intervenes. Jacob was not under severe environmental stress at the time, but he was isolated and escaping the wrath of his brother. The vision occurred in transit from his family camp to the tents of Laban, where Jacob intended to find a wife. The dream, therefore, marked Jacob's transition from son to the head of a household.

Interestingly, Jacob noted the special nature of the site after he had the dream. He was afraid after he awoke and said: "How awesome is this place; it is none other than the house of God and the gate of heaven" (Gen. 28:17, *WS*). The place was a "nonplace" when Jacob lay down. He perceived nothing unusual about the site.[4] The presence of the Lord gave the site special meaning, and Jacob erected a pillar to commemorate the event.

After Jacob had married and fathered eleven children, he had a second isolated encounter with the Divine while crossing a ford of the Jabbock river, a tributary of the Jordan. Jacob sent his family across ahead of him and remained alone on the far side of the wadi (a ravine or canyonlike stream bed). A "man" attacked Jacob and he struggled with him. The identity of the "man," who never gave his name, has been argued, and some interpretations suggest the shadowy combatant was a demon. Others suggest the "man" was an angelic messenger or the Lord himself.[5] Neither wrestler won the contest, but, before the unknown assailant fled, Jacob forced a blessing from him. The attacker both blessed him and renamed him Israel ("the one who strives with God") for Jacob had "struggled with God and man and [had] prevailed" (Gen. 32:29, *WS*). Jacob named the place Peniel for he had "seen God face to face" (Gen. 32:31, *WS*) and had survived.

Both these narratives of Jacob have the theme of seeing God, and both contain a promise or a blessing. In both cases, Jacob was in transition, and it was necessary for him to be alone. In the first case, Jacob was fleeing Esau, while in the second he was going to meet him under hostile circumstances. The wrestling match stands unique among the "wilderness encounters" in Genesis. In all the other examples,

God acted, and the individual responded. At wadi Jabbock, Jacob took the initiative and asked for a blessing. When his assailant responded, he gave Jacob a name. Jacob, now called Israel, was to become the father of twelve tribes and of a holy nation.

We can now extract some major themes in these passages about the beginnings of the Hebrew people. First, these isolated encounters in wild sites were associated with theophanies—appearances of God or of other-worldly beings. The form of the appearances varied, but in several of the cases, the messengers called from or were seen in heaven. These appearances carried promises and blessings. All the promises concerned God's faithfulness and the future of the people. In all cases, the content of the encounter implied it was necessary for the people involved to be by themselves. Isolation was a repeated and important factor.

Throughout the texts, seeing and hearing God was of great concern—great enough that both the child of the wilderness, Ishmael, and several of the sites were named for "seeing" or "hearing." The type of wild or isolated site, however, varied. Abraham sought God on the mountain, while Hagar found God unexpectedly in the desert. Jacob's most important spiritual transition was at a river crossing. His dream occurred at a site lacking distinctive natural topography but graced with the gates of heaven.

It should be noted that in these very early stories, natural objects served as sanctuaries. A tree might make a location a sacred place. When Abraham travels to the terebinth tree of Moreh, for example, the Hebrew indicates the tree is either the place of the oracle (a person making prophetic statements) or perhaps even, from pre-Hebrew religious beliefs, that "the tree itself gives the oracle."[6] Westermann suggests that "the patriarchs looked for places that had long since been sacred and lay outside the walls of settlement,"[7] when establishing their sanctuaries. The spring on the way to Shur, where Hagar saw the angel, may well have already had status as a religious site. These sites, then, preceded the establishment of the cult as sacred places and speak to the great age (and possible animist roots) of the wilderness traditions with which they are associated.

Both Hagar and Jacob were dealing with human antagonists. Abraham was directly engaging God, but was under a great burden because of the impending sacrifice of his son. In all cases, God's mercy was very visible in the wild. The Lord provided not only a spring and ram, but also the strength to go on and receive the promises.

In each case, the Lord was a presence. This is most obvious in the case of Jacob, to whom God says: "I am Yahweh the God of Abraham your father, and the God of Isaac." In each case, the Lord was also a protector and a provider—of water, of the ram, and, for Jacob, of protection wherever his lineage goes, even into exile.

In each case, the Lord was the generator of the glorious future, in the blessings on the offspring and in the creation of the nations.[8] People in isolation in natural settings hear promises of multitudes of offspring and promises of national history.

Biblical scholars credit parts of the story of Hagar in the wilderness to the geographically disparate *J, P* and *E* sources (the letters denote different hypothetical documents), which may have been collected separately in the kingdoms of Israel and Judah.[9] The description of Hagar's first wilderness sojourn is not necessarily from the same prescriptural source as the story of Abraham and Isaac on the mountain. Yet, if these narratives lack common names for God, they have common plots and themes. This suggests "wilderness seeing and hearing" is one of the most ancient concepts of the Hebrew religion and arises from a very old oral tradition of narratives about meeting with the divine in isolation or at certain distinctive wilderness sites.

We have in these examples the *Genesis* or *foundational model* of wilderness spiritual encounter. Some characteristics of the foundational wilderness experience are

1. God is clearly seen or heard and also sees those in distress. "Seeing" includes theophanies and dreams.
2. God makes promises about the future of a divinely elected people—including Ishmael who is not of the chosen line. This serves as both prophecy and covenant.
3. Divine providence continues to operate and may be more apparent than elsewhere.
4. Stress is associated with divine encounter.
5. One may be drawn to the mountain by God's calling, e.g., Abraham, or God may appear unexpectedly in an isolated setting. God is not always sought in the wilderness *per se*; the encounters are not necessarily willed by the person who engages the divine.
6. Wilderness events are associated with spiritual transition, with assuming new roles, or with acquiring new futures.

Ears Grown Dull

My experience on the way to Goshen Prong was limited in comparison to the experiences described in Genesis. Neither God nor an angelic messenger appeared, rather the voice of God echoed deep in my heart. A detailed comparison is unfair, since the context of the Biblical promises to the patriarchs is not applicable. Yet isolation was as critical to what happened on the way to Goshen Prong as it was in the angelic encounters of Hagar and Jacob. In light of contemporary social conventions, a hiker is an unlikely participant in a spiritual event. The Bible, however, clears away cultural prejudices when a fleeing maid servant meets an angel in the wilderness, not once, but twice. I returned convinced of God's reality, and this was also an outcome of the Genesis stories.

The central issue is not what we expect to happen in the wilderness, but what we expect to happen in our spiritual lives as a whole. For many modern Christians, God does not enter the physical world. For the ancient Hebrew, God interacted in a very personal way. Abraham spoke to God frequently, had visions, and met messengers. For Abraham, both the present and the future lay in God. The messenger was not solely a presence in the tent—a mystical happening—but a connection in the entire chain of history and in God's purposes. Abraham's spiritual vision was spatially and temporally far more expansive than our own.

The story of Genesis is really a story of family and community, yet several of the most dramatic events occurred in the wilderness. God often chooses to speak to the individual. In Genesis, there are no right or wrong settings for interactions with the Lord, only best settings. Although we find Abraham traveling to the "mountain of Yahweh" and Jacob sleeping at the gates of heaven, God was not limited to these sites. Wilderness did, however, function as a place where one could more easily perceive God and come to understand the will of Yahweh.

3
The Water Became Sweet
(The Exodus)

The Burning Bush

The wilderness events we have investigated from Genesis all involved one or two people, rather than a larger group. The best known Biblical wilderness sojourn, however, brought the entire Hebrew people into the desert and became a repeated exercise in group dynamics. God's plan for the salvation of his chosen people centered in the Exodus—the story of the growth of the individual wilderness experience of Moses into an expedition for twelve tribes.

In the Exodus, wilderness themes found in Genesis are further developed, which is not surprising considering the common authorship and editing of the first two books of the Pentateuch. Whether one takes the conservative approach and considers Genesis and Exodus to be the work of Moses, or the modern form-critical approach and considers the two books to be edited versions of several documents or historical traditions, there can be little doubt the use of wilderness motifs in both works is strongly related and originates from common sources or from a consistent tradition. Exodus does develop some new themes, however, and emphasizes the response of the community over that of the individual. The Exodus becomes an exercise in "seeing" and "hearing," not just for Moses and other leaders, but for the people of Israel as a whole.

The book of Exodus begins with a report of Egyptian oppression of the Hebrews and the birth of Moses. Brought up in agricultural and urban Egypt, Moses hardly seemed—from his origins—to have the making of a wilderness leader. Yet, after killing an Egyptian, he fled to the land of Midian and ended up herding sheep in the desert:

> Now Moses was tending the flock of his father-in-law, Jethro, the priest of Midian, and he led his flock behind the wilderness, and came to the mountain of God, to Horeb. The angel of the Lord appeared to him as a flame of fire from the midst

25

of a bush. And he looked, and to his amazement the bush was burning with fire
without being consumed. And Moses said, " I must go across to see this marvelous sight, why the bush does not burn up." When the Lord saw that he went
across to look, God called to him from the midst of the bush, saying, "Moses,
Moses." He answered, "Here I am." He said, "Do not come any closer; take
off your sandals from your feet, for the place on which you are standing is
holy ground." Then he said, " I am the God of your father, the God of Abraham,
the God of Isaac, and the God of Jacob." And Moses hid his face, for he was
afraid to look at God.   (Exod. 3:1–6, *CHL*)

In Moses' first encounter with God, he was both in the desert and at a mountain.
The incident of the burning bush has several similarities with wilderness events in
Genesis. Moses had left Egypt fearing for his life and had become an outcast from
his people. He was alone in the desert and was not expecting divine action. At the
burning bush, Moses approached a dramatic form of theophany in order better to
see and understand why the bush was not burnt. In a proclamation very similar to
God's statements in Jacob's dream, an angel of the Lord appeared and announced to
Moses the presence of God. The content of the angel's speech, as well as the setting,
tied Moses to his ancestors and to the events of Genesis. The "mountain of God" described in these passages was not the mountain visited by Abraham. Moriah is presumably under the remains of the temple in Jerusalem (and crowned by a mosque),
while Horeb is in Sinai. Moriah became a place of sacrifice—the site of the temple
and the location of the crucifixion, while Horeb functioned as a site of revelation for
Moses, the children of Israel, and the prophet Elijah.[1] Exodus 3:7–4:17 continues
with God's call to Moses to lead Israel out of Egypt. The theophany of the burning
bush was similar to the theophanies seen by patriarchs, but the content of the message to Moses was quite different. When Yahweh spoke to Hagar, Abraham, and
Jacob, the message was personal and the commands were of an immediate nature,
such as the instruction to Hagar to return to Sarai. Moses, in contrast, received a
commission to take the message from God to others and to lead the Hebrew people
out of bondage. Hagar, Abraham, and Jacob returned to the family with an understanding of the person of God[2] and a promise for the future. Moses returned with
a mission.

The theophany at Horeb was a transition from the personal spiritual experience of the patriarchs to the prophetic call. With the transition came a change from
concern for a family to concern for a people. Reluctant, Moses protested that he
lacked authority and the gift of clear speech. In the Old Testament, the office of
prophet began as a struggle and continued as such. As Moses doubted his leadership, so did the people.

Important, although somewhat obscure, is the giving of God's name at Horeb:

> Then Moses said to God, "If I went to the Israelites and was saying to them, 'The God of your fathers has sent me to you,' and they asked me, 'What is his name?' what should I say to them?" God said to Moses, " I AM WHO I AM." And he said, "So you shall speak to the Israelites, 'I AM has sent me to you.'" And God spoke further to Moses, "So you shall say to the Israelites, 'The Lord (YHWH), the God of your fathers, the God of Abraham, the God of Isaac, and the God of Jacob, has sent me to you.' This is my name for ever, this my designation in every generation."   (Exod. 3:13–15, *CHL*)

God then directed Moses to assemble the elders of Israel and tell them God had appeared to him. The Lord declared that the Hebrews would enter a new land "flowing with milk and honey" and ordered Moses to go to the king of Egypt and say, "The Lord, the God of the Hebrews, has met with us. Now let us go a distance of three days into the wilderness, to sacrifice to the Lord our God" (Exod. 3:18, CHL).

Critics have questioned why Moses asked for the name of God, when Moses surely already knew it. The giving of the name may be an assurance that Yahweh, rather than some other deity, was selecting the prophet, or it may be a request for a reaffirmation of the relationship of Yahweh to Israel. For our purposes, we need to note that the wilderness again served as a site for the giving of the name of God and for the understanding of God's person. In the generations between Joseph and Moses, God's activity among the children of Israel had not been worthy of record. Then Yahweh spoke again and spoke very dramatically. God did not attempt to call Moses while in Egypt, but waited until he was in the desert. The theophany's instructions to Moses to lead the people of Israel into the wilderness for three days of worship implied that Egypt was an unsuitable site for worship of Yahweh—the wilderness was preferred. When God chose to reopen communication with Israel, God's voice was heard first in the wilderness, and God's call was to the wilderness.

A Strong East Wind

When Moses entered the wilderness again, he had escaped from Pharaoh and had the children of Israel in tow. According to the book of Exodus, God chose the route, not to test the people with environmental difficulties, but to avoid war with the Philistines:

> When Pharaoh let the people go, God did not lead them by the way of the land of the Philistines, although it would have been nearer, for God thought: "Lest the

people change their minds when they experience war, and return to Egypt."
Rather God led the people round about by way of the wilderness toward the
Reed Sea. The Israelites went up armed out of Egypt.

(Exod. 13:17–18, CHL)

As they approached the desert they were joined by a spectacular theophany:

> Setting out from Succoth they encamped at Etham on the edge of the desert.
> The Lord used to go in front of them, in a pillar of cloud by day to guide them
> along the way, and in a pillar of fire by night to give them light in order that they
> might travel by day and night. The pillar of cloud by day and the pillar of fire by
> night never moved from before the people.   (Exod. 13:20–22, CHL)

The Lord then turned the Israelites toward the Reed Sea[3] and hardened Pharaoh's heart so the king and his charioteers were foolish enough to pursue their
escaping slaves. The terrified Israelites complained to Moses saying they would
rather serve Egypt than die in the desert. Moses replied that the Lord would deliver
them.

As the children of Israel entered the Reed Sea, the pillar of cloud moved
behind them, as did an angel:

> Then Moses stretched out his hand over the sea and the Lord drove back the sea
> with a strong east wind all night, and turned the sea into dry land. The waters
> were split, and the Israelites went into the sea on dry ground, the waters forming
> a wall for them on their right and on their left. The Egyptians pursued after them
> right into the sea, all of Pharaoh's horses, chariots, and horsemen. At the
> morning watch, the Lord looked down upon the Egyptian army in the pillar
> of fire and cloud, and threw the Egyptian army into panic. He clogged the wheels
> of their chariots so that they moved forward with difficulty. Then the Egyptians
> said, "Let us flee from the Israelites for the Lord is fighting for them against
> Egypt."
>   Then the Lord said to Moses, "Stretch out your hand over the sea, that
> the waters may come back upon the Egyptians, upon their chariots and horse-
> men." So Moses stretched out his hand over the sea, and as morning broke,
> the sea returned to its normal course, and as the Egyptians were fleeing
> before it, the Lord shook the Egyptians into the sea.   (Exod. 14:21, CHL)

In the drowning of the Egyptians, we find both a mighty act of God and a
co-occurrence of natural events. "A strong east wind" drove the waters back, and
the bed of the sea, naturally very muddy, clogged the wheels of the chariots. The
view of the pillar of the cloud had caused a panic among the Egyptians without further divine action. The Israelites on their first excursion into wild nature found the
environment itself very friendly to their cause, as God worked, not only through

the water, but also through the marshy ground, to ensure their deliverance. The Egyptians, who had already experienced the force of environmental disaster in the seven plagues, found the Reed Sea no more friendly than the blood tainted, frog filled Nile had been. The walls of water on the right and the left imply supernatural intervention, but the book of Exodus also cites wind, the water, and the seabed in Pharaoh's demise.

The Sweetening Tree

After a victory celebration, the children of Israel entered the wilderness of Shur. Having traveled three days without water, they came to a spring at Marah, but found the water bitter.

> So the people murmured against Moses, saying, "What are we to drink?" He cried to the Lord and he directed him to a tree, and he threw it into the water, and the water became sweet.
> There he made for them a statute and an ordinance and there he put them to the test.
> (Exod. 15:24–25, CHL)

In this passage, we find a new motif, that of the wilderness murmuring. Although the people balked when crossing the Reed Sea, this text concerning Marah uses a particular Hebrew verb that means to grumble or complain, and originates in what Brevard Childs terms "a stereotyped language of complaints."[4] The camp was free from rebellion at this point, but the people were thirsty and unhappy. The children of Israel appear to have had short memories. Less than a week before, the sea parted, yet, when confronted with bitter water, they immediately assumed the trip had turned against them. Under the Lord's direction, however, Moses finds an environmental solution to an environmental problem. This difficulty, so quickly resolved, was hardly a test. The "test" in the story has instead to do with a "statute and an ordinance."

The "murmuring" motif has been the source of much of the negative commentary on Hebrew and Christian attitudes towards wilderness. The wilderness as a test, however, does not seem to be Yahweh's original reason for leading the people into the desert; Yahweh was actually avoiding a more serious problem with the Philistines. Their greatest difficulties lay not in the desert, but in occupied Canaan. Further, although wilderness conditions sometimes precipitate the "murmur," the "murmur" in the book of Exodus is always directed against Moses or against God. The Hebrews are not struggling with the environment nearly as much as they are struggling with their own prophet and with Yahweh.

The problem of testing appears again in the case of the manna and the quails. The children were hungry and voiced their usual "murmur" that they would have been better off staying in Egypt. Again there was divine provision, and ironically, the test lay in whether the people could follow divine instructions in utilizing it: "And the Lord said to Moses, 'I am going to rain bread from heaven for you, and the people shall go out and gather each day a day's portion, that I may test them, whether they will follow my instructions or not. On the sixth day, when they prepare what they have brought in, it shall be twice as much as they have ordinarily gathered." (Exod. 16:4–5, *CHL*).

In Exodus 16, both Moses and Aaron tell the Israelites that their murmurings are directed against God, and then, while Aaron was speaking to the whole community

> . . . they looked toward the desert, and suddenly the glory of the Lord appeared in the cloud. The Lord said to Moses, "I have heard the murmurings of the Israelites; say to them, 'Between dusk and dark you shall eat meat, and in the morning you shall have your fill of bread; then you shall know that I am the Lord your God.' "
> In In the evening quail came up and covered the camp and in the morning there was a layer of dew about the camp. When the layer of dew lifted, there appeared on the surface of the wilderness a fine flaky substance, as fine as frost, on the ground. When the Israelites saw it, they said to one another, "What is it?," for they did not know what it was. Moses said to them, "It is the bread which the Lord has given you to eat."   (Exod. 16:10, *CHL*)

The confrontation between Moses and Aaron and the people thus resulted in another theophany and in further provision via creation. The dew became bread, and quails appeared in the camp. The passages again mention testing, but not a test concerned with the wilderness itself; rather the test was in the statutes set by God. The practice of keeping the manna on the sixth day was a precursor of the Sabbath law.

It should be noted the book of Deuteronomy (8:2–3) does state that the wilderness sojourn and the lack of food was intended as a test of the children of Israel and thus gives a different impression than the Exodus account:

> And you shall remember all the way which the Lord your God has led you these forty years in the wilderness, that he might humble you, testing you to know what was in your heart, whether you would keep his commandments, or not. And he humbled you and let you hunger and fed you with manna, which you did not know, nor did your fathers know; that he might make you know that man does not live by bread alone, but that man lives by everything that proceeds out of the mouth of the Lord.

The purpose of God's delay in provision, according to Deuteronomy, was to teach the children to depend on him. The lack of physical sustenance in the desert became a means for God to force the children of Israel to turn to him and became a motive for God's mighty acts.[5]

Through the Exodus, the murmuring grew worse and became more and more a test of God. Yahweh, in return, continued to provide for the people, but set standards for faith. There can be no doubt the stress of the wilderness instigated the complaints—but the book of Exodus makes two things clear: (1) the murmuring was against God and God's mission; and (2) greater dangers lay ahead: the environment was the least of their problems.

The Call from the Mountain

The murmurings form an ironic precursor to the events at Sinai. The people continued to complain even though God provided the necessary food and water. Despite their spiritual inadequacies, the Israelites proceed further into the wilderness of Sinai, and then they camp "in front of the mountain" where the Lord calls to Moses saying:

> Thus you shall say to the house of Jacob, and announce to the children of Israel: "You have seen what I did to the Egyptians, and how I bore you on eagles' wings and brought you to myself. Now then, if you will really hearken to my voice and keep my covenant, you shall be my special possession among all peoples (for all the earth is mine), and you shall be to me a kingdom of priests and a holy nation."
> (Exod. 19:3–6, *CHL*)

Just as Moses received his call at a mountain in the desert, the entire people now receive theirs. They are to become "a kingdom of priests and a holy nation."

Following this community call to mission and ministry, we find one of the most spectacular theophanies reported in the Bible:

> The Lord said to Moses: "I am now coming to you in a thick cloud in order that the people may hear when I speak with you and then may trust you forever . . .
>
> "You must set boundaries for the people round about, saying, 'Beware of going up the mountain or even touching the edge of it. Whoever touches the mountain shall be put to death. No hand shall touch him, but he shall be either stoned or shot; whether beast or man, he shall not live.' When the horn sounds a long blast, they shall come up to the mountain."
>
> On the third day, when morning came, there were peals of thunder and flashes of lightning, and a dense cloud upon the mountain, and a very long blast of the horn, and all the people who were in the camp trembled. Moses led the people out of the camp toward God and they took their stand at the foot of the mountain.
> (Exod. 19:9, 12–13, 16–17, *CHL*)

Again, the purpose was hearing and speaking to God. The reception of the message was a combination of the individual—Moses as the primary contact with

Yahweh—and of the community. The entire nation prepared for the sacred event, and the entire nation saw the theophany.

The product of the interaction with the Divine at Sinai was far more than an affirmation of Yahweh's presence. The Lord gave Moses the Law, including the Ten Commandments, and established the priestly office and the pattern of worship for Israel. Out of the wilderness arose an ethical system and the framework for Hebrew religious practice. In the tradition of Abraham, the children of Israel made a covenant with God. They accepted the Law—"the book of the covenant"—in return for becoming a "holy people." Moses, to seal the covenant, built an altar at the base of the mountain, and "then Moses and Aaron, Nadab and Abihu [the sons of Aaron], and seventy of the elders of Israel went up [on the mountain], and they saw the God of Israel, and under his feet as it were a pavement of sapphire stone, like the very heaven in purity. Yet he did not lay hand on the leaders of the Israelites, but they beheld God, and they ate and drank." (Exod. 24:9–11, *CHL*).

According to the Old Testament scholar, Brevard Childs, the Hebrew verb translated "saw" means exactly that. He notes, however, that many interpreters have reduced the bluntness of the statement when translating, and have assumed the "seeing" was intellectual rather than literal.[6] The passage is similar to the visions of Isaiah and Ezekiel. Just as Isaiah only describes God's throne and never discusses the divine person directly, this text describes only the floor underneath God and avoids any direct discussion of the form of the deity. Although brief, the account speaks of glittering lapis lazuli, spanning a great area, and having the clarity of unpolluted desert skies. The description is also of a God whose majesty is incomprehensible and whose "seat" is beyond the reach of the particulars of a written report.

The text expresses surprise, that the elders, privy to such a sight, have not died. After the lightning of the first theophany in the cloud, the elders, on approaching Yahweh, have not found either destruction or terror, but rather enjoy a covenant meal. The image expresses acceptance by God and participation of the new nation in God's glory.

This is followed by yet another theophany, which was physically different in appearance from the manifestation of the sapphire floor. God calls Moses up to give the tablets of stone containing the Law, and so taking Joshua, while leaving the elders behind, Moses:

> . . . went up to the mountain and the cloud covered the mountain. The glory of the Lord settled on Mount Sinai, and the cloud covered it six days. On the seventh day he called to Moses from the midst of the cloud. Now the appearance of the glory of the Lord was like a consuming fire on the top of the mountain in the sight of the people of Israel. And Moses entered in the midst of the cloud

and went up the mountain, and Moses was on the mountain forty days and forty nights.

(Exod. 24:15–18, *CHL*)

The return of the fiery theophany was a reexpression of God's power and terrifying control over nature, while Moses' entry into the cloud was an expression of Israel's new relationship with God. The Lord gave Moses the tablets of the Law—tablets made by divine rather than human hands—as well as the plans for the tabernacle, soon to enclose the Holy of Holies.

We should recognize that God called the Hebrew people into the wilderness for a specific purpose—to worship at the mountain—and that this was, at the same time, necessary in order to initiate their new role as a nation of priests. The children of Israel had to "see" God before they could enter into this new ministry, thus the manifestations of the Divine were critical to their identity as a community and their transformation into an independent nation. We should also note that everyone, including women and children, saw the theophanies on the mountain and were therefore incorporated into this nation of priests. Although the Old Testament identifies spiritual leaders, such as prophets, and affirms a religious social class, the Levitical priesthood, the entire people saw the manifestations of Yahweh, and the entire people are therefore responsible for executing the will of God.

Moses Alone

When Moses tarried on the heights, the people grew impatient and called to Aaron to make them new gods. Ironically, while Moses remained forty days receiving instructions for the tabernacle, the site of true worship of Yahweh, the people were busy making the golden calf and participating in idolatry. After the glory of the theophany of the sapphire floor, the elders seem either to have forgotten the God they had encountered, or to have been too fearful of the people to halt the sacrilege. The Book of Exodus does not provide any evidence for environmental stress or other external troubles during this time interval—the people merely grew impatient waiting for Moses. Before Moses came down from the mountain, the Lord informed him of the infidelity of the nation. Moses pleaded with God and reminded Yahweh of the promises made to Abraham, Isaac, and Israel. Once again, in a wilderness setting, the covenant to increase the descendants of Abraham became a critical matter, but this time a mortal approached God about the covenant, rather than a divine being approaching a mortal. Moses reminded God of the promise to make Abraham's descendants "as many as the stars in the heavens." Despite his willingness to petition for the people, when Moses entered the camp, he became enraged and shattered the tablets "at the foot of the mountain" (Exod. 32:19, *CHL*).

The failure resulted in yet one further encounter with God. After the apostasy of the calf, God ordered Moses to take the children away from Mount Horeb and on to the land flowing with milk and honey (but occupied by the Canaanites, Amorites, Hittites, Perizzites, Hivites, and Jebusites). God also announced he would no longer accompany the people, lest he "consume" them on the way, due to their "stiff-necked" attitudes (Exod. 33:1–3). The people, recognizing their loss and seeking to repent, stripped off their jewelry and remained without ornaments from Horeb onward. Moses petitioned and requested not only that God's presence remain with the people, but also that Moses himself might see God. God was willing to grant both requests and gave Moses further access to the Holy Person and the ultimate theophany:

> Then he [Moses] said, "Let me see your glory." He answered, "I will make all my goodness pass before you and I will proclaim the name Lord before you. I will be gracious to whom I will be gracious and I will show mercy on whom I will show mercy. But," he said, "you cannot see my face, for no mortal man can see me and live." And the Lord said, "Look, there is a place beside me. Station yourself on the rock, and, as my glory passes by, I will put you in a crevice of the rock and shield you with my hand until I have passed by. Then I will remove my hand and you will see my back, but my face shall not be seen."
> (Exod. 33:18–23, *CHL*)

This wilderness passage gives us a detailed description of how one sees God. First, God is viewed in terms of attributes rather than a physical appearance. Moses "saw" God's goodness and God's glory. Second, a human can not see God's face and live. This refers not to a face, as a human countenance, but to God's essence and the source of God's Word. Third, seeing God is clearly related to the revelation of the name of God. When God's goodness passed, God's name was revealed. Fourth, both God's name and God's attributes are "vehicles of his essential nature, ... defined in terms of his compassionate acts of mercy."[7] The sentence "I will be gracious to whom I will be gracious and I will show mercy on whom I will show mercy" is in the same formula as "I am whom I am." In the wilderness, Moses encountered the depths of God's being. Moses requested and received a meeting with Yahweh and discovered not human features or flashing lightning, but mercy, goodness, glory, graciousness, and the Name of the One Who Is and Ever Shall Be.

The book of Exodus finishes with the reestablishment of the Law and the renewal of the broken covenant. Moses cut two new tablets for the Lord, and, as the Lord passed before him, the Lord proclaimed:

> "The Lord, the Lord, a God compassionate and gracious, long-suffering, rich in steadfast love and faithfulness, extending steadfast love to thousands, forgiv-

ing iniquity, rebellion and sin. Yet he does not remit all punishment, but avenges the iniquity of the fathers upon the sons and grandsons to the third and fourth generation." And Moses hastened to bow to the ground and he prostrated himself.

He said, "If I have gained your favor, O Lord, then may the Lord go in our midst, although it is a stiff-necked people, and pardon our iniquity and our sin, and take us as your own possession."

He said, "I hereby make a covenant. Before all your people I will do such wonders as have not been performed in all the earth or in any nation. All the people among whom you are living shall see how awesome is the Lord's work which I will do for you." (Exod. 34:6–10, *CHL*)

Again, the manifestation of the divine was linked to the understanding of what Yahweh is like. The covenant finished with God's promise to drive the people out of the land. Moses descended with the new tablets, and his face so shining the people could not bear to look at him.

The Book of Exodus ends with the execution of the tabernacle instructions. When the work was completed:

Then the cloud covered the tent of meeting and the glory of the Lord filled the tabernacle. Moses was not able to enter the tent of meeting because the cloud had settled upon it and the glory of the Lord filled the tabernacle. Whenever the cloud was lifted from the tabernacle, the Israelites used to set out on their journey, but if the cloud was not lifted, they would not set out until the day that it was lifted. The cloud of the Lord was over the tabernacle by day, and there was fire in it by night, in the sight of all the house of Israel through all their journeys. (Exod. 40:34–38, *CHL*)

At the end of Israel's sojourn at the foot of Horeb, the theophany moved into a new residence. The imagery of the cloud and the fire combines both the pillar of the cloud and the appearance of God on the top of the mountain. The text implies God had come down from the mountain to reside with Israel. God's being had combined with the mission of the cult.

Sprouts, Blossoms, and Almonds

From Horeb onward, the people continued their murmuring and repeatedly failed to execute the will of God. Before reaching Canaan, for example, the people again craved meat and complained to Moses that they had had much better food in Egypt. This time the Lord sent a wind bringing quails from the sea and let the aeolian forces deposit them two cubits deep around the camp. But, "While the meat was yet between their teeth, before it was consumed, the anger of the Lord was kindled against the people, and the Lord smote the people with a very great plague. There-

fore the name of that place was called Kibroth hattaavah, because they buried the
people who had craving" (Num. 11:33–34).

A short time later, the people reached the borders of Canaan, and Moses sent
spies to assess the land. When the spies returned they brought back both the rich
fruit of the well watered hills and discouraging reports of the great strength of the
peoples already inhabiting the region. The children of Israel began murmuring once
again and, thinking their enemies were too strong to overcome, rebelled and moved
to chose a captain to lead them back to Egypt. Moses pleaded with God to forgive
them, and the Lord did not destroy the people, but sent them back into the wilderness for forty years in order that the rebellious generation would pass away before
Israel entered the Land. Another plague followed (Num. 13–14).

For the remainder of their sojourn in the wilderness, the children of Israel
continued to struggle with the Lord and the Lord's ordinances. God's patience,
so evident at the onset of the Exodus, was replaced by a series of disciplinary
measures. When Korah and his sons doubted the legitimacy of the Levitical priesthood, "the ground under them split asunder, and the earth opened its mouth, and
swallowed them up . . ." (Num. 16:32). When the children of Israel traveled back
towards the Reed Sea and began to murmur once more about the quality of the food
and the lack of water, "the Lord sent fiery serpents among the people, and they bit
the people, so that many people of Israel died" (Num. 21:6).

In these passages, we find the environment unleashed against the children of
Israel, much as it was against the stubbornness of Pharaoh. Incidents beneficial to
the people, such as the miring of the chariots in the Reed Sea, the sweetening of
the waters at Marah, and the coming of the manna, have been replaced by earthquakes, inedible quails, plagues, and fiery serpents. The plagues began with the
incident of the golden calf and arose again with each new apostasy. Nature was
forever on the Lord's side even if the people were not.

Not all the environmental responses after Horeb were negative. At the height
of the conflict between Moses and Aaron and the congregation, nature spoke for
the authority of the Levitical priesthood through one of the most beautiful mighty
acts of God in the Scriptures. After the people had complained about the death of
Korah and being struck by a plague, the Lord instructed Moses to get each of the
heads of the twelve tribes to give Moses a rod to deposit in the tent of meeting.
The next day, when Moses returned, "the rod of Aaron for the house of Levi had
sprouted and put forth buds, and produced blossoms, and it bore ripe almonds"
(Num. 17:8), indicating God's choice of Levi for the holy office.

The Lord also provided water from the rock at Meribah (meaning "contention"). Notably in this case, the people really did lack water, as

they had at Marah, and the Lord produced it without any discipline for their complaining. In the Exodus, we can not, in fact, consider the wilderness evil if it was repeatedly executing Yahweh's will. One post-Enlightenment concept of wilderness is of an uncivilized and ungodly land that needs human presence to be "purified." The wilderness in the Exodus is almost the opposite. The people leaving Egypt are unwilling to obey God and under the continuing influence of an ungodly culture. The people do not need the wilderness *per se* to be purified, but they do need to meet God in the wilderness and at the mountain to better understand who their God is. Wild nature assists the Creator in driving Pharaoh back, in receiving the theophanies, and in providing physical evidence—such as the fiery serpents and the blooming staff—of the Righteous One's will.

Twelve Tribes in the Wilderness

The wilderness sojourns of Moses and of the children of Israel had much in common with the *Genesis* or *foundational model.* First, the "seeing" and "hearing" in the wilderness were very clearly repeated. God did not call Moses until he had left Egypt and traveled to Horeb. The people both see and hear God in the theophanies at Sinai where there was continued direct contact with the Divine. The communications of the wilderness may be summarized in Moses' day-to-day supervision of the tent of meeting:

> When Moses entered the tent, the pillar of cloud would descend and stand at the entrance to the tent while he spoke with Moses. When all the people saw the pillar of cloud standing at the door of the tent, all the people would rise and bow low, each at the entrance of his tent. The Lord would speak to Moses face to face, as one man speaks to another . . . (Exod. 33:9–11, CHL)

Second, covenants and promises were important in the wilderness setting. Wilderness became a site for vision into the future and for deepening relationships with Yahweh. Third, stress, from lack of food and water or hostile military encounters, precipitated God's providence, usually mediated through the environment. God provided water for the people when they entered the wilderness, just as the Sustainer had provided water for Hagar, generations before. Fourth, the Exodus produced spiritual transitions in the wilderness, first for Moses, on receiving his call at Horeb, and then for the people as a whole, as they changed from slaves of the Egyptians to a nation of priests.

The *Exodus* or *community model* of wilderness experience also has its own distinctive elements. Of importance are the theophanies, which were among the largest scale and the most integrated with natural phenomena of any such divine appearances described in the Scriptures. In Genesis, the messenger who spoke to Hagar in the wilderness of Shur probably appeared in human form, and Jacob's stairway to heaven was clearly recorded as a dream. In Exodus, in contrast, the pillar of cloud and the lightning on Sinai were visible to the entire nation. Seventy elders saw God through the sapphire floor. Moses' vision of God from the crevice in the rock was probably the most direct contact with the Holy One in all Hebrew history, prior to the time of Christ. For a brief period in the wilderness, the entire nation of Israel assumed a prophetic visionary role and was privileged to see and hear the Almighty God.

The expansive nature of theophanies may have originated in a need for the Lord to communicate with the entire community or in a need for Yahweh to communicate divine power to a people raised under the Egyptian Pharaohs. Much of the murmuring repeated the notion that things had been better for the children of Israel in Egypt. This was not, to the ancient mind, a mere environmental and economic statement of fact. Provision would have been viewed as a product of a deity's power. When the people complained in Num. 11:4–5: "O that we had meat to eat! We remember the fish we ate in Egypt for nothing, the cucumbers, the melons, the leeks, the onions, and the garlic . . . ," they were falling short, not just because of an uncontrolled craving, but because they refused to recognize the basic person of the God they had seen. Yahweh not only overpowered Pharaoh, Yahweh could control the Nile (when it turned to blood) or the Reed Sea (when the waters parted). The pillar of the cloud and the fireworks on Horeb were expressions of Yahweh's creative power, as was the falling manna. The children of Israel, however, could not raise themselves above material concerns. God's justice and goodness—as well as their own role in God's purpose—escaped them.

This leads us to a second difference with the Genesis narratives. When Hagar, Abraham, and Jacob encountered the Lord in the wild nature, they all accepted the promises made by God and praised God for what transpired. Wandering in the wilderness of Shur, Hagar showed a deep understanding of how gracious God had been to her. The children of Israel, being servants of an unkind master, just as she had been, showed no appreciation or insight when the Lord sweetened the water at Marah. Instead, they continued to complain. There is no evidence in the Pentateuch that the children of Israel had ever been close to death due to lack of water or that they were ever without food after the falling of the manna, yet they murmured constantly. Hagar prepared to watch Ishmael die without cursing God or Abraham, and Abraham walked faithfully up Moriah, knife in hand. As much as the wilderness sojourns of Genesis bred faith, the wilderness sojourn of the Exodus grew into a long, weary trip from the people's lack of it.

The story of the Exodus adds an important element to the concept of God's providence and God's mercy as found in the wilderness narratives of Genesis. Although the faith of Abraham wavered and Hagar attempted to escape from Sarai, the spiritual blindness, the complaints, and the rebellion of the Exodus are missing among the members of Abraham's household. In Genesis, God's mercy takes the form of springs of water, a ram provided for sacrifice, or the promise of a future generation. In Exodus, God's goodness and mercy are much more all-encompassing. This can be seen both in God's just and patient treatment of a rebellious and troublesome people and in the giving of the Law. God only asked from Israel what was good for them to do in any case. In the wilderness, we find God revealed as "merciful and gracious, slow to anger, and abounding in steadfast love and faithfulness . . . forgiving iniquity, transgression, and sin . . . "[8] In the wilderness, we can begin to see past the limits of God's interactions with faithful Abraham and into God's interactions with a people who were constantly rebelling and pulling away.

In Genesis, God's wilderness promises were made to individuals and primarily concerned the future of their offspring and relationships to God. In the Exodus, God's wilderness covenants and call were much more complex. Both Moses and the people were drawn into a cosmic mission, and both resisted the roles God assigned them. The interaction of Moses with the theophany of the burning bush was both a repetition of the Genesis theophanies and a new event—the prophetic call. In Genesis wilderness was primarily associated with assuming a new relationship to God and with promises, while in the Exodus wilderness was associated with preparation for mission.

The role of the wilderness itself is an important question here. There is no evidence from the early chapters of the Book of Exodus that God intended the wilderness itself to be a test, although the concept of shortage of food as a test is found in Deuteronomy. The stresses of the Exodus were always limited by God's immediate action. In a spiritual sense, stress is not necessarily a bad thing—if it provides an understanding of God's character and builds relationships with the Divine.

The element of testing must be balanced with God's choice of the wilderness, as a "safe" zone, free of both Pharaoh and the Philistines. The wilderness was also a place where Yahweh could be manifested to the entire Hebrew people without interference from Egyptian or Canaanite cultures. The Lord originally led Israel into the wilderness to thwart Pharaoh and to bring the people to the mountain of God to worship. Israel's lack of obedience to Yahweh slowly turned the wilderness into a series of difficult experiences. At the beginning of the Exodus, we see the wilderness of the loving Creator God. The sea parted, manna fell from heaven, water flowed in the desert, and the mountain burned with the fire of the Lord. As the children rebelled and refused to acknowledge the power and will of Yahweh, the wilderness became a place of testing. The testing, however, was centered on God's ordinances rather than on the environmental conditions. In every case of legitimate need, God provided. After Horeb, the wilderness became more difficult as the environment responded to the attitudes and actions of the nation. The earth opened, plagues swept through the camp, and fiery serpents bit and killed. Israel, called because of promises to Abraham, Isaac, and Jacob, found sin spreading through the community and resting on the people until the third and fourth generation.

The isolation provided by the wilderness seems necessary to God's purposes in the Exodus. The amount of spiritual baggage brought by the people from Egypt indicates direct encounter with God would have been impossible in the realm of the Nile. Free of social influences and other religions, the wilderness became an unclouded mirror for the spirit of the newly forming nation. In the numerous conflicts with God, the people had no one to blame but themselves. One might say the stress and barrenness of the desert were partially a platform for learning about God's providence, as they had been for Hagar, and partially a place where Israel could see how far they had strayed from the faith of Abraham. The material nature of Egyptian culture and of Egyptian gods had settled very deeply into the being of the twelve tribes. If the sojourn in the wilderness did not remove the idols, it at least made the problem of idolatry clear.

One of the most important themes of the Exodus is the liberation of the Hebrew people from the social and political oppression of Egypt. Yet, Pharaoh's direct political control ends after the crossing of the Reed Sea, and the children of Israel could have marched on to Canaan in short order. The extended wilderness sojourn was partially a product, not of Pharaoh's military might, but of the bondage created by the mental, moral, and spiritual models nurtured by the long period of slavery. The Exodus assumes that physical freedom is only half the battle. The person who is not spiritually free as well will foolishly want to return to settled life in Egypt the moment the "environment" becomes difficult or unpredictable. The person who is not spiritually free may also carry Egypt and the Egyptian gods into the Promised Land. Liberation begins when the people (or their prophet) first hear God's call to confront Pharaoh. It is a mistake, however, to assume liberation is consummated when the sea washes over the retreating army. Liberation is consummated when the people accept Yahweh's Law rather than Pharaoh's law and begin to believe they belong to the Lord rather than to Pharaoh.[9]

The Call and the Community

The book of Exodus is, in fact, an excellent study of the response of a group partially freed from its cultural and social underpinnings and locked in a difficult journey through unpredictable environments. On such a group expedition, the environment may be troublesome, but personalities can often become a primary source of disruption. Conversely, a group can provide mutual encouragement and can accomplish more than an individual. When a group is under stress in wilderness conditions, the response will depend not only on the interior strengths and weaknesses of its members, but also on how the group begins to operate as a "community." Fortunately, groups are more than the sum of their parts. Leadership makes a difference, as does the temperament of the assembled travelers. The sequence and types of stresses can be critical. As each barrier is overcome, the "community" gains knowledge and an ability to deal with similar situations. The challenges often lie not so much in the difficulties posed by the physical environment, but in the chance for human failure. In the wilderness, you can see people at their heroic best and their incompetent worst.

In a spiritual framework, the response of the community also differs from that of the individual. The ability of the community to see and hear, for example, often differs from that of its leaders. Within the community are the young, the inexperienced, the doubting, and the selfish. Arguments arise in cases where an individual acting alone could make an easy decision. The community, however, can be a source of strength and protection. When God has a major job to be accomplished, the community, more often than not, is the only appropriate vehicle. The Exodus displays the tensions, and the amount of struggle necessary to make the community of the faithful fully functional.

Raising up the holy nation, the giving of the Law, and the establishment of the cult may seem like events far removed from wilderness conditions. Many passages in the Law, in fact, concern settled agriculture and situations more appropriate to life in Canaan. The possible dates of portions of the text aside, God chose to provide the terms of the covenant while "alone" with Israel in the desert. The text gives the impression the religious system was generated by the Lord's instructions, rather than by imitating neighboring peoples. This enhances the concept of the special call of Israel. The giving of the Ten Commandments on the desert mount also suggests their divine origin and universal applicability. The law given in the wilderness was much more God's Law than human law.

The wilderness became a culturally neutral ground, where the Lord formed a "holy" nation. There can be no doubt that, when the people left Egypt, they had lost the heritage and faith of Abraham. They saw them-selves not as the people of Yahweh, but as slaves of Pharaoh. They cared more about cucumbers, leeks, and garlic than about service to God. The stay at Horeb reacquainted them with the God of their progenitors and gave them a new calling as a nation of "priests." God had promised Abraham on Moriah: ". . . all the nations of the earth shall find blessing through your seed . . ." (Gen. 22:18, *WS*). At Horeb, Yahweh attempted to implement this blessing. When God could not free the people from their stubbornness and lust for the Egyptian life, God let the older generation pass away in the wilderness and then took the new generation, who had known only the desert and the presence of God, into the land.

The fact that the wilderness served God well—and was probably the only suitable place for God to act—is not reflected directly in the vocabulary of the text, but rather must be inferred from the context of the events. Throughout most of the Pentateuch, the desert is referenced without further description, either positive or negative, although there are passages that use terms such as "vast and terrible wilderness" (Deut. 1:19, *JB*) to stress the physical difficulties of the Exodus. The ultimate purpose of these specific passages, however, is not to record the attitudes of the people towards their environment, but rather to record their attitudes toward God, in the context of narratives that demonstrate God's love and salvation. All interpretation of the "meaning" of the wilderness sojourn must consider the overall purpose of the texts.

A related issue—and a theologically difficult one—concerns the notion of holy places. Several religious traditions in the ancient Near East incorporated mountains as divine residences or other forms of sacred location. The god Baal, in Canaanite religion, lived on a mountain and descended in a cloud. Other traditions recognized temples or mountains as cosmic centers, where heaven and earth were linked together.[10] The Genesis and Exodus accounts mention two mountains of God—Moriah and Sinai (assuming Horeb and Sinai are the same place). The treatment of the holy nature of the site is much more strongly developed in the story of Exodus. The theophany of the burning bush informed Moses he was standing on holy ground. The preparations at Sinai included a warning to the people not to touch the mountain:

> You must set boundaries for the people round about, saying, "Beware of going up to the mountain or even touching the edge of it. Whoever touches the mountain shall be put to d e a t h . . . ." Moses came down from the mountain to the people and had the people prepare themselves and they washed their clothes. And he said to the people, "Be ready by the third day; do not go near a woman."   (Exod. 19:12, 14–15, *CHL*)

The implication, with the mention of washing and ritual purity, is certainly of a holy place. The question is: Did the Hebrews believe one could only see God at this particular location? The text has an interesting tension to it. The pillar of cloud was with the people from the exit from Egypt onwards and was strongly associated with Yahweh God as well as with an angel of God. Yet in order to see Yahweh and to worship, the people went to Sinai, where they also encountered a fiery theophany. It is as if the presence of the Lord went everywhere in the wilderness with the people, yet could be best comprehended at a specific isolated mountain. When the Lord threatened to leave the people, Moses petitioned for God's continued presence and the glory of the Lord entered the tabernacle.

Form critics might argue at this point, the two different presences are the product of two different accounts of the escape from Egypt. Even if this is the case, it does not explain why the spatial conflict was accepted by the editor of the text. To the ancient Hebrews, however, who were not given to Platonic thought or the logic of mutually exclusive cases, God could have been traveling with the camp and could also have appeared more dramatically at the mountain. One theophany would not have precluded another. Note that the transfiguration of Christ on the mount is also a case of God "who is always with us" assuming an exceptional other-worldly aspect at a specific time and location.

If one were to consider the Bible folklore, one might suggest the story of Sinai represents the transition from gods inhabiting specific earthly localities to gods who transcend the physical. This explanation is not adequate, however, to the complexity of interaction between the Divine and the human in both Genesis and Exodus. The God of the Pentateuch is universal and omnipotent. Yahweh spoke to Abraham throughout his wanderings and tracked down Hagar in the desert. Careful analysis suggests the landscapes of these stories refer to real places, but they are, at the same time, "spiritual landscapes." Certain types of things are more likely to happen in certain types of places, in certain types of social settings. Deserts and mountains are better places than cucumber fields and royal palaces to see God. This may be a carryover from very ancient religious traditions, but it may also reflect a human reality that isolation and natural settings favor reflection and unprejudiced spiritual vision.

A comparison of the wilderness adventures of Genesis and Exodus suggests the Genesis narratives are very ancient and contain the oldest wilderness motifs. The vocabulary of the Genesis texts retains traces of animism and the presence of sacred natural features and the outcomes of the narratives proclaim the power of individual wilderness sojourns and dreamings—themes found even in Stone Age cultures. The remarkable theophanies of the Exodus, although also very ancient, speak more to the concerns of "organized religion." The tremendous outpourings of fire and flame at Horeb infer power over other nations and suggest a demonstration of authority over competing gods. The Genesis wilderness sojourns, in contrast, present individuals interacting with the divine in an open and culturally uncluttered landscape. The first wilderness sojourns of Genesis were primarily visionary, whereas much of the Exodus was dedicated to cultural purification. Leadership in Genesis is "reproductive" and visions and promises are to individuals and family groups. Leadership in Exodus is "prophetic" or "priestly" and the visions and promises take the form of political events and the development of written law. The Exodus associates the Levitical priesthood with divine power working through nature, whereas in Genesis, divine power over nature is not associated with those in control of cultic functions.

In the Exodus, the nation behaves very much as an individual behaves in Genesis. We find Moses first encountering the burning bush by himself and then returning with the people, so that the entire community can share the visionary experience. Isolation and stress in the Exodus apply to the children of Israel as a whole, and things happen that could not have occurred in Egypt or even in the Promised Land. Although the Exodus brought the entire nation into the wilderness, it was still a solitary experience where other nations were excluded from participation.

We can summarize the characteristics of the *Exodus* or *community model* of wilderness sojourn:

1. The community could see and hear God in the wilderness, just as the individuals in Genesis could.
2. The wilderness was the site of the covenant, the giving of the Law, and the establishment of the Hebrew religion.
3. God's mercy was very evident in the wilderness, not just in supplying physical needs, but also in forgiving sin and transgression.
4. Prior to the time of Christ, wilderness theophanies provided some the best knowledge of the person of God.
5. The wilderness served as an environment for freeing the children of Israel from Egyptian desires and practices and for reintroducing to them the faith of their ancestors. The stresses of wilderness travel helped to clarify matters of faith and belief. Isolation or solitude was important to their understanding of their former dependence on Egyptian culture and their need for renewal as children of Abraham.
6. Community interactions with God in the wilderness in the Exodus were fraught with much more strife than the individual interactions in

Genesis. It is notable that Moses, as an individual leader, had far fewer problems conforming to the will of Yahweh than the people as a whole. The failures in the wilderness were not the result of the level of environmental stress, they were due to the inadequate response of the people. The wilderness became a mirror where the people could see themselves as they actually related to the divine.

7. The wilderness was the location of the first prophetic call, that of Moses, and of the transition of the children of Israel into a nation of priests, "dedicated to God's service among the nations."[11] The wilderness was thus the site of the assumption of new spiritual roles, or, more specifically, the contact with the divine that consummated these.

8. The relationship of the children of Israel to the wilderness environment was a reflection of their relationship to God. Although the Exodus narratives may be rooted in several sources, their portrayal of the environment is very consistent: wild nature performs the will of God. If the children of Israel have been obedient, nature will be kind; if not, the people encounter earthquakes and fiery serpents, which are equally the will of God. This is consistent with the Hebrew concept of Yahweh as Creator of all things. The cooperation between wild nature and Yahweh serves as a proof of who the God of Israel actually is and how far the power of Yahweh extends (outside the bounds of the land). Since the wilderness, by its very nature, has never been subject to human dominance, every natural assistance offered to the children of Israel must have been rooted in the divine. [12]

Then the Cloud Covered

From a contemporary perspective, a repetition of the divine manifestations of the Exodus is improbable, and the wilderness revelations of the Exodus do not need to recur. Because of the length and social structure of the Sinai sojourn alone, the Exodus model is, of all the spiritual models of wilderness experience, the most difficult to implement in the contemporary church. The Exodus model suggests, however, that wilderness experience may be very valuable for bringing the Christian community to terms with its faith. The stripping away of material support resulting in increased dependence on God, and the temporary withdrawal from "un-Godly" cultural influences that creates an opening for God's work are both potentially beneficial. The Exodus may be repeated in the sociopolitical realm, and in this case, the wilderness wanderings may critical to loosening the spiritual bondage of Egypt.

Perhaps the most important wilderness theme of the Exodus for contemporary spirituality is that of purification and stripping away idolatry. The Exodus suggests that removing the "faithful" from a confining cultural context and placing them at the mercy of the elements (or, more properly, where the mercy of God is most clearly visible) will result in a repeated showing of weakness and longing for the comforts provided by other gods. In a first-world context, the wilderness can help to provide the needed cultural distance to reflect on the bondage created by the "idols of the west" and "the gods of the industro-technical realms." In a third-world context, the wilderness and its uncertainties can help to break the internal and psychological bondage created by economic or political oppression. Although forty years in the wilderness might seem like a very long stay, in the life of a nation it was a short but critical interval, necessary to prepare the nation for the Promised Land. A major transition was accomplished in a single generation. Without this period of cleansing and renewal, the nation would certainly have been absorbed by Egyptian culture and would have slowly disintegrated into a lower social caste scattered among the temples of the Nile.

The Exodus was written for the Hebrew people who were settled in the Land, and perhaps drifting away from service to Yahweh. It is, however, about the Hebrews when they lived in a land not their own and served other gods. Whether we spiritualize the text and look at its directives about belief, obedience, and service to Yahweh, or we consider its socio-political message, the Exodus is about the love of God as demonstrated in the wilderness. The themes of the Exodus are applicable to both the individual and the community. Both need to see themselves as completely dependent on God, and both need the opportunity to explore the person of God without cultural interference. Caught in our contemporary passion for "book-learning," we sometimes neglect the spiritually essential confrontation with the self and with God unfettered by our preconceived notions of the Holy. The wilderness, with its freedom from the human constructs, is the traditional site for this engagement.

4

The Paw of the Bear, the Paw of the Lion
(David and Jonathan)

The Slippery Crag

Wilderness experience with spiritual impact is not necessarily "religious" wilderness experience. The Bible not only reports conversations with angelic messengers in the wilderness, it also describes more mundane activities, such as herding sheep or conducting military maneuvers, and relates these to the will and purposes of God. Time spent in wild settings may influence one's relationship with God through much more subtle interactions than theophanies and visionary dreams. The problem of having to survive in landscapes without developed human support systems, at the very least, builds character and can, in the right social framework, teach dependence on God. The skills and attitudes learned in the wilds carry over into other settings. Long months in the wilderness teach the extent of individual strengths and foster trust in God, both of which help when one must take the initiative to solve problems.

The exploits of David and Jonathan in 1 and 2 Samuel form one of the most extensive series of wilderness adventures in the Scriptures, yet not a single theophany appears. A major purpose of the book of Samuel is to discuss the qualifications for a king of Israel and to describe the relationship of that king to God. David and Jonathan, the heroes of the story, are men who learned to survive on the steep, arid slopes above the Jordan Valley and the Dead Sea. They both became faithful servants of Yahweh and outstanding leaders of the people.

As most of us know, David spent his boyhood as a shepherd in the hills of Judea. The wilderness adventures in 1 Samuel do not, however, begin with David. Jonathan, the son of King Saul, is the first to prove his prowess in wild

47

country. The wilderness narratives then continue through David's victory over Goliath and his long struggles with Saul, and conclude when David, as deposed king, defeats his own son Absalom in the forests of Ephraim. The exploits recounted in Samuel report neither sin nor failure in the wilderness, at least on the part of conquering David. Throughout the life of the greatest king ever to rule over Israel, rough terrain and isolation were on the side of God's anointed.

At the time the prophet Samuel selected Saul as king, the Philistines had military control over Israel. After defeating Nahash the Ammonite, who had besieged the Hebrew town of Jabeshgilead, Saul mustered forces to engage the Philistines. Jonathan initiated the conflict by defeating a Philistine garrison at Geba. In reaction, the Philistines gathered a large force. The people of Israel, fearing reprisal, hid themselves or fled. Saul did not retreat, but waited for Samuel to come and renew the kingdom at Gilgal, a holy site. When Samuel did not appear as expected, Saul grew impatient and offered the burnt offering himself. Since Saul was neither priest nor prophet, this sacrifice was profane. Samuel, arriving too late to prevent Saul's mistake, prophesied the reign of Saul would not continue (1 Sam. 12–13). Three Philistine companies invaded Israel. Saul's troops, badly outnumbered and armed primarily with sharpened farming implements such as axes and sickles, prepared to face the superior force.

At this desperate juncture, the book of 1 Samuel recounts the first of the "kingly" wilderness exploits. Jonathan, accompanied only by his armor bearer, approached a Philistine garrison camped on a high shoulder overlooking *wadi es-swenit,* a canyon extending down through the wilderness into the Jordan valley. Scripture describes the formidable terrain: "In the pass, by which Jonathan sought to go over to the Philistine garrison, there was a rocky crag on the one side and a rocky crag on the other side; the name of the one was Bozez, and the name of the other was Senah. The one crag rose in the north in front of Michmash, and the other on the south in front of Geba" (1 Sam. 14:4–5). The names of the rocks—Bozez, "the slippery one," and Senah, "the thorny one"[1]—emphasize the impassible nature of the landscape. Any approach to the camp, other than from the well-watched road to Michmash, would be extremely difficult. Even if Jonathan and his armor bearer did reach the camp, the chances of surviving the engagement with the larger force were next to none.

Approaching the pass, Jonathan and his armor bearer showed themselves to the enemy, and Jonathan declared that, if the men of the camp told them to come up—rather than declaring that they would go down to meet them—it was a sign that "the Lord has given them into our hand" (1 Sam. 14:10). The men of the Philistines hailed Jonathan and said, "Come up to us and we will show you a thing" (1 Sam. 14:12). Jonathan took this as an indication of God's will. Instead of attacking the enemy directly, the king's son and his armor bearer made their way down into the wadi and thus crept out of sight of the Philistines.[2] Jonathan chose a difficult route over a rocky crag and then "climbed up on his hands and feet, and his armor bearer after him" (1 Sam. 14:13). The two brave Israelites took the garrison by surprise, killed twenty men, and routed the remainder.

The theme of Samuel is of dual strengths. Jonathan was courageous, strong, and agile—a man able to ascend over precipitous terrain and accomplish the unexpected. At the same time, Jonathan was a man dependent on God. He waited for a sign from the Lord and, after pulling himself over the top of the cliff, sent the "fear of God" sweeping through the Philistine camp. Even with superior rock-climbing abilities, two men should not have been able to put the Philistines to flight, but God intervened. Jonathan's wilderness skills were not based on self-confidence; they were based on God-confidence.

Jonathan's victory over the Philistines was followed by a curious case of "wilderness provision." Saul, seeing the fleeing Philistines, consulted the priest, then rallied the people, and pursued the confused enemy force. He—perhaps trying to court the favor of God—laid an oath on the people, saying, "Cursed be the man who eats food until it is evening and I am avenged on my enemies" (1 Sam. 14:24). Jonathan, having been separated from the main force and not hearing the oath, was passing through the forest and came upon some wild honey. A good soldier, he did not stop, but dipped his staff into the honeycomb and ate some of the sweet substance. According to 1 Samuel, "his eyes became bright," indicating the honey had a positive effect and helped relieve his weariness. One of Jonathan's companions who had heard the oath told him of his father's curse. Jonathan replied: "My father has troubled the land; see how my eyes have become bright, because I tasted a little of this honey. How much better if the people had eaten freely today of the spoil of the enemies which they found; for now the slaughter among the Philistines has not been great" (1 Sam. 14:29–30).

The people of Israel pursued the Philistines and, faint from hunger, fell upon the spoil. The army was so famished, they ate the sheep and oxen with the blood, which was, to the ancient Israelites, a deep sin. Saul, distressed, built an altar to slay the sheep and oxen properly, and then inquired of God whether he should further pursue the Philistines. The Lord gave no answer, indicating further fault among the people. Through the mystical means of the Urim and Thummim, the lot fell to Jonathan, who confessed he had eaten honey while under his father's oath. Saul would have been willing to kill his own son, but the people, recognizing Jonathan's role in the victory, ransomed Jonathan (1 Sam. 14:31–46).

The tale is a strange one. Saul's motive for forcing his army to abstain is obscure. Although Saul's curse was probably spiritually motivated, he inhibited military action. The text implies Saul did not bring as complete a victory as might have been possible. Jonathan, a hero throughout, was not informed of his father's decision. The text gives the impression Saul had not only fallen from Samuel's and God's favor, but also the land itself was willing to "disobey" Saul. The forest provided Jonathan with strength for further pursuit of the Philistines when Saul had denied his army the blessing of sustenance. Wild nature was on the side of righteous Jonathan.

The Paw of the Lion

In the aftermath of Saul's failures, Yahweh sent Samuel to seek another king for Israel. God directed Samuel to Bethlehem to the household of Jesse, but none of the seven sons of Jesse present were satisfactory before God. Samuel asked if all were present, and Jesse replied that one was still keeping the sheep. Samuel requested that they send for him, and, when he came in, Samuel saw that "he was ruddy, and he had beautiful eyes and handsome. And the Lord said, 'Arise, anoint him; for this is he' " (1 Sam. 16:12). The biblical text introduces David in humility—he was almost forgotten—and in virility[3]—he is ruddy and handsome. His wilderness vocation of herding kept him away from a potentially important meeting, but the Lord ensured David was considered. The text also implies David was physically blessed, and his appearance was enhanced by his outdoor life.

In 1 Samuel 17, David's wilderness background comes to the fore in the confrontation with Goliath. Jesse sent David to the scene of the battle to resupply his brothers rather than to fight. David was, at this point in his life, still tending the sheep. He had left them with a keeper while he traveled to the encampment in the Elah valley. Goliath had already come out to challenge the men of Israel. David, standing by, inquired: "What shall be done for the man who kills this Philistine and takes away reproach from Israel? For who is this uncircumcised Philistine, that he should defy the armies of the living God?" (1 Sam. 17:26).

> Now Eliab his eldest brother heard when he spoke to the men; and Eliab's anger was kindled against David, and he said, "Why have you come down? And with whom have you left those few sheep in the wilderness? I know your presumption, and the evil of your heart; for you have come down to see the battle." And David said, "What have I done now? Was it not but a word?"
>
> (1 Sam. 17:28–29)

Eliab mistook David's intention and mocked him for his wilderness vocation, which shortly would prove to be his strength.

David went to Saul and proposed to fight Goliath. Saul protested that David was still a youth and no match for the skilled man of war.

> But David said to Saul, "Your servant used to keep sheep for his father; and when there came a lion, or a bear, and took a lamb from the flock, I went after him and smote him and delivered it out of his mouth; and if he arose against me, I caught him by his beard, and smote him and killed him. Your servant has killed both lions and bears; and this uncircumcised Philistine shall be like one of them, seeing he has defied the armies of the living God." And David said, "The Lord who delivered me from the paw of the lion and from the paw of the bear, will deliver me from the hand of this Philistine."   (1 Sam. 17:34–37)

This passage claims the destruction of two wild predators which were, by Iron Age standards, risky to subdue. The text does not state that David used a bow or a sling to accomplish this and suggests close combat without mentioning the use of a metal sword or spear. A brown bear or a lion would be very difficult to kill with a club or a shepherd's staff. The lack of detail in this text thus accentuates the element of divine dependence.

David turned down the offer of Saul's armor and chose instead five smooth stones for his sling. When the giant saw David, "a youth, ruddy and comely in appearance" (1 Sam. 17:42), he was disdainful and threatened to "give [David's] flesh to the birds of the air and the beasts of the fields" (1 Sam. 17:44). Undaunted, David confronted the giant and proclaimed:

> You come to me with a sword and with a spear and with a javelin; but I come to you in the name of the Lord of hosts, the God of the armies of Israel, whom you have defied. This day the Lord will deliver you into my hand, and I will strike you down, and cut off your head; and I will give the dead bodies of the host of the Philistines this day to the birds of the air and to the wild beasts of the earth; that all the earth may know that there is a God in Israel, and that all this assembly may know that the Lord saves not with the sword and spear; for the battle is the Lord's and he will give you into our hand.   (1 Sam. 17:45–47)

David did as he promised, first striking Goliath with a stone and then cutting off his head.

The exploit parallels Jonathan's attack on the Philistines in several ways. First, both David and Jonathan were able to move against the superior force when Saul was not. Second, both David and Jonathan displayed wilderness skills. In Jonathan's case, past training was a probable element in climbing the crags. In David's case, his own statements credited his wilderness experiences with preparing him for the engagement with Goliath. David had, at a very young age, overcome both lions and bears. Third, both warriors relied on Yahweh and gave the Lord of Israel credit for the victory. Despite their maturing masculinity and their own intrinsic strengths, both David and Jonathan were God-dependent.

It is also worth noting that Goliath's threat to feed David to the birds and beasts speaks in a subtle way of who will be defeated in a contest for the ground under their feet. David committed the Philistine host to the earth, thus countering the Philistine claim to the land, and stated that, through his victory, "all the earth" would know of the God of Israel. David comprehended the importance of faith in God, not just in the land of the Israelites, but on a universal level.

Wildgoats' Rocks at En-gedi

The lives of Jonathan and David converge after the slaying of Goliath. 1 Samuel reports "the soul of Jonathan was knit to the soul of David, and Jonathan loved him as his own soul" (1 Sam. 18:1). Jonathan gave David his robe, his armor, his belt, and his bow and made a covenant with David. Ironically, David was already anointed to take the kingship from Jonathan, Saul's heir. As the story progresses, Saul begins to see David as a threat to his kingdom. Possessed by an evil spirit, Saul tried to strike David with his spear. "Saul was afraid of David, because the Lord was with him but had departed from Saul" (1 Sam. 18:12).

The end of 1 Samuel is a long chase: David fleeing Saul, and Saul unable to catch him. Jonathan disobeyed his father when he was asked to kill David and, instead, assisted David, first by warning him of Saul's intent and then by speaking of David's worth before Saul. Jonathan's petitions did little good, however, and Saul's dislike for David changed from political concern to obsessive hatred. Jonathan's love, however, did not waver. Meeting David in a field to warn him of impending danger, Jonathan promised to inquire of his father's real intentions towards David and to inform David of them. Jonathan then requested that David not cut off Jonathan—if he should live—or his house, and "Jonathan made David swear again by his love for him; for he loved him as he loved his own soul" (1 Sam. 20:17). When Jonathan returned to David with the bad news—Saul wished David dead—"David rose from beside the stone heap [where he was hiding] and fell on his face to the ground, and bowed three times; and they kissed one another, and wept with one another, until David recovered himself" (1 Sam. 20:41). Nowhere else in the Old Testament is such a deep love between two friends shown.

Throughout a long series of military movements—both to avoid Saul and to continue to battle the external enemies of Israel—David moved from region to region, always falling back to the desert, caves, or the forest. David fled from the field where he met Jonathan to Nob and thence to Gath. He then returned to Judah and hid in the caves of Adullam, where he gathered four hundred men to him. The prophet Gad warned David that his stronghold was not safe, so David withdrew into the forest of Hereth. Saul continued to pursue his young adversary, and

David, after defeating the Philistines at Keilah, "remained in the strongholds in the wilderness, in the hill country of the Wilderness of Ziph. And Saul sought him every day, but God did not give him into his hand" (1 Sam. 23:14).

Appropriately, Jonathan's last meeting with David was at the stronghold in the wilderness:

> And Jonathan, Saul's son, rose, and went to David at Horesh, and strengthened his hand in God. And he said to him, "Fear not; for the hand of Saul my father shall not find you; you shall be king over Israel, and I shall be next to you; Saul my father also knows this." And the two of them made a covenant at Horesh, before the Lord; David remained and Jonathan went home.   (1 Sam. 23:16–18)

Jonathan, in an ultimate act of friendship, assured David that he would be king over Israel. The son of Saul also proved his faith in God by throwing his support to David, God's anointed. Jonathan's fate was, from this point, tied with Saul's despite Jonathan's submission to Divine Will. The wilderness covenant between the two friends did not prevent Saul from bringing his entire house to ruin. David did not see Jonathan alive again after this meeting.

When the Ziphites betrayed David to Saul, he moved on to the wilderness of Maon:

> And Saul and his men went to seek him. And David was told; therefore he went down to the rock which is in the wilderness of Maon. And when Saul heard that, he pursued after David in the wilderness of Maon. Saul went on one side of the mountain; and David was making haste to get away from Saul, as Saul and his men were closing in upon David and his men to capture them, when a messenger came to Saul, saying, "Make haste and come; for the Philistines have made a raid upon the land." So Saul returned from pursuing after David, and went against the Philistines; therefore that place was called the Rock of Escape. And David went up from there and dwelt in the strongholds of Engedi. (1 Sam. 23:25–29)

The wilderness of these passages was not completely desolate, but incorporated the open pastureland on the slopes of the hills above the Dead Sea. The text mentions wilderness, mountain, rock, and hill country again and again to show how difficult David's position was[4] and also to show how able David was to handle strenuous circumstances. The wilderness always seemed to be on David's side. Even the covering of the rocks must occasionally fail, however, in which case God intervened to protect David by having Saul called home. The dual strengths were again present: David's stamina and familiarity with the country, which, in the end, had to be supplemented by the Lord's grace and providence.

While David was in En-gedi, Saul fell into his hands, but David did not take complete advantage of the circumstances. The text mentions a specific location—"the Wildgoats' Rocks"—again emphasizing the wild nature of the territory. Saul walked into the cave where David and his men were hiding and proceeded to relieve himself. David, rather than killing Saul, cut the skirt off Saul's robe, an embarrassment to any warrior. He walked out of the cave after Saul and called to the king to show him that he had had the chance to kill him—but did not take it. Saul in his reply to David acknowledged David would be ruler over Israel (1 Sam. 24). The wilderness, again, was to David's advantage—the place where David was always the overcomer and Saul was never able to win a decisive victory.

David went on the wilderness of Paran (probably actually a return to Maon), where he met Abigail and took her for his wife, and then returned to the wilderness of Ziph for another confrontation with Saul. David, accompanied by his nephew Abishai, managed to sneak into Saul's camp at night and, after refusing to kill Saul, absconded with the spear and water jar sitting by the head of the sleeping king. In this example of "wilderness skill," David must have negotiated his way into the camp in low light and then, in almost complete silence, moved among Saul's men until he came to their leader. Divine providence also played a role in David's success, as "a deep sleep from the Lord" had fallen on all Saul's men. David finished the incident by going to stand "afar off on top of the mountain," but in view of the camp. He called to Abner, Saul's commander, and to Saul to retrieve the spear (1 Sam. 26). The act was not just a demonstration of valor, it also served a prophetic function. David had refused to kill Saul on the grounds that the Lord would eventually take Saul's life. This would indeed soon come to pass.

In his last adventures before the death of Saul, David went to reside with the Philistines and not only secretly killed Philistines successfully, but also raided the other hostile neighboring peoples. Excluded by the Philistine king from going to battle against Saul and Israel, David went home to Ziklag, a temporary residence, to find that the Amalekites had burned the town and captured the families of all his followers. David conducted a campaign in the Negeb desert and regained his own two wives, all the families, and all the herds that had been lost. Meanwhile, the Philistines fought against Israel, and Saul and three of his sons, including Jonathan, were killed (1 Sam. 27–31).

When the news of the Israelite king's death reached David, rather than celebrate his release from persecution, David used the imagery of wild nature to eulogize the passing of his friends:

> Thy glory, 0 Israel is slain upon thy high places!
> How are the mighty fallen.
>
> . . . .
>
> Ye mountains of Gilboa
> let there be no dew or rain upon you,
> nor upsurging of the deep!

For there the shield of the mighty was defiled . . .
. . . .

Saul and Jonathan, beloved and lovely!
In life and death they were not divided;
they were swifter than eagles,
hey were stronger than lions.
. . . .

Jonathan lies slain upon thy high places,
I am distressed for you, my brother Jonathan;
very pleasant have you been to me;
your love to me was wonderful,
passing the love of women.   (2 Sam. 1:19, 21, 23, 25)

David, in his grief, called for a curse in the form of a drought upon the highlands where his friends have died and, in the process, ties the environment to the tragic event. David compared the slain leaders to the swiftest and strongest untamed creatures. Jonathan, who won his most important victory on "the heights" above the crags, had fallen on Mount Gilboa. David's love for his deceased comrade took precedence over the impending political victory, and he lamented rather than rejoiced.

The Thick Branches of a Great Oak

After the passing of Saul, David went to Hebron where the men of Judah made him king. With the land still fraught by war, David's fortunes turned, and he was able to defeat both Saul's heirs and Israel's neighboring foes. David became king over all Israel and expanded the boundaries of his empire to include much of the surrounding country, including the wilderness locations he had visited in his flight from Saul and his service to the Philistines. He moved to the king's house and turned his concerns toward transferring the Ark of the Covenant to Jerusalem, in order that both spiritual and political power would be centered in the Holy City. Throughout most of his reign, there is little mention of wilderness. 2 Samuel instead recounts the internal turmoil in David's household, as David covets Uriah's wife Bathsheba and his son, Ammon, forces his sister Tamar to lie with him. The core of 2 Samuel concerns sexual sin, revenge, prophetic reproof, and political intrigues in Jerusalem.

When David's rule is threatened, however, wilderness again becomes important. Absalom, his son, grew impatient and wished to become king in his father's stead. He gathered the men of Israel to him in Hebron, and David, on hearing of the impending coup, fled Jerusalem and took his followers into the wilderness. Absalom, receiving some bad advice from a counselor secretly loyal to his father, did not pursue immediately, but allowed his father time to safely cross the Jordan and organize his faithful troops. The counselor who discouraged Absalom suggested: "your father and his men are mighty men, and . . . they are enraged, like a bear robbed of her cubs in the field" (2 Sam. 17:8). He also suggested David was already making ready to fight his son and, when the first of Absalom's followers fell: "Then even the valiant man, whose heart is like the heart of a lion, will utterly melt with fear; for all Israel knows that your father is a mighty man . . ." (2 Sam. 17:10). The images of the lion and the bear return, this time not to convince a doubting king to let David fight, but to discourage the pretender to the throne from immediately engaging the well-proven David.

Men of the neighboring peoples, the Ammonites and the Gileadites, brought supplies and food to David's followers, saying "The people are hungry and weary and thirsty in the wilderness" (2 Sam. 17:29). The rations were well chosen for a fighting force, lacking both perishables and delicacies, but including "honey and curds" (perhaps symbolic of the milk and honey of the land).[5] His troops refreshed, David mustered his army and sent them out against Absalom. The battle was fought in the forest of Ephraim. According to 2 Samuel 18:7–8: ". . . the men of Israel were defeated there by the servants of David, and the slaughter there was great on that day, twenty thousand men. The battle spread over the face of all the country; and the forest devoured more people than the sword." The area was covered not only by thick forest, but also by heavy undergrowth and rocks. The dense cover put Absalom's troops at a disadvantage and caused many casualties.[6]

The forest was yet of further help to David's cause: just as Absalom chanced to meet some of David's followers, he rode his mule "under the thick branches of a great oak, and his head caught fast in the oak, and he was left hanging between heaven and earth . . ." (2 Sam. 18:9). One of David's commanders and his armor bearers killed the king's son against David's orders. Again, the wilderness seemed to favor David; the land itself supported God's anointed king.

From Cave to Crown

In this *Davidic* or *leadership model* of wilderness experience, we find major differences with both the *foundational (Genesis)* and *community (Exodus) models*. First, David and Jonathan were much better able to cope with the wilderness environment than were Hagar or the children of Israel. Scripture implies the young men had learned basic survival skills and could move easily across difficult terrain. Both David's and Jonathan's problems centered around military and political adversities, not around difficulties with finding food or water. In the story of David, there were no miraculous appearances of springs or manna, only

God's repeated deliverance, usually mediated through human action. David's ascension to the throne through the turmoil of Saul's rule was itself a mighty act of God.

Second, there are no theophanies in the wilderness portion of the story of David. God speaks to David's heart. The theophany has not disappeared from the history of Israel at this point—in fact, some of the most dramatic theophanies are yet to come—but most of the future angelic appearances and voices from heaven in the Old Testament concern the prophets. In 1 and 2 Samuel and related material in Kings, the role of the national leader partially separates from the role of the prophet. Moses, for example, assumed both prophetic and civil duties. He wrote down the law, spoke to God, and directed peoples' day-to-day activities. During the time of the judges, the head of the people was also a religious leader and, in the case of a prophet such as Deborah, led the people into battle. In 1 Samuel, these roles diverge. Saul spends a brief period among the prophets (and proves himself ill-adapted to the vocation). Much of Saul's failure as king was, in fact, a result of his poor understanding of God's will and priorities. Biblical critics often call David a prophet, but his role in society is quite different from that of Gad (who advises him to flee Saul) or Nathan (who comes to rebuke him for his behavior concerning Bathsheba). David's heart knowledge of God is, by implication, nurtured in the wilderness, just as the prophet Elijah's understanding of Yahweh is deepened by his trip to the desert mount of Horeb (see chapter 6). Yet while Elijah literally hears God, David relies on the inner, inaudible voice. In the lives of David and Jonathan, the verification of God's presence or God's purpose is found in the young warriors' statements acknowledging the role of Yahweh in their actions.

Third, the role of the environment is different in the story of David than in the Pentateuch. In Genesis, the wilderness and the mountain serve primarily as platforms for divine action. One sees God protecting Hagar in the wilderness and providing water—the wilderness *itself* does not protect or provide. In Exodus, the role of the environment becomes more complex. God worked a series of mighty acts, and the environment itself also seemed to protect the people, as in the miring of Pharaoh's chariot wheels. After Sinai, however, the environment worked increasingly against the people and became a means of divine disciplinary action. For David and Jonathan, arid lands, cliffs, caves, and forests offered protection and routes to victory or to safety. The interactions with nature were framed much less in terms of the miraculous and much more in terms of the actual physical characteristics and geographic locations of the sites. The modern reader, recognizing the historic nature of the writings, may overlook the subtle implications about the role of the land. David and Jonathan rambled about primarily on the property the Lord intended to give to the children of Israel. When the Forest of Ephraim aided David, the king was outside his home territory of Judah. Further, David was the Lord's anointed. The land itself, wild and cultivated, was always on his side and helped him win victories over the Philistines, Saul, and his own son.

Fourth, in the foundational and community models there is limited implication that the wilderness experience helped to build personal strength or leadership, at least in terms of environmental contact. Moses' first experience at Horeb, for example, was much more an encounter with God than with the wilderness. Genesis does suggest Ishmael's success at living in the wilderness was related to his skills as a bowman and warrior. We find this a much stronger theme, however, in the stories of David and Jonathan. David, on confronting Goliath, credits his boyhood experience as a shepherd with developing his expertise. It is quite clear, as he flees from Saul, that his youth in the wilderness has helped to save his life.

Fifth, the narratives of David and Jonathan and of David and his followers present wilderness as a setting for friendship and comradeship. There is little evidence from Genesis or Exodus that wilderness helps to build or maintain interpersonal bonds. The development of personal dependence—on others and on God—is, in contrast, a major theme of 1 and 2 Samuel.

We can summarize the most important features of the Davidic or leadership model, which has a grounding in the foundational model, but diverges in several key ways:

1. Wilderness experience built strength, stamina, and leadership and survival skills.
2. David's and Jonathan's success was established on God-confidence. Wilderness experience could, in the right social setting, help to develop this God-confidence.
3. Wilderness experience could produce heart knowledge of God by means other than direct encounter with a divine being.
4. Wild nature favored and protected the leader who was in right relationship with God.
5. Wilderness experience could produce camaraderie and group inter-dependence, which, at its best, was mediated by God.

Inner Spirit, Outward Bound

The Davidic model of wilderness experience is similar to many types of contemporary wilderness leadership training. One of the best examples of these modern programs is Outward Bound, which began as a wartime survival school and had as its original purpose to develop the will and the skill to survive in young British sailors who might be torpedoed during World War II. Through the years, Outward Bound has grown to include training centers in a variety of geographic locations and environmental settings. The programs use wilderness to not only teach technical skills, such as rock climbing and white water canoeing, but also to help participants "develop their inner strengths and resources, to recognize and dispel self-imposed limitations, and to learn to work cooperatively within a group for the benefit and service of others."[7] The participants find the programs valuable for improving personal conditioning, developing wilderness skills, learning to overcome challenges or hazards, learning to work as a team member, and developing an increased sense of the needs of others. Sociological studies have found the programs increase self-esteem, self-concept, self-empowerment, and physical fitness. Although most Outward Bound activities are organized for patrols of eight to twelve people, the longer courses also include a two to three day solo, where the participant is left alone (with food and water) in the forest, desert, or other wilderness environment.[8]

The Davidic model of wilderness experience is also partially based on learning or using wilderness skills under stress. As a shepherd David spent long hours by himself. He had to account for his livestock. He climbed over rocks looking for his lost charges and improved his hand and eye coordination by using his sling. He developed courage as he took a sheep from the jaws of a lion. He learned to navigate in rough country and to supply himself with food and water. The major difference between the Davidic model and the Outward Bound model is that Outward Bound attempts to develop only strength, skill and self-confidence; in the case of David, these are always supplemented by God-confidence. Both David and Jonathan give credit to Yahweh for all their victories. They lived somewhere between the physical and the spiritual. For both, the strength in their arms and the sharpness of their eyes were necessary to their role among the people of God. Caught up in the progress of holy history—rather than just pursuing individual achievement—their spiritual heritage and their heart knowledge of God guided their actions.

One wonders how David, left out in the wilds to tend the sheep, became a king with such a strong orientation toward God. The Scriptures introduce David in the context of his family and their relationship to the land of Judah and describe his service as a young musician in Saul's court. David was a product both of his family heritage and of his wilderness vocation. For spiritual development, the wilderness experience was valuable, but not by itself adequate. David grew in the midst of Hebrew culture and his worthy family. There is no evidence he had any formal religious training. Somewhere between the household of Jesse and the rocky hills of the wilderness, he developed a heart knowledge of Yahweh.

The modern Outward Bound model does not deal with these sorts of basic spiritual and social ties, although it may, in some wilderness therapy programs for troubled youth, attempt to compensate for the lack of them. For David, the skills he used in defeating Goliath were learned in the wilderness. His arrival in the valley of Elah, however, was at his father's bidding, and, in going to the site of the battle, he was following his brothers. Physical stamina and courage do little good if they are not applied in the right place, at the right time. The leader must also have the discernment to enter the battle on the Lord's side, or he or she should not enter it at all.

A last comparison with the Outward Bound model lies in the development of camaraderie. Despite the association of wilderness and isolation, programs like Outward Bound try to both change self-perception and foster better relationships among people. The challenges of the wilderness are supposed to teach participants to expect more of themselves and to expand their concept of their own personal limits. Working in groups "helps to develop bonds of mutual trust and an understanding that accomplishment requires cooperation." Through the courses, the participants not only gain strength and insight, but also "lasting friendships."[9] In the stories of David and Jonathan we find numerous cases of this type of active interdependence. It begins when Jonathan's armor bearer follows him on to the crags and continues as Jonathan gives David his armor and seeks David out at the risk of his own life. We see in David a leader conscious of the needs of his troops, while, in the case of Saul, we see less concern for his soldiers and less willingness to care for them.

When Jonathan gives David his armor, the mutual identification between two young men in their physical prime is one element in the relationship. Their submission to the will of God is another. Jonathan and David are very much alike. They have the same skills and the same interests. Through it all, however, it is not the challenge that builds the relationship and carries it through the final tragedy, it is the love of Yahweh entering into the human bond. The challenge and the stress help the relationship to develop, but they are not what tie it to eternity.

Wilderness training should thus be quite appropriate to the building of Christian leaders, and the techniques and goals of secular wilderness schools are generally applicable to the task. The Christian leader, however, needs to develop a sense of self relative to the person of Christ and must consider God's will in selecting any course of action. The God-dependence so critical to David's rise to power has to be nurtured in Christian community as well as in the wilds. The Christian can not simply enter the wilderness, get tough, and return to society to assume a military command, a political office, or a managerial position. Somewhere in the process, the Spirit of God needs to enter the human relationship and supply the anointing. Then—and only then—is wilderness leadership training complete.

5

The Mountain Haunts
of the Leopards
(Writings)

Emerald Lake

We had started out from Estes Park, Colorado, with a clear view of the mountains, but, as we drove up the road to Bear Lake and gained elevation, the sky grew gray, covering the peaks in a dense cloud. Coming around a sharp curve, we encountered a light glaze of ice on the road. I slowed and tried to keep my eyes on the vehicle ahead while continuing to admire the white dusting of snow on the gold of the aspens. The weather changes quickly in the Rockies in the fall.

Just as we got to the parking area, the storm began in earnest. Lightning flashed overhead, and precipitation smacked the windshield.

"It's hailing," I reported to Sandy, "and I think we ought to wait for the lightning to quit before we head for the lake."

"Are you still up for a hike?" I asked after a doubtfilled pause. I had just met Sandy, a visiting South African, at a wilderness conference and had no idea what she thought of setting out in zero visibility.

"It's magnificent," she muttered, looking out at the storm. She turned towards me and replied with conviction, "I'd love to go walking in the snow. I haven't had a chance to do this in months."

We waited about fifteen minutes for the thunder to move off to the east and then strolled up to Bear Lake, which turned out to be only a few hundred feet away. The cloud, swirling across the surface of the water, would lift and expose the conifers on the far shoreline and then would settle again, completely obscuring the forested slopes. Sandy and I positioned ourselves in the shelter of a fir and observed the cloud's movements as it coated boulders, tree branches, and our rain suits with wet snow. I looked at Sandy, trying to determine what she thought of

63

the damp and the cold. She gazed across the choppy water into the cloud. "It's wondrous," she said with an open smile.

"Let's head toward Dream Lake," I suggested, adding my frosty breath to the saturated atmosphere. "It sounds like an appropriate place to visit in the fog."

We trudged up a well-worn path and encountered at least a dozen hikers, some of them without rain gear or jackets, rushing back towards the parking area. As we climbed, we could catch an occasional glimpse of a nearby rock face or a stream just below the trail, but most of the landscape remained lost in the thick, moisture-laden air. Slowly, as the storm spent its energy, the snowfall broke into flurries. By the time we reached Dream Lake, the snow had stopped.

We made our way to some boulders near the water's edge and sat down to absorb the setting. The lake, warmer than the air above it, was producing a white wispy "smoke," which rose and dissolved. Seen through the mist, everything was shades of gray. A tree jutting out from a ledge fifty yards away was nothing more than a dark outline. We leaned quietly back on the rocks and let the shifting cloud change the scenery around us. The tree disappeared and reappeared again.

"You hike by yourself a lot, don't you," said Sandy.

"It's true," I replied. "How did you know that?"

Sandy didn't answer me. The ceiling was lifting, and the dark tops of a hundred trees suddenly rose across the water. A beam of sunlight shot through the haze, illuminating the gnarled branches of the tree in front of us. We were beginning to see flashes of green. The cloud rolled gently upwards, revealing the length of the lake, the height of the ridges, and the complexity of the boulder-strewn shore. I hadn't realized how far into the glacial valley we had come. I hadn't been able to see it.

"It's wondrous," said Sandy.

"Want to go on?" I asked. "There is one more lake at the end of the trail." Sandy didn't need any further prompting. We ambled through the small wet grassy patches along the edge of the shore and then scrambled up the stony trail that took us higher yet.

When we reached Emerald Lake, the cloud still obscured the peaks, but was allowing enough light to penetrate its milky interior to turn the water in the tarn bright green. We could see the cliffs at the head of the narrow valley rising out of the talus at their bases, but owing to the cloud, we couldn't view their full height. The misty, wet veil swirled against the stone faces, fought the sheer walls, and then pushed itself up over the peaks. As the cloud dissipated, two thousand feet of snow crowned granite suddenly towered above us and touched a very blue Rocky Mountain sky.

"It's wondrous," sighed Sandy for the umpteenth time.

"Wondrous . . . ," I concurred. Staring up at the tremendous blocks of stone made me contemplative. What sort of God would make both this deep green lake and those awesome gray walls? What sort of God would place the bright blue above them? What sort of God would allow me to see both the emerald and the azure and to enjoy every breath I took of the crisp mountain air? I let my eyes travel slowly from the snow beneath my feet, across the shining water, over the angled talus, up the raging cliffs to the unreachable summit. Majesty . . . beauty . . . power . . . order . . . A God who had enough of all these to create the mountains.

Sandy and I glanced at each other for a moment and then sank back in our own thoughts. She had the good graces to leave me wandering my own internal pathways. I threw off the feeling the cliffs were eternal and immutable and plunged into geologic time. In comparison to these tremendous blocks of granite, I was ephemeral. Yet the great rock slabs at the base of the massive stone faces betrayed the action of ice, water, and gravity. The soft snow I had been striding through, stacked miles high and miles long, could shave the ridges of the highest sierra and clear away the most massive landforms on earth. In the hands of the Creator, the delicate snow flake could become a cosmic hammer. Before the Almighty God, these hugh towers of stone were a passing beauty. The slow lifting of the cloud had made me even more aware of how fleeting is the glance we get of the cosmos. The window opens, and then it closes. Here, in a brief viewing, God's time intersected with mine.

I nodded to Sandy, and we both stood up. Being mere creatures, we were about to be driven back down the trail by falling temperatures and failing sunlight. As I took a parting gaze at the mountains, I could think of nothing but God's majesty. I have experienced the reflection of divine splendor in the wilderness many times, and each new vision, each touch of grace, each glimpse of otherworldly glory gives me a reason to travel further and higher. The wilderness always humbles me, something which I badly need and, at the same time, leaves me very pleased with the small place I have in the whole.

* * *

Today we strongly associate wilderness and environmental spiritual experience with aesthetic experience, yet the biblical materials we have reviewed thus far contain very little discussion of the beauty or wonder of nature. Not the wild setting but the theophanies and visions have been colorful, immense, or awe inspiring. Beauty rests in the lapis lazuli floor below the throne of God, not in Horeb itself. The Bible does not, however, ignore beauty in the wild portions of the created universe. Some of the older biblical books show a sensory appreciation for untamed beasts and landscapes. Most of the relevant passages are not, however, found in historic texts, but are located in the sacred writings, including the psalms and the wisdom literature, and in the works of the later prophets. These works enhance the historic themes of the Pentateuch by describing God's relationship to the entire natural order. The biblical writers did not, therefore, just appreciate wild nature for its own sake. They thought it informed humankind about the person of God—a Creator who takes a joy in all creation, a Master of the Heavens who forms the mountains and feeds the wild beasts with infinitely loving hands.

A Lyre Belonging to a Lion

Having discussed David as a warrior in the wilderness, we should not neglect David as a poet and a musician. In contemporary western culture, we have little difficulty imagining a great military commander as a fine national leader—a man envisioning building programs and seeking alliances with neighboring countries. We find it harder to cast this man of valor as a devout servant of God and even more difficult to see him as world class composer. Today a military vocation is one thing, a religious vocation another, and an artistic vocation a third. In the biblical times, both kings and prophets were poets.

David not only knew how to use a sling, he also knew how to play the lyre. When King Saul had fallen from God's grace and was tormented by an evil spirit, he sent for David on the grounds that he was a skillful musician. David won the heart of the stricken ruler, for when David played: ". . . Saul was refreshed, and was well, and the evil spirit departed from him" (1 Sam. 16:23). In his first assignment in the royal court, David's "military" skills were secondary—gentle sounds pouring forth from his instrument "conquered" the adversary. The soul soothing qualities of David's music distinguished him as God's anointed.[1]

The Scriptures do not mention where David learned to play, but it is usually assumed he spent part of his time in the wilderness practicing. Even today, Middle Eastern shepherds entertain themselves with flutes and other instruments. If David's boyhood on the isolated hills allowed him to develop a strong arm and a sharp eye, it also allowed him to develop nimble hands and an ear for the Lord. According to tradition, he authored many psalms himself and helped to develop this art form so full of natural imagery.

Artistic expression was also found in David's son Solomon, who was of a more scholarly bent than his father:

> And God gave Solomon wisdom and understanding beyond measure, and largeness of mind like the sand on the seashore, so that Solomon's wisdom surpassed the wisdom of all the people's of the east, and all the wisdom of Egypt. For he

was wiser than all other men, wiser than Ethan the Ezrahite, and Heman, Cal-
col, and Darda, the sons of Mahol; and his fame was in all the nations round
about. He also uttered three thousand proverbs; and his songs were a thousand
and five. He spoke of the trees, from the cedar that is in Lebanon to the hys-
sop that grows out of the wall; he spoke also of beasts, and of birds, and of the
reptiles, and of the fish. And men came from all peoples to hear the wisdom
of Solomon, and from all the kings of the earth, who had heard of his wisdom.
(1 Kings 4:29–34)

Solomon's wisdom expressed itself both in an understanding of nature and
in literary ventures. He had a touch of the classifier and administrator and, perhaps,
even of the scientist. The passages do not, however, indicate Solomon's products
possessed the spiritual power so central to his father's work. The book of 1 Kings
implies his forays into wisdom had a worldly cast and were used to win an international political competition. Among all his amusements and duties, the prag-
matic king had time to write one thousand and five songs.[2]

In the household of David, we find a *poetic model* of wilderness experience.
Wild nature inspired the psalmist and enhanced the gentle singer's knowledge of
God. The historic narratives tell us little of how this functioned, but they clearly
declare the role of poet may be combined with other spiritual callings. Out of the
poetry grew prophecy and out of prophecy grew the history of Israel.

Our interest in the poetic books and wisdom literature will be twofold. First,
what do these texts tell us about wild nature and God's intent for it, and second,
since our interest is also human spirituality, what do wild nature and wilderness
tell us about the person of God and our relationship to the divine?

Rivers Raise Their Thunders

Perhaps no one ever understood better than the psalmist God's role as Creator
and God's continuing care for creation.[3] Psalm 104, which is based almost entirely
on natural imagery, begins with a series of poetic descriptions of the original
creation:

> You fixed the earth on its foundations,
> unshakable for ever and ever;
> you wrapped it with the deep as with a robe,
> the waters overtopping the mountains.
>
> At your reproof the waters took to flight,
> they fled at the sound of your thunder,
> cascading over the mountains, into the valleys,
> down to the reservoir you made for them;

you imposed the limits they must never cross again,
or they would once more flood the land.   (Ps. 104:5–9, *JB*)

The psalmist not only credits Yahweh with the creative act, he also associates
God with the eternal and unchangeable in nature—the unshakable foundations.
God has a basic power over creation, having established its limits. The psalmist recognizes the great volume and pounding force of the waters and declares
the thunder of God causes them to flee. Yahweh controls all nature and natural
forces.

Psalm 104 then moves into the present, preventing the reader (or hearer) from
seeing the life-giving acts of the Creator God only in the distant past:

You set springs gushing in ravines,
running down between the mountains,
supplying water for wild animals,
attracting the thirsty wild donkeys;
near there the birds of the air make their nests
and sing among the branches.

. . . .

The trees of Yahweh get rain enough,
those cedars of Lebanon he planted;
here the little birds build their nest
and, on the highest branches, the stork has its home.
For the wild goats there are the mountains,
in the crags rock badgers hide.   (Ps. 104:10–12, 16–18, *JB*)

These passages praise God for continued provision of water and for the establishment of appropriate habitats for the wild animals. God's providence extends to
the desert and to crags. Water from Yahweh's fountains grows vegetation, which
in turn provides nests for birds. The creative act continues in the form of God's
loving providence. The waters of the previous passage are now tame enough to
nourish the trees and the wild asses. The Lord has tied creation so tightly together that the raging flood splashes down as quiet rain and produces the foliage
which hides the small birds and their offspring.

The psalmist then reports the perfection of God's timing:

You made the moon to tell the seasons,
the sun knows when to set:
you bring darkness on, night falls,
all the forest animals come out:
claiming their food from God.

The sun rises, they retire,
going back down in their lairs,
and man goes out to work,

and to labor until dusk.
Yahweh, what variety you have created,
arranging everything so wisely!
Earth is completely full of things you have made:

among them the vast expanse of ocean,
teeming with countless creatures,
creatures large and small,
with ships going to and fro
and Leviathan whom you made to amuse you.
All creatures depend on you
to feed them throughout the year;
you provide the food they eat,
with generous hand you satisfy their hunger.

You turn your face away, they suffer,
you stop their breath, they die
and revert to dust.
You give breath, fresh life begins,
you keep renewing the world.   (Ps. 104: 19–30, *JB*)

The Psalm praises God's skill in ordering the creation and lauds Yahweh's deep compassion in the continuing renewal of the world. Without the mercy and the love of God, the cosmos would instantly cease to function. Our contemporary tendency is to see "the love of God" in acts of altruistic human giving. The psalmist sees the love of God in the diversity and order of the nonhuman portion of nature. Streams splash down the mountain, badgers scurry into crevices, and lions roam by God's care. The God of Psalm 104 is a God much wider and deeper than either the frail and false Baal or our modern God in a box, whose best expression is an occasional act of charity or a special telecast for the hungry.

The passage about "leviathan," possibly a whale or other large sea creature, has been variously translated, "Leviathan which thou didst form as a plaything," "Leviathan, which you formed to frolic there," or "Leviathan which thou didst form to sport in it."[4] In any case, the passage speaks of God's joy shared with all creation. This aspect of God is difficult to see in the Exodus with its continuing tension over the obedience of the people, or in the wanderings of Hagar with her desperate physical needs. Yet in Psalm 104 we see a God who delights in the natural world and who has made creatures who can rejoice in their own existence.

The Psalm concludes:

Glory for ever to Yahweh!
May Yahweh find joy in what he creates,
at whose glance the earth trembles,
at whose touch the mountains smoke!

I mean to sing to Yahweh all my life,
I mean to play for my God as long as I live.
May these reflections of mine give him pleasure,
as much as Yahweh gives me!   (Ps. 104:31–34, *JB*)

The result of the psalmist's contemplation of creation is glorification of God. The psalmist wishes a blessing on God when he sings: "May Yahweh find joy in what he creates" (Alternate translation: "May the Lord rejoice in his works," from the *Revised Standard Version*). The joy in the psalmist's own spirit rises out of his understanding of creation and how he sees the Lord interacting with wild nature. The omnipotent God of the Exodus, who makes the mountains smoke, has time for the small birds and the stork in the cedars and for the frolicking sea monster.

The themes of creation, providence and joy are found in other psalms using natural imagery. In Psalm 65, the strength of God as Creator is tied to God as righteous and God as Savior.

Your righteousness repays us with marvels,
    God our savior,
hope of all the ends of the earth
    and of distant lands.

Your strength holds the mountains up,
    such is the power that wraps you;
you calm the clamor of the ocean,
    the clamor of its waves.   (Ps. 65:5–7, *JB*)

In the modern mind, the relation between salvation and creation is indirect. Salvation is something experienced by country folk singing in small churches and suburbanites running out of alternatives and pushed to the edge and has nothing to do with raising the continents above the seas. In observing nature, the psalmist ties the power that lifted the mountains to the power that spreads hope across the earth. God's righteousness is as much a source of "marvels" as God's creative essence.

Psalm 65 ends with a series of images of providence in nature including:

You crown the year with your bounty,
    abundance flows where you pass;
the desert [wilderness] pastures overflow,
    and hillsides are wrapped in joy. . . .
                                (Ps. 65:11–12, *JB*)

Hope, salvation, righteousness, and providence: all are part of the same divine being, and all pour forth from the same creative well. Today we do not think about being able to detect evidence of God's love in nature because we think of expressions of God's love as being confined to human beings. Looking at the natural world, the psalmist understood how universal God's love is.

In Psalm 98, the psalmist records not only the joy, but also the majesty of God in relation to righteous and justice:

> Let the sea thunder and all that it holds,
>   and the world, with all who live in it;
> let all the rivers clap their hands
>   and the mountains shout for joy.
>
> at the presence of Yahweh, for he comes
>   to judge the earth,
> to judge the world with righteousness
>   and the nations with strict justice.   (Ps. *98:7–8, JB*)

The psalmist understood how the various elements in God's character are related and could see creation responding to virtues—such as justice—which we humans disassociate from the cosmos outside human society. There is a gentle irony in the rivers and mountains rejoicing at the arrival of the righteous judge, especially when one suspects much of contemporary religious culture would be terrified by the same event.

And in Psalm 93, the rivers and the sea are used to prove the majesty and transcendence of God. The poet observes the tremendous power in untamed nature and then counts Yahweh's power greater:

> Yahweh is a king, robed in majesty;
> Yahweh is robed in power,
> he wears it like a belt . . .
> . . . .
> Yahweh, the rivers raise,
> the rivers raise their voices,
> the rivers raise their thunders;
> greater than the voice of ocean,
> transcending the waves of the sea,
> Yahweh reigns transcendent in the heights.
>
> (Ps. *93: 1–4, JB*)

The psalmist understood the majesty and power in nature and, thus, had a glimpse of the majesty and power of God. A great temptation in David's time, as well as in our own, was to seek a God who could be manipulated and controlled. The psalmist understood God's transcendence and, in the process, recognized Yahweh as the ultimate ruler, the sovereign whose will could not be bent by human desires. Without the recognition of transcendence, the power of God as deliverer and savior slips by us. Only the Transcendent One can unlock the chains of human history.

Hiding Among the Reeds in the Swamp

Some of the same themes treated in the psalms reappear in the sacred writings of later periods. The Hellenistic "wisdom" literature, for instance, does not discuss wilderness sojourns *per se,* but draws illustrations of various spiritual principles from wild nature. The author of Ecclesiastes, much more cynical than any psalmist, explains his philosophical dilemma in the cosmic pattern of rivers continually running to the sea and the seas never filling, and in the winds blowing from all cardinal directions and never blowing out. Nothing is new, he complains. The author of the apocryphal book of Wisdom, more of an optimist, finds God's glory displayed in the cosmos and uses imagery similar to that of Psalm 104. The best examples, however, of poetic motifs from wild nature may be found in the Book of Job, which ironically concerns theodicy (the problem of determining the role of a loving God in allowing evil in the world), and in the Song of Songs, a book which is difficult to date and is unlike any other literature found in the Hebrew canon of Scripture.

In the Book of Job, God allows Satan to exercise disruptive and destructive power in the life of God's most faithful human servant, Job. Job's friends, assuming that the evil calamities are the result of some sin, come to visit Job and attempt to induce him into confession and repentance. Job refuses and continues to claim he has done nothing wrong. In Job 38–41, the Lord himself speaks from a whirlwind in response to the commentaries of Job and his friends on Job's sufferings. Beset with trouble and personal disaster, Job has been justifying himself and challenging "the credibility of God."[5] God, arriving in a whirlwind (another theophany), condescends to answer Job. God's lengthy speech emphasizes, first, God's power in creation and humankind's inability to serve as Creator. God asks Job if he was present for the laying of the foundations of the universe, then poses a series of questions about the continued functioning of the world:

> Have you ever in your life given orders to the morning
>   or sent the dawn to its post,
> telling it to grasp the earth by its edges
>   and shake the wicked out of it, . . .
>
>   . . . .
>
> Have you ever visited the place where snow is kept,
>   or seen where the hail is stored up, . . .
>
>   . . . .
>
> From which direction does the lightning fork
>   when it scatters sparks over the earth?
> Who carves a channel for the downpour,
>   and hacks away for the rolling thunder,

so that rain may fall on lands where no one lives,
    and the deserts void of human dwelling,
giving drink to the lonely wastes
    and making grass spring where everything was dry?
Has rain a father?
    Who begets the dewdrops?
What womb brings forth the ice,
    and gives birth to the frost of heaven,
when the waters grow hard as stone
    and the surface of the deep congeals?

                     (Job 38:12–13, 22, 24–30, *JB*)

God's challenge—ironically—never really answers Job's complaint. Although God makes the position of humanity in the universe very clear, the Lord does not directly accuse Job of wrong doing. Using a series of natural examples, God articulates the extent of divine rights. The primary thrust of the original text was to deal with suffering among the righteous and the unexplainable elements in God's interactions with humankind. For our purposes, we should note Yahweh's speech declares that the Creator is continually active in natural processes and that humankind will never be God's match. Yahweh's concern and activities extend to the desert, where no humans live, and to the grassy springs, even where there are no herds or flocks. Far more than the psalms, the Book of Job integrates God's role in nature with God's relationship to the human. When Job overstepped the bounds of his knowledge, Yahweh responded with a series of examples of divine roles as the Master Planner, Master Builder, and Eternal Provider of the universe. God's speech is intended to give Job a better idea of who God is and how humankind relates to divine powers and prerogatives.

In the process of explaining God's powers, the Book of Job describes God's care for wild animals in detail, and makes it clear God intends to provide for them:

Do you find prey for the lioness
    and satisfy the hunger of her whelps
when they crouch in their dens
    and lurk in their lairs?

Who makes provision for the raven
    when his squabs cry out to God
    and crane their necks in hunger?
Do you know how the mountain goats give birth,
    or have you ever watched the hinds in labor?
How many months do they carry their young?
    At what time do they give birth?

> They crouch to drop their young,
>   and leave their burdens fall in the open desert;
> and when calves have grown and gathered strength
>   they leave them never to return.
>
> (Job 38:39, 39:1–4, *JB*)

The point of these passages is not that it is impossible for humans to feed lions or to see a hind giving birth, but that God's providence reaches beyond anything human beings could provide or organize. God maintains all of nature, even in the wild and isolated places.

Wilderness itself is part of the divine order:

> Who gave the wild donkey his freedom,
>   and untied the rope from his proud neck?
> I have given him the desert as a home,
>   the salt plains as his own habitat.
> He scorns the turmoil of the town:
>   there are no shouts from the driver for him to listen for.
> The mountains are the pastures that he ranges
>   in quest of any type of green blade or leaf.
>
> (Job 39:5–8, *JB*)

God claimed to have given the wild ass his "freedom" as well as his natural habitat. God clearly affirmed the existence of creatures who do not serve humans and, in fact, established "wildness" as a divine prerogative. Again the point is not that a wild ass cannot be tamed—a young ass can be easily captured and trained. By God's mandate, however, a large part of creation is not under human control and is not intended to be.

The use of wild examples in God's discourse with Job is not coincidental. "Job, in his speeches, has challenged God's right, one might even say his freedom.... Job has improperly and 'without understanding' interfered in God's affairs."[6] God's statements set limits to human dominion and clarify divine rights through the example of creatures other than privileged Adam. Our contemporary tendency is to assume God's relationship to human beings should always be presented in human or divine terms. Job 38–40 discusses humankind's relationship to God in terms of God's relationship to wild nature. Job could not understand the real presence and power of God without understanding the extent of God's role as Creator.

> Does the hawk take flight on your advice
>   when he spreads his wings to travel to the south?
> Does the eagle soar at your command
>   to make her eyrie in the heights?
> She spends her nights among the crags
>   with an unclimbed peak as her redoubt,

> from which she watches for prey,
>> fixing it with her far-ranging eye.
> She feeds her young on blood:
>> wherever men fall dying, there she is.

(Job 39: 26–30, JB)

Job could not tell the hawk where to fly, and neither can we. To further the irony, when humans fall in battle, their bodies become food for wild creatures. In the end, humans are little different from the rest of creation, something humans too often forget.

In the psalms, creation frequently speaks praise to God. In Job this is also the case, but praise of the divine becomes a minor theme against what creation says to humankind. In this "self-witnessing of creation," Job encountered a flood of difficult questions about the mysteries of the cosmos and the role of the divine in maintaining it. Throughout God's entire discourse in Job, God lets creation speak for the Creator's prerogatives. In one of the longest biblical discourses attributed directly to Yahweh, God points to creation to explain divine motives, rather than presenting some otherworldly form of wisdom. Job must ultimately put the entire world back in the hands of God, as he recognizes that the earth exists for the Lord, and it is God alone who supports and maintains it.'

The modern scientist might naively ask: but don't we now know a great deal more than Job? What seemed like wonders to Job and his friends are now thoroughly understood. We can watch mountain goats giving birth or visit the clouds where snow and rain are kept. The Book of Job is actually discussing a deeper type of knowledge than this. We might, for instance, be fooled by the discussion of Behemoth:

> Now think of Behemoth;
>> he eats greenstuff like the ox.
> But what strength he has in his loins,
>> what power in his muscles!
> His tail is stiff as cedar,
>> the sinews of his thighs are tightly knit.
> His vertebrae are bronze tubing,
>> his bones hard as hammered iron.
> He is the masterpiece of all God's work,
>> for his Maker threatened him with the sword,
> forbidding him the mountain regions
>> where all the wild beasts have their playground.
> So he lies beneath the lotus,
>> and hides among the reeds in the swamps.
> The leaves of the lotus give him shade,
>> the willows by the stream shelter him.

> Should the river overflow on him, why should he worry?
> A Jordan could pour down his throat without his caring.
> So who is going to catch him by the eyes
> or drive a peg through his nostrils?   (Job 40:15–24, *JB*)

Many commentators suggest that Behemoth here is not only a hippopotamus, but also a mythical symbol of evil. Whatever the allegorical meaning of the passage, the whole is a very accurate description of the species' morphology and habitat. But can't modern man capture Behemoth?

Yes, we can capture a hippopotamus, and humankind, through its history, has destroyed thousands. We can also divert the water away from the hippopotamus's home river bed and make his habitat run dry. We do not, however, hold the sword which keeps Behemoth from the mountain, nor can we put him under the lotus or make him to withstand the river's floods. Interpreting this passage as a description of a real animal or a mythical beast, we are equally helpless. We didn't make the hippopotamus what he is, and we can't attempt it now. We can destroy and dislodge him, but we cannot control him the way God can. Our scientific knowledge does not provide creative power of the sort described in Job. We can barely rehabilitate an ecosystem with all the elements still present, much less create one from nothing. We can only think of ourselves as better than Job if we confuse the power to create with the ability to destroy.

The theophany in the whirlwind is also as the master architect of the universe, a Creator who imparts wisdom and order to creation and rejoices and delights in it. God uses wild nature to challenge all humankind: Who are you? Where are you? Were you there when the world was formed? Do you understand how all nature came to be?[8] Like Job, we have to stand in awe of God's works and continuing care, and, like Job, we are humbled by God's power and majesty.

The Book of Job uses images from wild nature in a more sophisticated way than Psalms. The speeches about nature are, for example, no longer in the mouth of the author, they are in the mouth of God. The questions being asked are more difficult, and the replies of the All-Powerful are indirect. The issue is not just who created the earth, but why creation is managed as it is. The psalmist proclaims his righteousness, while Job struggles with deviations from the supposed rewards of a God-fearing life. The wild beasts of the psalms glorify God by taking their place in the desert and the forest. The wild beasts of Job speak of a God with inscrutable motives and inherit modes of being beyond the perspective of limited humankind.

Among the Lilies

The Book of Job is not the only model for human relationship with God. Job approaches the issues very philosophically, and asks very deep "Whys?" The

Scriptures also deal with the divine-human relationship in a more emotional framework and use wild nature as a model for our perceptions of spiritual beauty. Despite the controversy over its symbolism and origins, the Song of Songs, a poem built around a set of splendid natural images, has inspired many who were looking for a model of love between the Lord and the faithful. Conservative Christians tend to see the entire book as a spiritual allegory of the relationship between Christ (as the lover) and the church (as the bride). This type of interpretation is also found in the Talmud of Judaism, in which Yahweh is the lover and the Jewish nation is the bride. At the other extreme are those who see it only as a love poem, possibly of secular origin, and an almost accidental inclusion in the Bible. Since the poem is of pre-Christian origin and presumably served some religious purpose in the ancient Hebrew cult, the original usage was almost certainly more than a glorification of human romance. The Song may have been part of the liturgy of a spring or fall festival. It could thus carry both images—of the union between the divine and the human and of earthly marriage and fertility.[9]

Whatever the original intent, the poem uses nature to represent the spiritual and personal characteristics of the lover and the beloved. Motifs from agriculture mingle with those from the wild. The beloved calls her lover (in religious interpretation, Yahweh or Christ) "a cluster of henna blossoms from the vineyards of En Gedi" (Song of Songs 1:14, *NIV*), and says, "his appearance is like Lebanon, choice as its cedars" (Song of Songs 5:15, *NIV*), while he calls her, "my dove in the clefts of the rock, in the hiding places on the mountainside" (Song of Songs 2:14, *NIV*). She refers to herself as "a rose of Sharon, a lily of the valleys" (Song of Songs 2:1, *NIV*). Her beloved says: "Your two breasts are like two fawns, like twin fawns of a gazelle, that browse among the lilies" (Song of Songs 4:5, *NIV*). The beloved, in describing their union, says:

> My lover is mine and I am his;
>   he browses among the lilies.
> Until the day breaks
>   and the shadows flee,
> turn, my lover,
>   and be like a gazelle
> or like a young stag
>   on the rugged hills.   (Song of Songs 2:16–17, *NIV*)

The lover calls to his bride:

> Descend from the crest of Amana,
>   from the top of Senir, the summit of Hermon,
> from the lions' dens

and the mountain haunts of the leopards.
(Song of Songs 4:8, *NIV*)

And with a mixture of the wild and cultivated, the lover paints yet another portrait
of the bride:

You are a garden locked up, my sister, my bride;
you are a spring enclosed, a sealed fountain.
Your plants are an orchard of pomegranates
with choice fruits,
with henna and nard,
nard and saffron,
calamus and cinnamon,
with every kind of incense tree,
with myrrh and aloes
and all the finest spices.
You are a garden fountain,
a well of flowing water
streaming down from Lebanon.
(Song of Songs 4:12–15, *NIV*)

The song not only uses the imagery of wild nature as a setting for a maturing love,
it also uses natural imagery to describe how the lovers see each other. The poem
equates the natural with the object of godly love and alternates between wild and
cultivated nature. The Song of Songs does not differentiate between the beauty of
a gazelle among the lilies and "a flock of goats descending from Gilead" (Song
of Songs 6:5, *NIV*).

The beauty seen in wild nature draws our attention to a pure form of love.
The knowing of the lover requires all the senses—seeing and hearing, touching
and tasting. The poem implies even the most intense beauty in nature does not
reflect what the lovers perceive, yet the natural images are the most provocative
available. In the closing verses, the lover calls to his beloved who is in a garden
"with friends in attendance," (Song of Songs 8:13, *NIV*) and she replies to him:

Come away, my lover,
and be like a gazelle
or like a young stag
on the spice-laden mountains.   (Song of Songs 8:14, *NIV*)

In observing wild nature, we slowly begin to recognize only God's unselfish love
could have produced such beauty.

The Still Waters

The sacred writings present wild nature in one further role—aiding or encouraging spiritual restoration. In oft-memorized Psalm 23, the wilderness wanderer sings: "The Lord is my shepherd, I shall not want. He maketh me to lie down in green pastures, he leadeth me beside the still waters. He restoreth my soul." Although the psalmist attributes the spiritual restoration, which may include comfort and healing, directly to God, the pastoral landscape becomes part of God's method. The verdant grasslands, which were probably in the biblical wilderness, and the quiet of the stream aid the establishment of what we might currently term "inner peace." The Psalm, itself intended to provide comfort and encouragement, suggests natural features can reflect or perhaps provide a channel for the regenerative powers of Yahweh.

Another wild landscape image associated with restoration is the "mountain of God." In Psalm 3:4, for example, the "virtuous man under persecution" calls to Yahweh, who answers "from his holy mountain" (*JB*). Psalm 43:3, a lament, requests that God: "Send out your light and your truth, and let these by my guide, to lead me to your holy mountain and to the place where you live" (*JB*). Although such passages are associated with Jerusalem on the mount, they also invoke images of wilderness theophanies and direct experience with the Divine. On the holy mountain all sorrows flee and are replaced by rejoicing. In an image that extends far beyond what modern romanticism projects for natural restoration, Psalm 72:3 proclaims the coming of the Lord by requesting: "Let the mountains and hills bring a message of peace for the people. Uprightly, he [the promised king] will defend the poorest, he will save the children of those in need, and crush their oppressors" (*JB*).

We should note that in the writings wild nature is sometimes used to portray a turbulent or distressful spiritual state, as in Psalm 42:7 where the poet cries: "Deep is calling to deep as your cataracts roar; all your waves, your breakers, have rolled over me" (*JB*). In this case, wild nature and the troubles are still considered to be in the hands of God, which gives the psalmist further hope for overcoming them.

Who Sees the Glory?

In the *poetic model* of wilderness spiritual experience, descriptions of God's relationship with wild nature inform us about the person of God. The psalmist sees beyond the structural and functional aspects of creation into the divine. The *Poetic model* of wilderness and wildness:

1.   Emphasizes God as Creator: by observing wild nature, the psalmist gained an understanding of the creative power of Yahweh and of the magnificence of the cosmos.

2.   Praises God's providence: by showing the extent of God's care for wild nature, the psalmist demonstrates our total dependence on God.

3.   Teaches that God takes joy in the wild creatures and the rest of the cosmos: the psalmist knows God's care extends to all creatures, from the depths of the sea to the furthest reaches of the deserts.

4.   Presents God as righteous, majestic, just, transcendent, omnipotent, omniscient, and loving. The psalmist, without theophanies, had a broad understanding of the person of God.

5.   Proves wildness itself is part of the divine order and a prerogative of God.

6.   Teaches human dominion is restricted and can not interfere with the divine prerogatives. Creation itself teaches human limitations: we do not have the power to create from nothing nor can we exercise God's wisdom in caring for the creation.

7.   Suggests peaceful or beautiful natural settings might aid in spiritual restoration.

8.   Demonstrates the beauty God bestowed on wild nature can be a model for spiritual relationships and spiritual vision. The aesthetic aspects of wild nature are a mirror of God's unselfish love.

Today, when we observe nature, we often assume a scientific mindset. What is it? How big is it? How does it work? What does it look like if you take it apart? The poetic literature of the Bible has little interest in such material questions. The psalmist observed the stork in the cedars and asked not, "What does this tell me about what storks do?" but, "What does this tell me about who God is?" Observations of wild nature, accompanied by an understanding of God's creative role, brought the authors of the wisdom books face to face with an all powerful Lord and a frail humanity bent on overstepping its intended limits. The singer of the Song of Songs found that the more one enjoys the beauty of nature, the more one understands the love of God, and vice versa. For the psalmist the appreciation of the wild was one more means of glorifying God. In the *poetic model* we find, not just "seeing" and "hearing," but also "touching," "tasting," and "smelling."

The psalmist wasn't confronted with an angel or a voice from heaven, but with the countenance and song of creation. Contemporary Christianity, preoccupied with personal salvation and personal blessings, can't seem to hear the wilderness when the wilderness praises God.

In the Bible, the poet or the composer of prophecy in poetic form may be a shepherd alone in the hills or a warrior waiting for an enemy attack, a king with a special gift of wisdom or a woman dancing on the shore of the Reed Sea, a member of a priestly family or a poor tender of sycamore trees. A poetic understanding of wild nature is not limited to the full-time musician or writer, but is available to anyone desiring a heart knowledge of Yahweh.

The poetic model should be as accessible to us today as it was in David's time and is, perhaps, the easiest way to participate in "wilderness spirituality." This model does not require long journeys into the desert or fearful confrontations with large, carnivorous animals. Athletic calf muscles and a tolerance for blustery weather are helpful, but not essential. What is necessary is a willingness to see God's person and God's actions displayed in the nonhuman portions of the cosmos. An afternoon spent exploring a desert spring, appreciating a mountain meadow, or watching feeding whales can provide joyful inspiration. The diminutive has as much to teach us about God as the spectacular or grand. Squirrels gathering nuts, flowers blooming on a road edge, or dragonflies zipping over a farm pond: all speak to us about the Creator's pleasure in diversity, and care for the least to the greatest. The poetic model forces us to move outside our usual sphere of very human (and often very self-centered) preoccupations and encourages us to look at God's interests on a much more sweeping scale than our own focused problems and desires usually allow.

The poetic approach to wild nature can also encourage spiritual restoration and emotional or even physical healing. Experiencing the beauty of peace of God in Nature is not a substitute for direct interaction with the regenerative powers of the Creator, but, as with the other characteristics of God, the mending and binding so necessary to heal our stress-filled lives may flow through creation. For the spiritually oppressed or the socially injured, a pleasing or quiet natural environment can help provide spiritual release. Resting by a clear, free-running river or sitting on a sunny slope in blooming desert grassland can bring peace and joy into very clouded souls.

The poetic model can provide sound spiritual exercise for novice wilderness sojourners. Christian pastors and educators sometimes worry about nature appreciation growing into idolatry. In a culture dominated by material possessions and very erotic concepts of beauty, nature "worship" is a possibility, especially if the enjoyment of nature is limited to mere aesthetic pleasure (and thus develops as *eros*). Contemplation of nature in the light of the psalms and other relevant biblical literature will, however, discourage materialism since it demonstrates the limitations of humankind and proves the Creator's love, power, and providence. Cast against our favorite idols including rock singers, sex appeal, technological innovation, automobiles, body building, professional advancement, high sales, and television evangelists, wild nature seems like the least of our problems. Short sessions of prayer, praise, and reading the Bible in natural settings are relatively safe and, according to biblical example, conducive to spiritual growth. The question is not whether one may study nature, but how one should study it.

6

Fed by Ravens
(The Former Prophets, Elijah and Jonah)

Under the Palm of Deborah

The role of wilderness in building the careers and skills of military and political leaders has been examined; it is now appropriate to investigate the role of wilderness in the lives of the spiritual leaders of the Hebrews—the prophets. Despite God's appointment of the tribe of Levi as priests who cared for the tabernacle, the tent of meeting, and, eventually, the temple, the real voice of God in the Hebrew tradition was usually the prophet. Prophets might or might not come from priestly families, were of both sexes (unlike the priests), and were of varying social backgrounds. From the time of entry into the Promised Land until the coming of Christ, prophets not only brought the word of God to the people, but also advised on military campaigns and matters of state.

As the Hebrews attempted to take up their role as a holy people and priests to the nations, the prophets became the conscience of Israel and Judah. (The nation split into two kingdoms after the time of Solomon.) The prophet heard and spoke the Word of the Lord, yet was still sensitive to evil. When the people failed to do the Will of God, the prophet informed them of their shortcomings. The prophet, thus, was an "iconoclast, challenging the apparently holy, revered, and awesome,"[1] always pursuing the highest good, no matter what personal difficulties this precipitated. The prophet was austere and accusing, willing to identify sin and speak of destruction. Yet, at the same time, the prophet was compassionate, understanding the love of God and the needs of the people.

Because of the nature of the prophetic mission, the "forthsayer" was frequently lonely or under stress from the weight of divine vision. Driven away by threats from a king, or ostracized by his own people, the one who told God's truth often became an outcast. Prophets were welcomed by those in political power as long as they said what the leaders wanted to hear. When the leadership fell away from God, the prophet became unwelcome in Jerusalem—or anywhere else in the kingdoms.

The prophets had an association with outdoor settings, in both wild and cultivated landscapes. The prophetess Deborah, who was also a judge over Israel before the time of Saul and the monarchy, "used to sit under 'the palm of Deborah,' between Ramah and Bethel, in the 'Mountain of Ephraim,' and the people of Israel came to her for judgment" (Judg. 4:5, SG). On numerous other occasions, the prophets were found sitting under trees or on hills or mountains. In 1 Kings 13:14, when an old prophet from Bethel pursued a man of God driven from the altar at Bethel, he found him sitting under an oak. God caused a plant to grow up over Jonah (Jonah 4:6–8) to protect him from the sun. In 2 Kings 1, an officer and his soldiers found Elijah sitting on top of a "hill," or a "mountain" (possibly Carmel).[2] When the Shunemite woman came searching for Elisha to raise her son, she found him on Mount Carmel (2 Kings 4:27).

Their association with natural settings tied the prophets to very ancient visionary traditions and undergirded their occasional confrontations with the cult and with the kings. The prophet represented the pursuit of God free from cultural concerns and in the faith traditions of the forebearers.

Fed by Ravens

The association of the prophets with natural objects or unpopulated locations had several motifs. One element was purposeful withdrawal. The story of the interactions between Elijah, "the settler," and king Ahab and his wife Jezebel, incorporates several rounds of confrontation and retreat. Elijah, in fact, introduced himself to Ahab by predicting there would be a drought in the land until Yahweh said otherwise, and:

> Then the word of Yahweh came to him [Elijah]: "Go hence and head eastward. And you shall conceal yourself in the wadi Kerith which lies on the opposite bank of the Jordan. Now it is arranged for you to drink from the wadi; and I have appointed ravens to provide for you there." So he went and did according to the word of Yahweh: he stayed in the wadi Kerith which lies on the opposite bank of the Jordan. And the ravens would bring him bread and meat in the morning, and bread and meat in the evening. And he drank from the wadi.
>
> (1 Kings 17:2–6, DeV)

Elijah, whose name means "Yahweh is my God," appeared abruptly to Ahab and disappeared just as abruptly. The purpose of the drought was not to burden the people, but rather it was to prove the power of the Lord, over both Ahab and Baal, the god of Jezebel. Baal was supposed to be the God of rain and dew and to ride on a thunderstorm; the drought, therefore, was a direct affront to Baal's supposed power.[3]

The text does not indicate Ahab threatened Elijah at this time, but Jezebel had already opened an attack on Yahweh's prophets and replaced them with prophets of Baal. God, presumably for the prophet's own protection, sent Elijah to an isolated locale "to conceal himself." Elijah was probably a settler from Gilead, which was "wild, forested, and large unsettled at this period,"[4] suggesting that Elijah was already adapted to the outdoor life. The purpose of his wilderness sojourn at wadi Kerith, however, was to cut Elijah off from all supplies of food and to make him completely dependent on God. Elijah, willing to make himself vulnerable and to rely on the Lord, found ample water in the brook and ample provision in the bread and meat the ravens brought.[5] The story is a parallel to the manna and the water supplied by God during the Exodus. As in the case of Moses and the children of Israel, wild nature protected the beloved of God. Elijah, however, unlike Israel in the Exodus, remained within God's will. Thus when the brook finally ran dry because of the lack of rain, God sent Elijah to the house of a widow in Zarephath in Sidon, where a jar of flour and a cruse of oil were miraculously never depleted.

The theme of withdrawal evolves into "hiding" and "flight." In 1 Kings, Elijah encountered a companion of Ahab's named Obadiah on the road. The Scriptures describe Obadiah as "exceedingly reverential toward Yahweh" and a man who, in the face of Jezebel's campaign to destroy the prophets, had hidden one hundred of them in two caves and had provided them with bread and water (1 Kings 18:3–4). This passage indicates that prophets other than Elijah had also withdrawn to wild sites—in this case, clearly because of the danger of physical harm. The prophets in hiding, however, were an interesting contrast to Elijah, who went off to speak to Ahab and then challenged Ahab to call the prophets of Baal to Mount Carmel (1 Kings 18:17–19). The prophet who had withdrawn at the direction of God and then, acting on God's instructions, had returned to confront Ahab successfully opened the battle against the false gods and their worshipers.

In the series of narratives about Elijah, withdrawal and confrontation become countermotifs. Not only the "flights," but also the "fights" occurred in wild natural settings. After Elijah returned from wadi Kerith, Elijah had Ahab call 450 prophets of Baal, 400 prophets of Asherah, and all the people of Israel (the northern half of the divided kingdom) to Mount Carmel. Elijah put the terms of the meeting very directly: "If Yahweh is God, go after him, and if Baal is, go after him" (1 Kings 18:21, *DeV*). Elijah directed the prophets of Baal to lay out an ox on a pile of logs. The prophets of Baal called on their god, but he did not answer. They leapt around their offering, but nothing came of it. Elijah followed their effort by repairing a damaged altar of Yahweh, laying out an ox on it, and building a ditch around the altar. He had the people fill jars with water and soak the ox and the wood, until the water ran all around the altar and filled the ditch. Elijah then offered oblation acknowledging Yahweh as God of Israel and "Yahweh's fire fell and consumed the sacrifice and the logs, the stones and the dust; and it even licked up the water that was in the ditch. And all the people saw it. And they fell on their faces and said, "Yahweh, he is God! Yahweh, he is God!" (1 Kings 18:38–39, *DeV*). The prophets of Baal were not only soundly beaten by Yahweh, the people seized them, took them down to the River Kishon, and dispatched them. Temporarily impressed by the mighty act of God in consuming both ox and altar, the people acknowledged Yahweh as Lord.

Not satisfied with proving the point by merely embarrassing false prophets, God, working through Elijah, also brought an end to the drought. Elijah warned Ahab it was going to rain. Then the prophet went up on Carmel and sent his lad to look for a cloud from the sea. When the cloud finally appeared, Elijah ordered the boy to

> "Rise up, say to Ahab, 'Hitch up and get on down, lest the rainstorm bog you in!'" And it did happen that here and there the sky began to grow dark with clouds and wind. And pretty soon there was enormous rainfall, with Ahab riding to get back to Jezreel. And the hand of Yahweh came upon Elijah. And, girding up his loins, he ran in front of Ahab as far as the approach to Jezreel.   (1 Kings 18:44–46, *DeV*)

After hearing the sound of rain which was not yet visible, Elijah climbed to the summit of Carmel in order to pray and to await the promised storm. The call to Ahab to return to Jezreel before the rain is ironic and served as a further sign of Ahab's lack of control. Ahab needed the rain badly, yet would not petition Yahweh for it. Now that the rain was coming, it threatened to trap Ahab, just as the Reed Sea trapped Pharaoh. Striking a final blow to Ahab's feeling of control, Elijah ran ahead of Ahab's chariot going into Jezreel.

Just as the Lord made himself visible on Sinai, the Lord made himself visible to Ahab and to Israel on Carmel. Yahweh did not show his face, but rather demonstrated two personal characteristics—power and omnipotence. First, by sending fire, Yahweh became the God above all gods, specifically Baal and Asherah. Second, by sending rain, Yahweh became the God of the elements and intruded directly on Baal's prerogatives. Elijah's role in both cases was to clarify Yahweh's control in these events. This was especially important in the coming of the rain, since, without the prophecy of Elijah, the prophets of Baal might have taken the credit.

The mount itself was doubly symbolic. First, Carmel provided a parallel to the experience of Israel on Sinai. The mountains were places where the Lord descended in a cloud and in fire. Second, Carmel was an "impressive topographical feature . . . referred to in Scripture as the very epitome of vegetative fertility, no doubt because it is crowned by a dense and beautiful forest."[6] Carmel was, therefore, claimed by Baal, the god of fertility, rainfall, and new growth. To challenge Baal on Carmel was to challenge the god on his own favored territory. An altar of Yahweh had been placed on the mount, but had been damaged. Elijah not only rebuilt the altar, he displaced Baal and began to rebuild the people of God.

Not in Wind, Earthquake, or Fire

After Elijah's victories on Carmel, Ahab told Jezebel about the destruction of her prophets, and she in turn sent a message to Elijah to tell him she intended to have him killed.

> And he was frightened; he arose and went for his life, arriving at Beersheba, belonging to Judah. And he dismissed his lad there, while he himself went on into the desert for a day's journey. And he came and sat down beneath a certain broom tree. And he asked for his life, so that he might die; and he said, "Enough now, Yahweh, take my life, for I am no better than my forefathers!" Then he lay down and slept beneath the broom tree. And behold, someone was prodding him and saying to him, "Get up and eat!" And he looked, and behold, near his head was a cake made on heated stones, as well as a cruse of water. So he ate and drank, then went back to sleep. And the angel of Yahweh came back a second time and prodded him, saying, "Get up and eat, for the trip is too far for you." So he arose, ate and drank, and walked on in the strength of that food for forty days and forty nights, as far as the mountain of God. (1 Kings 19:3–8, DeV)

The prophet who had so bravely reprimanded the king was now fleeing from the queen. The geography and the route taken into the desert represent the prophet's relationship to his mission. Elijah headed first for Beersheba, the outer edge of Yahweh's land. When he left his servant there and headed into the Negeb desert, he was not only entering the wilderness, he was also leaving the realm of royal political control and the realm of his own ministry. The withdrawal in this case was symbolic of Elijah's complete despair and abandonment of his prophetic call.

Ironically, he retreated because of fear for his life, but, on entering the desert, sat down under a broom tree and sought death.

The theophany in these passages was very like the angelic encounters of Hagar in desperation in the desert. The angel made simple provisions, a warm cake and some water. Elijah was so "burned out," the angel had to speak to him and feed him twice. Surprisingly, after this strengthening, the angel did not send Elijah back to Israel, but rather sent him deeper into the wilderness—all the way to Horeb and the mountain of God. What followed would come to be considered the classic "prophetic experience." The Word of Yahweh came to Elijah and asked why he was there, and, when Elijah reported that he alone among the people was still faithful to the God of Israel, the classic "prophetic experience" followed:

> And he said, "Go out and stand on the mountain before Yahweh." And be-hold, Yahweh was passing by, and a wind great and strong was tearing up the mountains, shattering the rocks before Yahweh. Yahweh was not in the wind. And after the wind came an earthquake; Yahweh was not in the earthquake. After the earthquake came a fire; Yahweh was not in the fire. And after the fire came a gentle little breeze. And so it happened that, when Elijah heard it, he wrapped his face in his mantle and going outside, he stood at the entrance to the cave. And behold, a voice came to him and said, "What is it with you here, Elijah?" And he said, "I have been furiously zealous for Yahweh, God of Hosts; for the Israelites have forsaken thy covenant, thine altars they have over-turned, and thy prophets they have slain with the sword. And I survive—I alone—but they are attempting to take my life."   (1 Kings 19:11–14, *DeV*)

Yahweh then ordered Elijah to return through the wilderness to Damascus. His mission was to anoint new kings over Syria and Israel, as well as to anoint Elisha as a prophetic successor. The Lord also informed Elijah that he was not alone, there were seven thousand in Israel who had never served Baal (1 Kings 19: 15–18).

In this wilderness sequence, Elijah withdrew about as completely as it is pos-sible for a man to withdraw. He traveled forty days into the desert and then sat in a cave. The voice of God asked him what he was doing there, and the prophet poured out his troubles and frustrations before God. Elijah felt that he was alone, the only one of the faithful left in the land, and that he too ultimately would die at the hands of the wicked Jezebel.

The Lord sent Elijah out on the same mountain where Moses and the chil-dren of Israel had seen God, but the theophany experienced by Elijah was quite different than the theophanies of the Exodus. Previously Yahweh had appeared in the wind, the fire, and the earthquake. This time these phenomena passed in front of Elijah, and Yahweh was not in them. In the famous "still small voice," or "gentle little breeze," Elijah encountered the person of God in a remarkable way. The prophet who had caused the fire of God to fall on Carmel and had prayed until a terrific thunderstorm drenched Ahab engaged a God beyond the force of the elements—and beyond the great energies of creation. Elijah was not allowed to grasp or control the being of God in the way the prophets of Baal attempted to do. Elijah experienced God in his infinite holiness and boundless majesty. "In this drastic moment, Elijah becomes freshly aware that it is Yahweh and none other and not himself, who is God. All his fears are set in their proper place. There is only one to fear, the same one who must be obeyed."[7] At Carmel, Yahweh showed himself more powerful than Baal, but the battle had been conducted in terms the fertility-oriented followers of Baal would understand. On Horeb, the God of Israel showed Elijah that the person of Yahweh could not be comprehended by degrees of power—or by comparison to the person of other gods. Yahweh was one, and Yahweh alone was God.

The voice of God also told Elijah that he was not the only faithful person left and that his prophetic tasks had not, by any means, ended. As with the Exodus, the isolated encounter in the desert was not an end in itself, but the beginning of a renewed ministry. Elijah was transformed from a depressed and failing servant, with no hope for himself, the nation, or God's mission, into a man who would bring about major change in the political scene and whose ministry would be followed by that of another great prophet. In the cave and on the mountain, Elijah presented his misconceptions to God, and God refuted them. Elijah had begun to think too much of his own strengths and weaknesses and his own role in ministry and had lost sight of the Lord. The sojourn in the wilderness deepened his knowledge of God and thereby erased his false perceptions of himself. He left the wilderness with renewed spiritual authority and renewed zeal for the difficult tasks before him.[8]

Taken up in a Whirlwind

In contrast to his wavering retreat into the Negeb, the end of Elijah's earthly ministry was marked by a purposeful crossing of the Jordan, near Jericho. Accompanied by Elisha, Elijah took his garment and struck the waters, and they parted. On the other side, "a fire chariot and fire horses" came between the two, and Elijah went up in a whirlwind to heaven (2 Kings 2:4–12). The fiery chariot here may well be a form of theophany, but neither angelic messengers nor the Lord appear to have been present. The whirlwind was a natural phenomenon, found elsewhere in the prophetic literature and often symbolic of judgment.[9] The passages do, in any case, describe a mighty act of God, similar to the crossing of the Reed Sea.

In this narrative, Elijah again left Israel via a wilderness route, but this time he departed as victor. The similarities to the Exodus were purposeful and demonstrated to the sons of the prophets, waiting on the far bank, the continuing action of God in Israel. The incident also provided the opportunity for the transfer of power and spiritual leadership to Elisha, who asked for a double portion of Elijah's spirit. Elisha proved his new role when he took Elijah's garment, struck the waters of the Jordan, and returned on dry ground to the sons of the prophets.

The tale includes the theme of transformation, both in the transfer of prophetic responsibilities and power and in the ascension of Elijah to heaven. His fears gone and his tasks complete, Elijah entered the whirl-wind. In this ultimate journey into wild nature, Elijah rose—bypassing death—to meet Yahweh once more. Elijah's triumphant departure was very much in contrast to Elijah entering the Negeb and asking God to let him die.

Swallowed by a Great Fish

The wilderness adventures of Elijah present an interesting comparison to the life of a later prophet, Jonah. God originally called Jonah to an urban mission—preaching repentance to the citizens of Nineveh. Jonah responded by fleeing to the west, finding a ship, and trying to get away from God as quickly as possible. God pursued Jonah and sent a great wind to stir up the oceans and the fear of the ship's crew. When they identified Jonah as the person who had fallen into God's disfavor, the crew promptly threw him overboard. Whether Jonah was swallowed by a great white shark, a whale, or some other sea creature is not the central issue. The enormous animal both saved Jonah from drowning and trapped him in the depths of wild nature. The recalcitrant prophet found himself on a very uncomfortable "wilderness sojourn."

Jonah needed time to think about things, and this was arranged in the wildest province of the greatest wilderness—the ocean. After offering a prayer containing a patchwork of verses from the psalms, Jonah declared "Deliverance belongs to the Lord" (Jon. 2:9), and the fish vomited him out. Complicated interpretation is unnecessary here. Jonah ended up in the digestive tract of a large marine organism because he was running from God.

After returning to dry land, Jonah carried the message to Nineveh, and the city repented. When God decided to call off the threatened destruction, Jonah became angry, rather than rejoicing, because his predictions were no longer accurate. Jonah went outside the city to wait and see what would happen:

> And the Lord God appointed a plant, and made it come up over Jonah, that it might be shade over his head, to save him from his discomfort. So Jonah was exceedingly glad because of the plant. But when dawn came up the next day,

God appointed a worm which attacked the plant so that it withered. When the sun rose, God appointed a sultry east wind, and the sun beat upon the head of Jonah so that he was faint; and he asked that he might die, and said, "It is better for me to die than to live." But God said to Jonah, "Do you do well to be angry for the plant?" And he said, "I do well to be angry, angry enough to die." And the Lord said, "You pity the plant, for which you did not labor, nor did you make it grow, which came into being in a night, and perished in a night. And should I not pity Nineveh, that great city, in which there are more than a hundred and twenty thousand persons who do not know their right hand from their left, and also much cattle?"   (Jonah 4:6–11)

In these passages, the rise and demise of the prophet's "tree" becomes a counterexample of God's true concerns. God made both the plant and the people of Nineveh. Jonah's preoccupation with the plant is compared to God's interest in this non-Jewish people. The plant or tree for shade became symbolic of the prophet's role, and Jonah looked more to his Hebrew righteousness than to God. The message to the postexilic Hebrew community was God's concerns for humanity were far beyond their limited nationalistic view.[10]

The story of the plant is a mirror image of the story of Elijah under the broom tree. In the case of Jonah, whose heart is in the wrong place, God takes away the tree and sends the east wind (the wind is on the side of the faithful prophet, such as Moses) to show Jonah his lack of understanding. In the case of Elijah, an angel provided food and water to strengthen Elijah for a difficult journey. In both cases, the prophets were shortsighted, and, in both cases, God used wilderness experience and natural features to teach lessons. Jonah, who continued to argue with God and to try to force human values on the divine, ended up lying in the heat and dust outside a foreign city. Elijah responded to God's voice and went on to further ministry.

A Lion by the Corpse

The more negative interactions of prophets with the wilderness environment extended past the provision of contemplative interludes and natural examples of God's will. A number of passages about the prophets portray wild nature as the avenger—or as a destructive force. In 1 Kings 13, for example, in a story about true and false prophecy, a prophet who had disobeyed Yahweh was killed by a lion: "And his corpse lay tossed in the road while the ass [he had been riding] remained standing alongside him; and the lion continued to stand alongside the corpse" (1 Kings 13:24, DeV). The fact that the lion had neither eaten the corpse nor attacked the ass were taken as signs of God's action in the incident.[11]

In a similar case, Elisha "was going up on the way [to Bethel], some small boys came out of the city and jeered at him, saying, 'Go up you baldhead! Go up, you baldhead!' And he turned around, and when he saw them, he cursed them in the name of the Lord. And two she-bears came out of the woods and tore forty-two of the boys. From there he went on to Mount Carmel and from thence he returned to Samaria" (2 Kings 2:23:25). This incident seems brutal to the modern reader and needs to be put in its Hebrew context. Elisha had just taken the ministry from Elijah at this point and was retracing Elijah's route before the elder prophet had crossed the Jordan. Elisha had also just performed a mighty act and "healed" the undrinkable water at Jericho. The boys did not understand the power or role of the prophet and came against him, much as the civil authorities had been doing. The narrative provides a contrast between Elisha putting salt in the water and blessing it so that the water might become clear and potable and Elisha cursing the boys and the she-bears descending on them. The text again makes it clear that the conflict was not with the wilderness in any way, but with the Lord. In the Hebrew Scriptures, all the Law and the ordinances of Yahweh apply to all the people, regardless of age (there are some special instructions concerning virgins) and any other social factors. The passage speaks to the absolute spiritual authority of the prophet and the risks of mocking God's chosen. As any Hebrew shepherd would have known, ordinarily two bears could not attack and maul forty-two boys. The bears prove prophetic authority is awesome and inscrutable, not to be treated lightly by those of immature vision.[12]

It should be noted that the concept of large wild animals responding to or acting for God continues late into the activities of the prophets. Daniel, for example, was a captive in Babylon when the king had him placed in the lions' den. After a long night in the den, Daniel reported to the king: "My God sent his angel and shut the lions' mouths, and they have not hurt me, because I was found blameless before him; and also before you, O king, I have done no wrong" (Dan. 6:22). In this case, God acts not by inspiring the animals to behave in a particular way, but by sending a messenger to physically interfere with the lions. The angel was not present, however, because God could not have remotely quieted the lions, or because the lions would not cooperate with God, but because a divine emissary was needed to make the king understand the event was an act of God. At the end of the story, the king has Daniel's accusers and their families thrown to the lions, and, executing prophetic justice, the big cats "overpowered them and broke all their bones in pieces" (Dan. 6:24).

The Gentle Little Breeze

Many of the elements in the *prophetic model* of wilderness experience are also found in the *Exodus model*. In the prophetic model, wild nature was very much on God's side and became an important agent of God's will. The wilderness was a site for revelation and theophanies, usually in isolation. The stress of the wilderness was associated with divine providence, and divine provision, such as the feeding by ravens, could be through natural agents. Wilderness remained a place for transformation. Elijah came back from the Negeb with a new purpose and a new mission. Jonah reconsidered and headed for Nineveh (although this was a false transformation in terms of a true understanding of God's will). Even Elijah's departure from earth was in a whirlwind.

The story of Jonah is a "counterwilderness" tale, with strong parallels to the Exodus. Jonah took an accidental isolated trip, presumably prayed his way out, but, in the end, still did not have a heart understanding of the will of Yahweh. The stay in the belly of the great fish transformed Jonah to the point at which he was willing to go on his prophetic mission, but not to the point at which he could avoid an argument with God over a merciful act. In Jonah, the story of the Exodus is replayed by an individual. If the people could be stubborn, so could a prophet.

The prophetic model does add some new concepts, however. Elijah's stay in wadi Kerith was a divinely ordained retreat, which did not include stress or struggling during the sojourn—or any rebellion or argument on the part of Elijah. The text suggests the prophet could enter the wild site and patiently wait on God.

Second, the "still small voice" was a new understanding of the person of God. After calling fire down on Carmel, Elijah was expecting the God of Moses on the mount. The "gentle little breeze" was the voice of a God who could operate on the inner person without all the power of the thunder and smoke. Elijah's encounter at the cave revealed not only God's will, but also Elijah's mistaken concept of it. The conversation with God caused a spiritual and psychological transformation that brought a far more obvious personal change in the prophet than did Moses's view of God from the crevice in the rock. The God of the gentle little breeze is also a very different god than Baal, whose person relies primarily on power over nature. The God of the gentle breeze is a god: (1) who is of all scales and concerns and all levels of action; (2) who moves in the spirit of humankind as well as in nature; and (3) who can bring human souls to fruit in the most arid places on earth.

We can summarize some of the characteristics of the prophetic model:

1. Prophets were frequently associated with wild sites or natural objects. Although they might enter the temple or tend the altars, they were also found on mountains and under trees. Their walks on the wild side partially freed them from the social order surrounding the royal courts and the official religious sanctuaries.
2. Prophets encountered God in the wilderness and entered deeper understandings of their missions in isolated sites.
3. The prophetic wilderness encounter with the divine could be deeply personal. The theophanies of Genesis and Exodus and the meetings

with angels still occur, but there are also the conversations with "the quiet voice of God."

4. Prophets entered the wilderness under a number of different circumstances including divine call, flight, and forceful ejection.

5. Prophetic experience was not always stressful or marked by rebellion. Elijah was "in tune" with the environment at the wadi Kerith.

6. The action or reaction of wild nature reflected the will of Yahweh in the lives of the prophets. This was especially noticeable in the case of interactions with large animals. The Scriptures report cases where animals killed prophets, captured prophets, acted on behalf of prophets, and did not harm them at close quarters.

7. Wilderness was an important location for spiritual transformation. The prophet who went into the wilderness returned with new strength for ministry, a new word, or in the case of Elisha, twice the spiritual power of his mentor.

8. Prophetic wilderness sojourns were often followed by a return to confront the problems of earthly kingdoms.

Looking back on the prophets, we recognize their mission was social. The wilderness, however, provided isolation—not only as protection from antagonists, but also as a time for personal reflection. Today, we have a need for the same type of strengthening Elijah experienced and for the same intimate knowledge of God. We may need to remove ourselves from conflicts with others, or to divest ourselves of the religious notions of our culture, in order to move towards God.

Elijah lived in an era when physical and political power dominated. His "gentle little breeze" would have made no sense to Ahab. Today, we have more concern for our spiritual and psychological state and perhaps expect the "still small voice" rather than thunder and lightning. The transformation available to us in the wilderness is limited not so much by an inability to hear God in the quiet as by our belief that God will depart from us when we leave the mountain. Both Elijah and Jonah found God stayed with them after they returned to their ministries. In our contemporary world we tend to confine religion to certain places and circumstances. If we seek God in the wilderness, we can't quite believe we will find the divine also back in civilization. The better seeing and hearing in the wilderness are not owing to a greater presence of God, but result from the removal of human pressures and clutter. Once a deeper understanding is reached, it should stay with us on the trip home. Once we have seen something clearly in the desert, we should be able to retain the vision as we return to community.

The importance of wilderness in the lives of the prophets presents a challenge to modern Christians. Wilderness spirituality hardly characterizes today's western church leadership. Are we having problems with false gods (i.e., Mammon) because isolated contact with God is rarely sought by those who presume to speak for the Lord? After the Exodus, the wilderness sojourn was not a common experience of the Hebrew people, but repeatedly fell upon the civil and spiritual leadership of the nation (although not always by choice). The Bible implies that wilderness fosters dependence on the divine, vastly improved spiritual vision, and the drive for new ministries. A rather short stay of one to forty days can change the course of holy history. Should our leaders also seek the cave or the mountain?

7

Jackals in Her Palaces
(Later Prophets)

The Storm Surge [1]

Taking an evening walk along the beach at Cumberland Island National Seashore, I diverted from the soft sand and strolled down to the moister, firmer substrate between the tide lines. The waves had left a wide line of shells after yesterday's storm. I crunched along on top of them, stopping to pick up an exceptionally large pen shell and admire its iridescent interior.

Meandering southward, I encountered a graveyard of scattered whelks and turned over a few to make certain the high water hadn't stranded any living animals. Each time I saw a purplish gray snail withdraw into her calcareous house, I took a healthy swing and tossed the storm survivor back into the surf.

The great treasure trove of shells extended the whole length of the island, from the wide sandflats at the south end jetty, past the glistening white dunes at Dungeness, to windswept Long Point. Yesterday's storm had washed up over the dune-stabilizing sea oats and had left marine life and driftwood littered everywhere. The sea had moved boxes, logs, and fragments of boats.

In my studies of Georgia Sea Island history, I had read of the great hurricanes in the nineteenth century. A storm in 1898 had thrown a twenty-two-foot storm surge on top of an eight-foot high tide over nearby St. Simons and Sapelo Islands. The water had been almost as high where I was now standing. I imagined thirty feet of ocean sweeping over me, over the high back dunes, over the marsh, and on to the main land. "It would have covered the whole south end of Cumberland," I realized. I pictured the island after a major hurricane—dunes flattened, beaches shifted, sand piled in trails, flotsam in the surviving shrubs.

My mind wandered back to St. Simons, where today summer homes stand right at the edge of the beach. I saw the storm surge breaking second story windows and carrying furniture away. I saw new dunes forming where the buildings had been. In 1898 after the water subsided, the survivors found human bodies tangled in the branches of live oak trees. I couldn't quite bring myself to project the fate of twentieth-century residents foolish enough to test the power of the storm.

"We've lost our respect for wild nature," I concluded as I strolled down to the granite jetty. "We think because we can place some rocks where they accumulate sand or can use heavy equipment to dredge out a channel, we can do anything . . . or withstand anything. No plantation owner would ever have situated permanent dwellings next to the spring tide line. Enough people drowned with the buildings on higher ground. It's as if we have no memory, or feeling for what the past tells us about the future. The islanders knew the sea will always have the final say about where the beach would be."

* * *

Cumberland Sound was covered with late afternoon fog as I headed down the River Trail from Sea Camp to Dungeness. In recent years, a series of northeasters had raised tides that cut into the low bluff at the edge of the Intercoastal Waterway. Their roots undermined by the erosion, mature hardwoods and loblolly pines had fallen into the water. The waves had already eaten away a road constructed in colonial times and were now working on the edge of the trail. I climbed out on a live oak, tilting laterally from the bank, and found an outlook point. On this quiet weekday, with little boat traffic, the drizzle muted every sound except the slap of the waves on the mud flat below my tree. I dangled my legs out over the diminutive breakers and rested silently, enjoying the movement of the water.

Suddenly a huge black shape loomed out of the salty mist. Taken by surprise, I almost fell from my perch. In the light rain and low light, the threatening object appeared as an undefined dark shadow heading my direction at a respectable speed. It was too tall to be a shrimp boat or a barge and sported neither masts nor rigging. I grabbed an oak branch and strained to see through the gray atmosphere, vaguely hoping the mysterious thing was a sea monster. As the distance closed, the outline of the hull and the conning tower slowly consolidated—it was a Poseidon submarine.

I had never seen a submarine underway before, and the vessel was much longer than I had expected. It seemed out of place in the sound—too large and too military for the Georgia salt marshes. The National Park Service had, in fact, been wrestling with the Navy over the plans to deepen the channel for an even bigger sub, the Trident. The increased dredging could further erode shoreline and cause changes in the tidal flow in and out of the estuaries. Shellfish and shorebirds make their living along the channel. Manatees, large vegetarian marine mammals, forage leisurely on the marsh grasses growing on the banks of the tidal creeks. Little white egrets step carefully through shallow water in search of small fish. Peaceful creatures, ill-prepared for the industrial age, they can't defend themselves against the dredges.

As the Poseidon turned towards Kings Bay Naval Submarine Base, I weighed the vulnerabilities and the strengths of nature—and of human beings. If the least terns and the snowy egrets did not like the changes in water quality, they would have no choice but to leave. A sub or even a tug could easily kill a manatee with its propeller. The sea, however, is more difficult to displace or to predict. A serious hurricane might close the channel and block the subs out of the Navy base. And then there is all the power in the dozen or so missiles inside those silent dark craft. Can we really hold nature by the tail and have her do our bidding without having her turn around and devour us? I thought about the biblical prophets and their respect both for the will of God and for the power of God expressed in wild nature.

Jeremiah (4:23–28) wrote, in regard to fallen Jerusalem and the demise of God's holy nation:

> I looked on the earth, and lo, it was waste and void;
>   and to the heavens, and they had no light.
> I looked on the mountains, and lo, they were quaking,
>   and all the hills moved to and fro.
> I looked, and lo, there was no man,
>   and all the birds of the air had fled.
> I looked, and lo, the fruitful land was a desert,
>   and all its cities were laid in ruins
>   before the Lord, before his fierce anger.
> For thus says the Lord, "The whole land shall be a desolation;
>   yet I will not make a full end.
> For this the earth shall mourn, and the heavens above be black;
>   for I have spoken, I have purposed;
>   I have not relented nor will I turn back."

Jeremiah obviously thought the fate of the birds and the fate of human beings were linked together in the hands of God.

As the Poseidon disappeared into its harbor, I contemplated our disinterest in the prophetic understanding of nature and our ignorance of the extent of human failing and human weakness before God. I imagined the wrath of all that is mighty turned full force in the power of creation.

Jeremiah stood at the edge of the ocean wilderness shaking his head. The sky brightened and then darkened, and the ocean roared. Egrets flew up in panic, and the gulls responded with terrified, hoarse cries. The waves leapt and crashed in a dozen directions, driven by earth tremors, rather than by the predictable pull of the tides. As the waters of the Sound drew back, the green marshes suddenly burst into flame and withered away, charred and barren. The soft mud of the shoreline turned into a fine dust blowing over the yellowish skeletons of thousands of stranded sea creatures.

Providence fled.

Chaos invaded.

I found myself—no longer myself—tangled in the branches of a blackened oak tree, above a dry, desolate shoreline. All I heard—all I perceived was the voice of Jeremiah. He sang of the failure of God's people and the destruction of the cosmos. His lament rose and fell, matching the cadence of the waves as the lunar tides slowly returned, and the ocean joined his sorrowful wail. Deep within, from some ancient understanding, I heard the sea participating in God's grief. Her song was high pitched and fractured, rising above the low moans of the wounded Sound. And I knew she was mourning . . . mourning for those who loved her and loved the Creator . . . mourning for the manatee, the egret—and me.

*   *   *

In our contemporary treatment of Christian wilderness traditions, we tend to shy away from wrathful passages and from those dealing with God's displeasure. In our battles to preserve wild lands and free roaming fauna, we want only positive, productive images of wilderness. In the biblical texts, however, wilderness association with wrath is linked with God's blessing and a prophetic understanding of the frailty of humankind. A careful reading of the relevant passages should make us thoughtful, rather than causing us to feel threatened. The words of the prophets do not disparage the wild, but rather display nature as fully responsive to the will of God.

Let Justice Roll Down like Water

In the prophetic tradition after the time of Elisha, we find dozens of images from wild nature—portraits of deserts, mountains, lions, wolves, and cedar trees—woven into the visions of the past, present, and future of the people of God. Although some of the later prophets, such as Amos of the shepherds of Tekoa, were clearly from rural backgrounds, the prophetic books such as Amos, Isaiah, and Ezekiel do not emphasize wilderness sojourns of the authors. They discuss the Exodus, however, and predict new wilderness wanderings having equal importance in the history of God's action on earth. They also discuss wild nature as an instrument of God's will, both for taking vengeance and for providing blessings.

As with the *poetic model,* the prophets do not always present actual events in these writings, but may use wild nature and wilderness allegorically. In some cases, the prophet may indeed be describing a real occurrence, but even then the prophet employs these images from wild nature for *emphasis,* rather than to provide historic detail.

The later prophets come from the time of the divided kingdom, when the twelve tribes had split into two nations, Israel (the northern kingdom) and Judah (the southern). The ministries of the later prophets are to be understood in relation to two major political events, Assyria's invasion of the northern kingdom and the consequent dispersion of Israel's populace to other kingdoms, and the fall of the southern kingdom and its population's exile in Babylon, which occurred over a hundred years later. The spiritual fidelity of David had faded from the land, and several of Israel's and Judah's larger neighbors were attempting to become world powers at the expense of smaller or less well-armed nations. The northern kingdom had engaged in officially sanctioned idol worship, although, if one believes Jeremiah, false gods had also spiritually captured the city of Jerusalem (which is in Judah) well before King Nebuchadrezzar conquered it militarily.

The later prophets were preoccupied with the sins of the people and God's impending judgment. Amos, the first to speak out against the religious and political condition of the northern kingdom, chastened the Israelites for their injustice and their exploitation of the poor. In the writings of Amos, we find the beginnings of three important themes that will appear and reappear in this continuing story of a nation's falling away from God, followed by divine judgment and reunion with God.

The first of these themes is the integrated relationship between God's role as creator and God's role as judge and savior. This concept is also found in Psalms, but it takes a far more powerful form in the later prophets. After he has listed the transgressions of Israel and the surrounding nations, including Tyre and Moab, Amos, speaking for Yahweh, declares environmental disaster has fallen upon Israel in the form of drought, blight, locusts, and pestilence; and yet Israel has not changed her ways. Amos 4:12 extends the famous warning: "prepare to meet your God, O Israel,"[2] and then, in verse 4:13, describes the coming avenger as "he who forms the mountains, and creates the wind, and declares to man what is his thought; who makes the morning darkness, and treads on the heights of the earth." This motif is repeated when Amos rebukes Israel for unrighteousness and then threatens:

> He who made the Pleiades and Orion [constellations],
> and turns deep darkness into the morning,

> and darkens the day into night,
> who calls for the waters of the sea,
> and pours them out upon the surface of the earth,
> the Lord is his name;
> who makes destruction flash forth against the strong,
> so that destruction comes upon the fortress. (Amos 5:8–9)

God's power in creation is displayed as the source of God's power in judgment and also as a source of God's right to judge. A nation worshiping false gods will not recognize this, nor will a nation ungratefully consuming the fruit of the land without providing for Yahweh's beloved poor and dispossessed. In the poetic language of the prophet, God's justice takes the form of creation when the prophet petitions: "let justice roll down like waters, and righteousness like an everflowing stream" (Amos 5:24).

The second theme found in Amos is the participation of wild nature, or the nonhuman portion of the cosmos, in the Creator's judgment. Amos uses wild nature to portray the state of the nation and to display the response of the Lord to apostasy. Attacking idolatry, Amos 1:2 states: "The Lord roars from Zion, and utters his voice from Jerusalem, the pastures of the shepherds mourn, and the top of Carmel withers." The roaring of the Lord reminds the reader of a lion or a thunderstorm, in this case, bringing drought across the face of the land. The imagery of predators is repeated in 3:8: "The lion has roared; who will not fear? The Lord God has spoken; who can but prophesy?" And again in 5:18–19 in a dire prediction of the fate of the nation:

> Woe to you who desire the day of the Lord!
> Why would you have the day of the Lord?
> It is darkness, and not light;
> as if a man fled from a lion,
> and a bear met him;
> or went into the house
> and leaned with his hand against the wall,
> and a serpent bit him.

Just as the lion killed the erring prophet in 2 Kings, the Lord and the day of the Lord both become predatory beasts and consume the erring people. Although these texts portray lions and snakes as dangerous animals, the predators are neither evil nor unrighteous. The wild beasts execute the judgment of God and are not party to the sins of the nation. In threats of divine action—on a national or international scale—the imagery of wild nature is used to display the power of Yahweh.

The third theme (presented in a very basic and undeveloped form in Amos) is that the Exodus in the wilderness was to have been a time of proper worship ("Did you bring to me sacrifices and offerings the forty years in the wilderness, 0 house of Israel?" [Amos 5:25]) and of grace:

> Yet I destroyed the Amorite before them,
> whose height was like the height of the cedars,
> and who was as strong as the oaks,
> I destroyed his fruit above,
> and his roots beneath.
> Also I brought you up out of the land of Egypt,
> and led you forty years in the wilderness,
> to possess the land of the Amorite.   (Amos 2:9–10)

The prophet reminded Israel of Yahweh's previous care and suggested the Israelites were no better at avoiding idolatry in his time than they were in the time of Moses. The nations still did not value God's love and protection.

I Will Allure Her, and Bring Her into the Wilderness

Hosea, assuming his ministry shortly after Amos and also directing his efforts towards the sins of the northern kingdom, was just as hard on the people and as critical of their spiritual state as his predecessor. In the work of Hosea, however, we find a type of wilderness romanticism barely developed in Amos. Hosea expanded the theme of "God's protection of wandering Israel in the desert" into a desired return to the wilderness to commune with God. In this "new Exodus," the Lord would call Israel back into the wilderness to purify the nation. Hosea 2:3 depicts the nation of Israel as an unfaithful wife, whom the Lord threatens to "strip . . . naked and make her as the day she was born" and to "make her like a wilderness [desert] and set her like a parched land and slay her with thirst." The reference to arid lands is not to an existing wilderness, but to a threatened drought. The sins of the people will effect the state of the land.[3] The prophet continues to speak for Yahweh: "And I will lay waste her vines and her fig trees, of which she has said, 'These are my hire, which my lovers have given me.' I will make them a forest, and the beasts of the field shall devour them" (Hos. 2:12). Creation again participates in God's judgment against Israel.

These negative images are countered by Yahweh's promise:

> "Therefore, behold, I will allure her,
> and bring her into the wilderness,
> and speak tenderly to her.

> And there I will give her her vineyards,
> and make the Valley of Achor a door of hope.
> And there she shall answer as in the days of her youth,
> as at the time when she came out of the land of Egypt."
> "And in that day," says the Lord, "you will call me
>   'My husband,'
> and no longer will you call me, 'My Baal.'
> For I will remove the names of the Baals from her mouth,
> and they shall be mentioned by name no more.
> And I will make for you a covenant on that day
> with the beasts of the field, the birds of the air,
> and the creeping things on the ground;
> and I will abolish the bow, the sword, and war from the land;
> and I will make you lie down in safety.
> And I will betroth you to me forever;
> I will betroth you to me in righteousness
> and in justice, in steadfast love, and in mercy.
> I will betroth you to me in faithfulness;
> and you shall know the Lord."   (Hos. 2:14–20)

In these passages, Hosea used the wilderness as the place where the lover (God) can be alone with his beloved (the people) and bring her devotion back to him. The Valley of Achor is at the edge of wilderness near Jericho, for example; Hosea called it "a door of hope."

The prophet interpreted the wilderness of the Exodus as a place where the children of Israel had a more perfect relationship with the Lord. Living in agricultural Canaan had turned them to idol worship and to abandonment of true religion. This concept was repeated in Hosea 11:1-3, in which the prophet portrayed Israel in Egypt as a beloved child and Israel in Canaan as following the Baals.

According to Hosea, God could best restore the nation to her former status in the desert wilderness. God's covenant would not only bestow peace on the nation, creation would be brought into harmony with Israel, and the wild animals would no longer execute part of the divine judgment. The Lord would again be betrothed to Israel, and the new covenant would be maintained by God's righteousness, justice, steadfast love, mercy, and faithfulness. The Hebrew word *hesed,* translated "steadfast love," implies an undying relationship between the Lord and the people and an everlasting covenant.[4] Hosea desired the entire nation to return to the desert and to a simple, more godly life.

The impending peace with creation—once the people have returned to their true Lord—becomes the fourth theme of the later prophets of concern to us. If the wild creatures can carry violent judgment in their jaws, they can also carry olive branches. The creation has to respond in harmony to the justice of the Creator.

A Fire Devouring Every Tree

The four themes appear in a number of well-developed variants in the proclamations of the prophets following Amos and Hosea. Micah, who lived at the time of the Assyrian threat, but prophesied the later fall of Jerusalem, portrayed wild nature as a witness in a covenant lawsuit. God accuses the people of wrongdoing, and presents the case to the mountains and hills:

> Now listen to what Yahweh is saying:
> Stand up and let the case begin in the hearing of the mountains
> and let the hills hear what you say.
> Listen, you mountains, to Yahweh's accusation,
> give ear, you foundations of the earth,
> for Yahweh is accusing his people,
> pleading against Israel.
> (Mic. 6:1–2, *JB*)

Wild nature, as a great work of the Creator and part of the cosmos, was worthy enough to participate in a hearing over Israel's sin. This passage also implies the spiritual state of God's holy people is of interest to the entire creation and a matter of universal importance.

Wilderness images finally came to full flower in the writings of two of the prophets dealing with the exile in Babylon. Ezekiel, who was taken into exile with other citizens of Jerusalem in 597 B.C., spoke of a day when the Land would be restored and a new temple would be built.[5] Ezekiel was undoubtedly familiar with the sayings of Hosea and modified some of Hosea's motifs. The image of Israel as a woman found by her lover, Yahweh, in the desert becomes transformed into the image of the city of Jerusalem found as an outcast child lying in a field.[6] Ezekiel also picked up the theme of the "new Exodus." Trapped in Babylon, just as the people had been stuck in Egypt, the nation needed to review the wilderness wanderings and develop a new understanding of them. Ezekiel reversed Hosea's concept of a glorious new wilderness experience for the people and concentrated on God's wrath. Yahweh had forgiven Israel many times over, but Israel had continued in apostasy. Ezekiel, speaking for God, first described God's mercy in the deserts of the Sinai:

> I thought I would pour out my wrath upon them in the wilderness, to make full end of them. But I acted for the sake of my name, that it should not be profaned in the sight of the nations, in whose sight I had brought them out. Moreover, I swore to them in the wilderness that I would not bring them into the land which I had given them, a land flowing with milk and honey, the most glorious of lands, because they rejected my ordinances and did not walk in my statutes, and profaned my sabbaths; for their heart went after their idols. Nevertheless my

eye spared them, and I did not destroy them or make a full end of them in the
wilderness.   (Ezek. 20:13–17)

This passage was followed by a description of the failures of the second generation and of God's threat, made in the wilderness, to scatter the children of Israel
among the nations (Ezekiel 20:23-24). The implication of these interpretations of
Israel's history is that Yahweh will extend the same grace to the people in
Babylon that was extended to the children of Israel in the wilderness. The prophets prior to the exile had made it quite clear the people had fallen from God. The
same was true during the wilderness wanderings, and God spared them.

The prophet went on to describe the wandering of the people in a wilderness,
not of dry sands and rugged mountains, but "of the peoples":

> As I live, says the Lord God, surely with a mighty hand and an outstretched
> arm, and with wrath poured out, I will be king over you. I will bring you out
> from the peoples and gather you out of the countries where you are scattered,
> with a mighty hand and an outstretched arm, and with wrath poured out; and
> I will bring you into the wilderness of the peoples, and there I will enter into
> judgment with you face to face. As I entered into judgment with your fathers in
> the wilderness of the land of Egypt, so I will enter into judgment with you, says
> the Lord God.   (Ezek. 20:33–36)

God then threatens to purge the rebels in an obvious parallel to the removals of the
rebellious during the Exodus. In this allegorical use of wilderness, the Lord takes
them through a wilderness of the peoples, presumably the nations where they were
scattered as well as the deserts standing between the dispersed Hebrews and Palestine.[7] As in the first Exodus, God's judgment will fall on the apostate. This does
not mean, however, that the rebels will die in a literal desert, but implies that the
political exile will remove the unworthy through unspecified means.

The prophet added the final touches to the predictions of God's wrath by
picturing the demise of the southern kingdom as a raging wildfire:

> Son of man, set your face toward the south, preach against the south, and proph-
> esy against the forest land in the Negeb [south]; say to the forest of the Negeb
> [south], Hear the word of the Lord: Thus says the Lord God, Behold, I will kin-
> dle a fire in you, and it shall devour every green tree in you and every dry tree; the
> blazing flame shall not be quenched, and all faces from south to north shall be
> scorched by it. All flesh shall see that I the Lord have kindled it; it shall not be
> quenched.   (Ezek. 20:46–48)

Ezekiel associated the first exodus into the desert wilderness and a coming
exodus into the cultural "wilderness of the peoples" with both God's wrath and

God's mercy. Wilderness was a place of judgment, not through the arid environment, but by the hand of God. Despite the negative tone, Ezekiel is actually encouraging the people to remain faithful to their God and hope for a return to Jerusalem. Just as the nation survived Egypt and the wilderness, they will survive Babylon.

Not to end the discussion of Ezekiel on a wrathful note, Ezekiel was also one of the later prophets who used some of the traditional "spiritual landscapes" for visionary experience. The book of Ezekiel frequently mentions "the Spirit" falling on Ezekiel and initiating prophecy. The Spirit (or the "hand of God") also "lifts Ezekiel and takes him off 'in visions of God.' "[8] In Ezekiel 8:2–3, an angelic messenger put out a hand that grabbed Ezekiel by the hair, and the prophet reported "the Spirit lifted me between earth and heaven, and brought me in visions of God to Jerusalem." In Ezekiel 40:2, the hand of Yahweh takes Ezekiel and sets him down on "a very high mountain" where he has a vision of the temple. Mount Horeb and Moriah are combined into the mystical peak where the holy city and the great temple sit. This combines urban imagery with Moses' desert mount. Although the probable intent of the text was to describe a literal journey from one place to another, it has never been clear if the experience of translocation was actually part of the vision. The mountain has become not just a literal site for vision, but also a symbolic site, and a precedent has been established for mystically entering "wilderness" landscapes rather than making a long trip on foot. These patterns will continue into the New Testament.

Forests and Mountains Singing

Ezekiel did not have the last word on the future of the people in the wilderness, and his interpretations must be balanced with those of Isaiah. The authorship of the book of Isaiah is controversial, as are the date or dates of the text. Conservative scholarship attributes the entire book to the prophet Isaiah who lived in Judah during a period in which the southern kingdom became part of the Assyrian empire. A majority of academic commentaries assume the first 39 chapters are older than the remainder (and written by Isaiah), while the rest is by other authors who wrote during or after the exile in Babylon. The book of Isaiah uses all four wilderness themes found in Amos and Hosea and surpasses all previous prophecy in the development of the new Exodus. In contrast to Ezekiel's somber predictions and dark colors, Isaiah associates wild nature and wilderness with both the coming Messiah and cosmic peace. Just as wild nature could be an instrument of judgment, peaceful coexistence with the wild characterizes the new kingdom of God.

The Book of Isaiah does present the wild as an instrument of God's wrath, both against Judah and her enemies. In Isaiah 56:9, in which the leaders of the people are rebuked, the Lord commands: "All you beasts of the field (wild beasts), come to devour—all you beasts in the forest." The intention here might have been the consumption of the bodies of men killed in battle, or it might have been the raiding of untended flocks. In Isaiah 18:6, in which the Assyrians "shall all of them be left to the birds of prey of the mountains and the beasts of the earth. And the birds of prey will summer upon them and the beasts of the earth will winter on them," the intent was clearly the consumption of the fallen. In either case, God set wild animals against men for human failure to be obedient to the divine will.

To bring an end to one of the most critical spiritual experiences of the Hebrew people, Isaiah prophesied that God would make Babylon—the great city filled with sin—the possession of the wild beasts:

> But the desert creatures will lie there,
>   jackals will fill her houses;
> there the owls will dwell,
>   and there the wild goats will leap about.
> Hyenas will howl in her strongholds,
>   jackals in her luxurious palaces.
> Her time is at hand,
>   and her days will not be prolonged.
>
> (Isa. 13:21–22, *NIV*)

The prophet's intent in describing empty buildings occupied by wildlife was not purely symbolic. The city, one of the centers of civilization, was to be abandoned and returned to the wild.

The first part of Isaiah introduces not just a new kingdom, but also a new king. The well-known Messianic prophecy of Isaiah 11 begins with a description of this son of David who will come to lead the people away from both political adversity and their sins. The Spirit of the Lord will rest on this "shoot from the stump of Jesse" (Isa. 11:1), and he will rule in complete wisdom and righteousness. His reign will bring justice to the poor and judgment to the wicked. When he comes, the good will of God will reach through all creation, and the entire world, natural and human, will enter a new era of peace:

> The wolf shall dwell with the lamb,
> and the leopard shall lie down with the kid,
> and the calf and the lion and the fatling together,
> and a little child shall lead them.
> The cow and the bear shall feed;
> their young shall lie down together;
> and the lion shall eat straw like the ox.
> The suckling child shall play over the hole of the asp,
> and the weaned child shall put his hand on the adder's den.

> They shall not hurt or destroy in all my holy mountain;
> for the earth shall be full of the knowledge of the Lord
> as the waters cover the sea.   (Isa. 11:6–9)

Creation, as always, is responsive to the will of God. Under the supervision of the Lord's anointed, just rule will extend past the boundaries of Israel and Judah to the entire earth and all its inhabitants, human and otherwise. The use of natural imagery identifies a cooperation intrinsic to all creation and not a merely temporary or superficial "human" peace. God's peace is not a peace of the spread of civilization, of the flowering of science, or of sophisticated treaty negotiations, but is a peace welling up through the foundations of the universe. Wild creatures not under human control lie down with children and domestic livestock. Under the righteous reign and the "seeing" and "hearing" of the root of Jesse, humankind returns to the Garden of Eden where they can call the wolf by name, and the wolf will come and sit beside them.

The second section of the book of Isaiah (chapters 40–55), also called Deutero-Isaiah, introduces yet another pattern for a new wilderness exodus. The prophet calls the people into the biblical wilderness not to experience God's wrath, but to experience God's joy:

> Comfort, comfort my people,
>   says your God.
> Speak tenderly to Jerusalem,
>   and cry to her
> that her warfare is ended,
>   that her iniquity is pardoned,
> that she has received from the Lord's hand
>   double for all her sins.
> A voice cries:
> "In the wilderness prepare the way
>   of the Lord,
>   make straight in the desert a
>   highway for our God.
> Every valley shall be lifted up,
>   and every mountain and hill be
>   made low;
>   the uneven ground shall become
>   level,
>   and the rough places a plain.
> And the glory of the Lord shall be
>   revealed,
>   and all flesh shall see it together,
>   for the mouth of the Lord has spoken."
>
>   (Isa. 40:1–5)

In an immediate sense, the passage was a declaration to the people that their sufferings in captivity were over. The highway in the wilderness was the way home from exile in Babylon, and the equivalent of the triumphal highways constructed in Babylon for the purpose of letting the gods enter the city.[9] Here Yahweh constructed a highway for himself and makes the way home easy. This should not be interpreted as a divine dislike of wilderness, but rather as a divine display of power. The passage implies that, just as the children of Israel saw Yahweh's glory on Sinai, all living beings will see the glory of God revealed in the wilderness. This coming glory is presumably not a theophany, but a presence of God which institutes a new kingdom of righteousness on earth.

Throughout Isaiah 40–55, the author included reminders of God's role as creator, and of God's creative power:

> Who has measured the waters of earth in the
> hollow of his hand
> and marked off the heavens with a span,
> enclosed the dust of earth with a measure
> and weighed the mountains in scales
> and the hills in a balance?   (Isa. 40:12)

The book of Isaiah suggests that the God who has the power to create the world also has the power to save the people. Just as God's role as judge is rooted in God's role as Creator, God as Savior relies on divine creative abilities and prerogatives to bring forth mighty acts to free the people from their captors and their sins.

The prophet related Yahweh's power to care for the poor to Yahweh's power to create:

> When the poor and needy seek water,
> and there is none,
> and their tongue is parched with thirst,
> I the Lord will answer them,
> I the God of Israel will not forsake them,
> I will open rivers on the bare heights,
> and fountains in the midst of the valleys;
> I will make the wilderness a pool of water,
> and the dry land springs of water.
> I will put in the wilderness the cedar,
> the acacia, the myrtle, and the olive;
> I will set in the desert the cypress,
> the plane and the pine together;
> that men may see and know,
> may consider and understand together,
> that the hand of the Lord has done this,
> the Holy One of Israel has created it.
>   (Isa. 41:17—20)

Note that Yahweh's creative act does not turn the wilderness into a cultivated landscape, but rather the availability of water fosters the growth of wild tree species. The presence of water would also make the way easier for the people as they returned home. The book of Isaiah uses water symbolically throughout, to show God's ability both to perform mighty acts and to heal.

The prophet then repeated the promise of water and easy way in the wilderness, this time providing for the wild beasts instead of the wild plants:

> Behold I am doing a new thing;
> now it springs forth,
> do you not perceive it?
> I will make a way in the wilderness
> and rivers in the desert.
> The wild beasts will honor me,
> the jackals and the ostriches;
> for I give water in the wilderness,
> rivers in the desert,
> to give a drink to my chosen people. (Isa. 43:19–20)

The provision made for the people would also care for the wildlife who can recognize Yahweh's power in providing water. These passages are similar to the proofs of God's providence in Psalm 104.

The prophet called for creation to praise God for mighty saving acts and expected the participation of the wild:

> Sing, O heavens, for the Lord has done it;
> shout, O depths of the earth;
> break forth into singing, O mountains,
> O forest, and every tree in it!
> For the Lord has redeemed Jacob,
> and will be glorified in Israel. (Isa. 43:23)

Just as Israel should praise God as Creator, all creation should praise God as Savior. From a contemporary perspective, it may seem strange to have the prophet call creation to respond to God's saving act directed towards a human community, yet, when the perfect will of God is done, creation should rejoice. Our current Christian perspective, which emphasizes personal salvation, tends to restrict spiritual events to the life of the individual and disregards events of importance to the community or to the entire cosmos.

The prophet ended the description of the new Exodus with a joyful call to both the nation and creation:

> For you shall go out in joy,
> and be led forth in peace;
> the mountains and the hills before you
> shall break forth into singing;
> and all the trees of the field shall clap their hands.
> Instead of the thorn shall come up the cypress;
> instead of the briar shall come up the myrtle;
> and it shall be to the Lord for a memorial,
> for an everlasting sign which shall
> not be cut off.   (Isa. 55:12–13)

These passages expressed the feelings of the captives in Babylon as they were released and began the journey home to the promised land. Creation joined in the celebrations of redemption, and the response of creation pointed to the redeemer who raised the myrtle and the cypress. Note, again, that cultivation did not displace wildlands, but rather forest trees displaced plants associated with land abuse and overgrazing.

The Return to the Wilderness

The later prophets all agree that God's role as Creator is central to both God's actions concerning the nation Israel and creation's response to God's concerns. Wild nature may either act as an instrument of God's judgment or serve as an indicator of God's peace. Several of the prophets also predict there will be a new Exodus into the wilderness. The prophets vary, however, on the circumstances and purpose of this wilderness sojourn. Ezekiel declared the coming wrath, while Isaiah called for joy and comfort. The old concept of spiritual transition in the wilderness becomes very sweeping and is disassociated from an actual desert region. We should recognize that, in the history of the Jews, any return of the widely scattered nation to the land could be called a new Exodus, as could any escape from apostasy and return to God. (One also had to cross deserts to move between nations in the Middle East.)

The return to the wilderness, although it has physical elements, becomes primarily a reinstitution of righteousness and a road to the re-establishment of the cult in Jerusalem. Hosea was concerned about the spiritual state of people still resident in the Promised Land and called them back to the desert to repent and learn again of God's wrath in return for the sins of the nation. He set his face against the people to try to get them to obey God and look forward to the day when the temple would once more be their nation's worship site. The book of Isaiah was certainly discussing the captivity in Babylon and the return to Jerusalem, but was also predicting the rise of a new David with a new kingdom marked by righteousness and peace.

In this *new Exodus model* of wilderness spirituality we find:

1. An extension of the prophetic relationship with nature, in which large predators attack or consume wayward prophets, to a wild nature attacking or consuming a wayward people. In some cases, this may be symbolic, and the Lord appears as a lion.

2. An association of peace between human beings and wild nature with the coming of the "shoot of Jesse" and the new kingdom of God.

3. The participation of wild nature in spiritual events on an international scale.

4. A conversion of wilderness into a symbolic or allegorical event, usually associated with repentance or renewal of a relationship with God.

5. An emphasis on the need for a new Exodus, which may be physical in some cases, but always is viewed primarily as a spiritual transition from a fallen to a renewed state. In both Isaiah and Hosea, this includes a "wilderness romanticism," whereby the time in the wilderness is seen as a great blessing for the people and a time of closeness with God.

Jackals in Her Shopping Centers

From a contemporary perspective, we may have trouble believing God's wrath could fall on us through wild nature. We are so confident of our human power to subdue, we find it difficult to picture jackals prowling through our shopping centers. The prophets suggest, however, that our relationship with God is reflected in our relationship with wild nature. Today we usually contemplate the spiritual future of the cosmos solely in terms of the human. Isaiah suggested living at peace with the wild will be an outcome of righteous rule. Do we believe this?

The tremendous growth of both agricultural and industrial technology has convinced us we have control. Even those few of us who deem God powerful enough to be angry—and to be able to do something about it—have relegated divine wrath to the afterlife. The prophets considered the outcome of human sin, particularly on a social or national level, to be more immediate. Is our disrespect for creation actually the result of a disrespect for God and for God's powers and prerogatives? Technological humankind asserts itself over creation and tempts "the storm." Do we tempt God as well?

The prophets have expanded the idea of the Exodus from a single formative event in the history of the children of Israel to a recurring phenomenon. The "new Exodus" can be a retreat into the wilderness for spiritual renewal, or an escape from bondage. It can also be a confrontation between God's wrath and human sin. The question for the contemporary Christian is: Should we take this literally? Is a period of withdrawal away from "Canaan" or "Babylon" valuable? Do individuals or Christian bodies sometimes need a new exodus?

Since the authors of the New Testament saw Christ as the "shoot of Jesse" and, therefore, the fulfillment of many of these prophecies, our next task is to determine if some of these *new Exodus* themes reappear in the Gospels or other New Testament books. As the prophecies of a coming Messiah and a New Kingdom are central to Christian interpretation of the Old Testament, we need to determine if Christ's ministry in any way served as a new Exodus into the wilderness.

8

And He Withdrew to an Isolated Place
(Intertestamental Times, Christ, and John the Baptist)

The Chattooga

I dropped the manuscript I was rewriting and put my head in my hands. I felt like every change was irrelevant. "I'm not doing a good job on this," I muttered as the telephone rang. It was a graduate student apologizing for missing an appointment. I gave her another time slot. As soon as I put the phone down, it rang again. It was my boss asking me to get some proposals ready for a meeting next week. He said he was sorry and he knew I didn't have much time to get the things done. I muttered that I already had a technical report due on a contract and then hung up. As I leaned back in my chair and tried to rein in my temper so that I could return to the manuscript, a mustached face peered through my door. "Oh no, another graduate student." Although he hadn't signed up for an appointment, I motioned for him to come in. He wanted advice on a job application. I gave it to him.

"I'm burning out," I thought to myself after he left. "I'm tired. I'm not physically tired . . . I'm tired tired." The secretary came in with some draft letters. I looked them over without thinking about the contents. Recognizing I was in no mood to chat, she retrieved the signed copies and backed out of the office. Then the phone rang again. . . .

* * *

That Saturday I was stuck working on the proposals. On Sunday I decided I wouldn't touch another scientific document or meet another social responsibility. The Cambodian refugees, whom I often took out for some local weekend adventure, could stay home this time. I went to an early church service and then shuffled through my Appalachian maps and hiking guides. I didn't want to spend too much time on the highway and just felt like being somewhere by myself.

The Chattooga River was attractive—not because it had the most spectacular trail system in the Blue Ridge mountain, but because it was close and I probably wouldn't see many people.

I arrived at the trailhead about noon, stuffed a Bible, a canteen of water, and a few granola bars in a pack, and began a leisurely walk up a path high above the river. In the first quarter mile I observed a few swimmers and white water rafters. As I moved away from the highway, however, I soon ran out of company. Crossing a side creek with a small cascade, I stopped to enjoy the shade of the early summer foliage and the splash of the water. As the minutes dissolved, I lost all desire to go anywhere or to do anything. Goalless, I started to relax.

Strolling further upstream, I began to enjoy each new view of the river as I rounded one bend after another. Between the dense forest cover and the river's winding course, it was impossible to see any great distance. Each rock, each gravel bar, each fallen log had to be discovered individually as I slowly made my way along an old overgrown logging road and dropped down here and there to the river bank. Finally, just as I was losing the trail, I found a little sandy beach with a smooth, dry log to lean on.

Long ago when I was working in the Great Smokies, I had developed the habit of taking a few hours off and going out into the woods to read my Bible and to pray. Sometimes I would go by myself; sometimes one or two friends would come along. We would rest under a dark green canopy of rhododendron in a forest of giant hemlocks or on a stream bank decorated with spring flowers, such as fringed phacelia and large flowered trillium, and take turns reading favorite passages of Scripture or exchanging prayers as we felt led. The mute, dignified forest seemed to inspire silent contemplation, and we would sometimes sit without speaking for close to an hour. Although we often used these forest devotions for serious prayer or seeking God, we rarely used them for formal Bible study. They were less of an academic exercise and more of a free spiritual interchange with each other and with Christ.

I find all my pleasant associations with forest devotions make it easy to return to them. Sitting by the Chattooga, I pulled my Bible out of my backpack and selected a few psalms. I prayed for a little while about business concerns and then went back to reading. I snuck over to the river and tried to observe the trout moving in a deep pool with a gravel bottom. I returned to my comfortable log and opened my heart to the Sustainer. The river rolled by me, sunlight flashing on the peaks of the small ripples. I pushed the clean white sand around with my hands and continued my conversation with the Holy. It wasn't a time for major spiritual transformation—or for deep revelation. I just rested on the beach, let the afternoon drift away, and praised God.

*   *   *

Despite the breadth of wilderness spiritual experience in the Hebrew Scriptures, the Old Testament does not report short, purposeful, non-stressful wilderness sojourns or journeys for rest or personal renewal. This type of wilderness does have a biblical basis, however. As we begin to investigate the New Testament we will see several changes in wilderness utilization, including a greater emphasis on purposefully entering the wilderness for spiritual reasons, some modifications in the duration of isolated stays in natural settings, and the introduction of withdrawal for rest. Although there is a strong continuation of Old Testament wilderness traditions in the New Testament, a change in the form of prophetic ministry becomes the driving force behind changes in wilderness spirituality.

Living on What Grew Wild

In the intertestamental period between the end of the Old Testament and the beginning of the New, social conditions changed greatly in Palestine. First Alexander the Great invaded, bringing with him Hellenistic culture. Then Rome became a world power, further advancing pagan philosophies, new technologies, urbanization, and international economic ties. Between the Testaments, Jewish culture in Palestine and elsewhere produced a variety of written works including compendiums of wisdom, narratives of righteous action such as "Tobit," and historical accounts of Jewish revolts against foreign invaders. References to wilderness and wild nature are scattered throughout this literature, but they do not produce any new "models" of wilderness spiritual experience. The Book of Sirach, for example, proclaims the glory of God in nature, in the tradition of the Psalms. 1 Maccabees discusses military movements into the wilderness, but emphasizes history rather than theology. Influenced by Greco-Roman models, it describes Jewish rebels hiding in the marshes of the Jordan, yet says little about the divine. The book of 2 Maccabees adventures into the supernatural,[1] but gives only limited attention to wilderness: "But Judas Maccabeus, with about nine others, got away to the wilderness, and kept himself and his companions alive in the mountains as wild animals do; they continued to live on what grew wild, so that they might not share in the defilement" (2 Macc. 5:27). For the leader of the Jewish resistance, hiding in the desert provided a freedom from food which was not ritually clean and the other unholy threats presented by Greek culture. When the Maccabees returned as conquerors to the Holy City: "they had been wandering in the mountains and caves like wild animals. Therefore bearing ivy-wreathed wands and beautiful branches and also fronds of palm, they offered hymns of thanksgiving" (2 Macc. 10:6–7). In the writings of the intertestamental period, wilderness became a haven for rebels against foreign powers and a route of military escape. Historically, the trend continued through the efforts of the Zealots to the demise of the Jews at the fortress of Masada in 68 C.E. Wilderness insurrection, however, was not the mission of the coming Messiah.

Locusts and Wild Honey

Wilderness comes to the fore, however, right at the beginning of the New Testament with the ministry of John the Baptist, a prophet reminiscent of his Old Testament predecessors. In a birth similar to that of Samuel, an angel announced John's conception to childless Zechariah (Luke 1). The Baptist's affinity for the desert wilderness marked him as an heir of Moses and Elijah. It is tempting to see John, in his garment of camel hair, simply as a return to the "forthtellers" of pre-Hellenistic times, but this analysis neglects subtle differences in the core of John's ministry.

First, Luke 1:80 notes that John "was in the wilderness till the day of his manifestation in Israel." Nowhere in the canonical books of the Old Testament do we find a prophet raised in the desert or spending long years there, except during the Exodus. Elijah's forty-day trip to Horeb was more typical, and both Elijah and Elisha were more often in residence on Carmel than in the wilderness of Judea or in the Negeb. Moses did spend most of his adult life in the arid lands between Egypt and Canaan, but he was raised in Pharaoh's domains. Coming to maturity in the wilderness was more clearly associated with David or Ishmael, both political leaders and men of war.

Second, Luke 3:2–3 reports "the word of God came to John the son of Zechariah in the wilderness; and he went into all the region about the Jordan, preaching a baptism of repentance for the forgiveness of sins," and Mark 1:4 says "John the baptizer appeared in the wilderness, preaching a baptism of repentance for the forgiveness of sins." In the Old Testament, there is little evidence wilderness residence was necessary to the Word of the Lord. The prophetic books are more likely to report the coming of the Word of the Lord by time than by place. Moses received his call to ministry at Horeb, but this was not a continuing pattern. Other than during the Exodus, preaching and calling to repentance were not wilderness ministries *per se*. The prophets tended to come out of the wilderness, engage the leadership of the nation, and then flee or retire back to the wilds.

Third, Matthew 3:4 describes John as wearing a garment of camel's hair and eating locusts and wild honey. Luke 7:33 says the Baptist ate no bread and drank no wine. The hair shirt and leather belt were traditional—Ahab recognized Elijah by similar attire—but long wilderness sojourns eating wild foods were not. The more common Old Testament repast was divine provision, such as the manna and the quails, or Elijah's ravens (who brought bread and meat). The Old Testament prophets certainly fasted, but there is no evident for this type of continuing diet. John relied on the Lord for food, as had the prophets of old, but his life-style was more completely ascetic.

Fourth, John arrives on the scene in Judea quoting from the book of Isaiah:

> . . . The voice of one crying in the wilderness:
> Prepare the way of the Lord,
> make his paths straight.
> Every valley shall be filled,
> and every mountain and hill shall be brought low,
> and the crooked shall be made straight,
> and the rough ways shall be made smooth;
> and all flesh shall see the salvation of God.   (Luke 3:4–6)

Old Testament prophets did not see themselves as fulfillments of previous wilderness prophesy or announce themselves as such. This prophecy of Isaiah referred originally to the return of the children of Israel from captivity in Babylon and the crossing of the wilderness to return to the Promised Land. Isaiah had prophesied the road from Babylon, unlike the path of Exodus, would "be made straight."[2] John's use of Isaiah not only identified John as part of the Messianic event, it also verified John's call for the repentance of sins. The road to the coming Messiah would be easier for those who repented and were ready to receive him.[3]

John the Baptist was more thoroughly a wilderness figure than any of the pre-Hellenistic prophets. His coming to maturity, his call, and his life-style were all rooted in the desert and unoccupied lands. The most important part of his ministry—the baptism—took the people to the transition between the Promised Land and the wilderness—the Jordan River. John was even more of a contrast to the social order of his time than Elijah was to his. Cast against the arts and letters and the economic wealth of Greco-Roman culture, John was not simply a throwback to another age, he was a spiritual protest against the priorities of the majority. Just as John's choice of habitats differed from that of the Herods, the social antagonism between John and the ruling powers was the natural outcome of deep differences in world view. The prophet needed to be different—to act in a distinctive fashion to make his spiritual point. John's version of wilderness ministry was extreme, but his wilderness lifestyle was an integral part of his message of repentance—and of the coming kingdom.

Waiting in the Desert

Was the intensification of the wilderness aspects of prophetic ministry original with John? According to the historic accounts of Josephus, there were others in the desert before the Baptist. Josephus mentions a man named Bannus "who lived in the desert, wore clothing made of tree bark, ate wild herbs and washed himself with cold water frequently during the day as well as at night for purification."[4] At least one apocryphal book, "The Ascension of Isaiah," records an ascetic wilderness retreat to the arid lands:

> And when Isaiah the son of Amoz, saw the lawlessness that was common practice in Jerusalem, and the worship of Satan and his wantonness, he left Jerusalem and settled in Bethlehem of Judah. And there too there was much lawlessness, so that he left Bethlehem and settled on a mountain in a desert place. And Micaiah the prophet, and the aged Ananias, and Joel and Habakkuk, and his son Josab and many of the faithful who believed in the ascension into heaven, also left where they were and settled on the mountain. They were all clothed with garments of hair, and they were all of them prophets: they had nothing with them, but were naked; and they all lamented with great lamentations because of the apostasy of Israel. And they had nothing to eat but only the wild herbs that they gathered on the mountains; and they cooked them and lived on them together with the prophet Isaiah. And they remained on the mountains and on the hills for two years.[5]

Unfortunately it is uncertain how much of this account of Isaiah's life originated in Jewish legend and how much resulted from later Christian additions. The extant manuscripts are from the sixth century C.E., and the original may postdate John. This narrative suggests, however, that Jews from the Hellenistic and Roman periods attributed ascetic behaviors to the prophets to an even greater extent than the Hebrew scriptures.

Of even more importance than these short literary accounts is the archaeological and historic evidence surrounding a sect called the Essenes and the community at Qumran, the site where the Dead Sea Scrolls were found. Josephus, Pliny, and Philo—dependable sources as ancient writers go—all mention the Essenes and their religious practices. A number of authors have noticed similarities between the Essenes and John, and some have suggested the Baptist grew up in an Essene community in the desert.[6] It is presently impossible to prove or disprove this theory, but we can easily verify desert asceticism preceding John. According to Pliny the Elder, the Essenes had at least one community on the shores of the Dead Sea.[7] (This may have been Qumran.)

The Essenes were separatists, refusing to sacrifice in Jerusalem and generally keeping company only with their own group. Like John the Baptist, they took the words of the prophets as a prediction of a coming kingdom arising in the wilderness, but in their "Manual of Discipline" (assuming the Essenes were the Qumran sect) they record Isaiah 40:1 differently than did the Baptist—in a version closer to the Hebrew original: "In the wilderness prepare the way of the

Lord, Make straight in the desert a highway for our God."[8] The Qumran community did not include the "voice crying" and emphasized preparation for the coming of the end times through the keeping of the Law. Convinced they were the only true interpreters of the Scriptures, "they devoted their exile in the wilderness to the study of the Bible."[9] They considered themselves to be the final remnant, under a "New Covenant" and the only "elect" among the Jews. They wore simple linen garments, strictly avoided unclean foods, and practiced ritual bathing for purification. At least part of the sect was celibate. The Essenes clearly has ascetic ideals.

John the Baptist's goals and ministry differed strongly from those of Qumran. The Baptist was not leading a closed sect, but was calling people from all occupations, including tax collectors and soldiers, to repentance. Although John's disciples apparently imitated his fasting and austere life, he did not ask those he baptized to follow him into the wilderness. According to the Scriptures, the Baptist made no exclusive or Messianic claims for himself.[10] While the Essenes were preparing themselves for the Last Days, John was preparing the people for the coming of Christ.

In their withdrawal to the wilderness, however, both John and the Qumran sect were reading the same Scriptures and may have been coping with some similar problems. Both John and Qumran thought the "new exodus" would begin literally in the wilderness. Qumran probably originated with a group in Damascus who were seeking a pure life under the Torah (Books of the Law). Social pressure and internal conflicts within the group may have forced them into the desert. For both Qumran and John, the isolation appears to have forwarded asceticism. Purity was an issue for Qumran and for the Baptist, and the wilderness helped to provide it.

Both John and Qumran were concerned with social evils. The Qumran sect was reacting to foreign "ungodly" cultures and to decadence in Judaism. John called the Pharisees and Sadduccees, the religious leaders of their time, a "brood of vipers" (Matt. 3:7). The wilderness offered them a freedom from other cultures and false religious notions, just as it had for Moses. On the slopes above the Dead Sea or on the banks of the Jordan, the high priest and the temple or the theological arguments of the Biblical scholars made little difference. The wilderness spoke neither of Romans nor Greeks, nor of the Jewish establishment, it spoke only of repentance and spiritual transition.

The Voice

John's appearance in the wilderness was neither an individual's copy of Qumran nor a mere attempt to add credibility to his preaching. John's life was, in many ways, a summary of the prophets who had gone before him. His ministry combined the environment of the Exodus, the lives of Moses, Samuel, and Elijah, and the preaching of Isaiah, Ezekiel, Amos, and Hosea, to produce, after a three-hundred-year silence, something simultaneously old and new in the land of Israel. The wilderness itself provided part of the message to the nation. In the prophetic books, Israel's time in the wilderness had been the time when the people were closest to God. The Exodus had been a period of transition when God freed the nation of Egyptian influence and returned it to service to him. The long sojourn in the wilderness had helped to build the first kingdom of Israel. In John's ministry, tradition and change combined at the shores of the river. John strode through the valley of the Jordan, announcing a new kingdom and the coming one who would baptize with the Spirit.

The first words of John in Matthew 3:2, "Repent, for the kingdom of heaven is at hand," associate wilderness with repentance. In calling the people to be baptized in the Jordan, John was drawing them back to the wilderness to truly become the children of God once more. John's message demanded a second Exodus—a crossing of the Reed Sea to a place where disobedience and rebellion could be left behind and the people could become "new" before God.[11]

John's ministry was certainly not an exact imitation of the first Exodus. Moses did not initially lead the children of Israel into the wilderness to repent; he took them there to worship according to Yahweh's command. The wilderness was intended to be the place where the people of Israel would see their God. Although the wilderness of the Exodus functioned as a site of testing, John does not subject the people as a whole to "physical hardship." The wilderness experience of the Exodus began, not with God's judging the people, but with God providing for their needs. Yahweh gave, first, freedom from Pharaoh, then food and water, then access to his holy throne, and then the Law and a covenant. Through the years of rebellion, God also poured out mercy and grace. John's offer of baptism parallels God's true intent for the original Exodus—freedom and new life for his people—but, unlike the first Exodus, it immediately identifies repentance and change as the primary issues.

A more direct association of wilderness with repentance is found in the later prophets. The use of desert imagery, however, varies with the context. Isaiah 40:3 and 48:20–21 portray the people passing along the straightened wilderness road after the Lord has called them to return to their homeland. In Ezekiel 20, in contrast, there is a long discourse on the sins of the Exodus, followed by a prophecy of God's wrath in the wilderness.[12] The preaching of John has both these elements: the way that is made straight, and thus easier, and the impending wrath, which will separate the lovers of God from the rest of humankind. John announced the one who is coming whose "winnowing fork is in his hand, and he will clear his threshing floor, and gather his wheat into the granary, but the chaff he will burn with unquenchable fire" (Matt. 3:12). God's judgment in the Exodus was usually mediated through the environment. The separation of the righteous from the unrighteous was in response to specific incidents and was meted out in terms of physical death. John followed the later prophets in predicting a more ultimate form of judgment, dealing with the final end of human spiritual life and of the cosmos.

The narratives about John imply the wilderness was necessary for his preparation, but there is no mention of a personal transformation like that of Moses or Elijah. Luke 1:39–45 reports that John recognized the presence of Christ while in his mother's womb. The transformation in the wilderness was not of the individual prophet, but, first, of the people and, second, of the spiritual "age." In the Exodus, the people are called to the wilderness to see themselves and God differently. As in the Exodus, a new kingdom was to be established. Unlike the Exodus, the new kingdom would not be an earthly kingdom of the children of Israel, but would be a spiritual kingdom of the Son of Man.

The Dove and the Wild Beasts

When Jesus was ready to begin his ministry, he, too, came to the Jordan to be baptized by John. At the river, the heavens opened and the Holy Spirit descended on Jesus "in bodily form, as a dove, and a voice came from heaven, 'Thou art my beloved Son, with thee I am well pleased' " (Luke 3:22). This text from Luke indicates *something* with a physical form descended and, presumably, was visible at the same time a voice came from heaven. Christ's entrance into ministry thus begins with a wilderness theophany in which the Spirit was seen and God the Father was heard. The message of the theophany did not concern the person of God the Father, but rather concerned the person of Christ and his relationship with God. The voice said Jesus was: (1) the Son, (2) beloved (the Greek form is *agapatos,* which implies spiritually loved), and (3) well pleasing or delighting to God.[13] Just as the Old Testament wilderness theophanies disclosed the characteristics of Yahweh, this theophany disclosed the spiritual characteristics of Christ.

It is not certain if anyone saw the theophany—other than Jesus and perhaps John—since none of the Gospel writers report any reaction from the crowd (although the wording in Matthew does imply the voice was speaking to the onlookers). In spite of John's protest that Jesus did not need to be baptized, however, we see Jesus joining with the people in a symbolic welcoming of the new kingdom. Luke indicates the descent of the dove took place after John had baptized all the people he was going to baptize—that is, Jesus' baptism was at the end of John's ministry. Unlike the people, Jesus did not need to repent of his sins, thus his baptism was an act of identification, rather than of personal transformation. The descent of the Spirit marked the beginning of Christ's ministry and the Messianic Age, and therefore, a transformation of the entire creation.

After the baptism, Mark 1:12–13 reports: "The Spirit immediately drove him [Jesus] out into the wilderness. And he was in the wilderness forty days, tempted by Satan; and he was with the wild beasts; and the angels ministered to him." Both Matthew and Luke add that he fasted—ate nothing for forty days (and nights)—and that at the end of that time he was hungry.

Jesus's confrontation with Satan would certainly fall into the class of a manifestation of the otherworldly, but it is the first such wilderness appearance we have reviewed where the otherworldly figure is clearly evil. Jacob's wrestling match with the unknown man at the banks of the wadi may have concerned a shadowy being, but the outcome of the match was an angelic message. Although Satan, the tempter, appears in the Old Testament, he is not part of the Hebrew wilderness tradition of encounters with the angelic.

The passages from the Gospels suggest other Old Testament parallels. Forty days was both the length of time Moses remained on the mountain of God, taking neither bread nor water, and the duration of Elijah's trip to Horeb where he ate nothing. Forty days could also be symbolic of the Hebrews forty years in the wilderness. Purposeful temptation by either God or Satan was not a part of Moses' and Elijah's experiences, however, and the narratives of Elijah actually suggest he did not need to eat after consuming divine provisions. The "testing" of the children of Israel was primarily through ordinances, although Deuteronomy 8:3 suggests God let Israel hunger so he could feed them with manna, that they might understand "that man does not live by bread alone, but that man lives by everything that proceeds out of the mouth of the Lord." In any case, there is no evidence Israel had fasted for forty days before the manna arrived.

Mark's comment about "the wild beasts" is reminiscent of the Old Testament passages where prophets interact with lions and bears. Remaining unharmed by large predators identified a true prophet in pre-Hellenistic times, and Isaiah had prophesied a child would be able to lead the most dangerous of beasts in the coming kingdom. The mention of "wild beasts" may have been intended to serve as a verification of divine protection and favor.

The Old Testament wilderness theophanies frequently included angels (usually one) who most often served as messengers. After the temptation, angels ministered to Jesus. The provision of food or water appears elsewhere in the Bible (Elijah, Hagar) in association with angels, but here we have the suggestion of an angelic host. The defeat of Satan was appropriately marked by the return of his holy counterparts.

As with the story of John, the temptation reminds us of several Old Testament encounters with God—yet it is not an exact match for any of them. The content of the temptation, as described in Luke and Matthew, concerns Jesus' obedience and relationship to God (just as the theophany at the Jordan had). Satan tempted Jesus first to turn stones into bread, and thus repeat the miracle of the manna to feed himself. Jesus answered from Deuteronomy 8:3 and, in obedience to the will of God, refused to break the fast. The Scriptures do not say that Christ's long fast was intended as a period of communion with God, but that was the case in Moses's forty days on the mountain and the Hebrews' long stay in the wilderness. If the fast had been broken, we might assume that communion with God would have been broken also. If Jesus had given in to Satan he would have repeated the impatience of the children of Israel and preempted the Father's ultimate provision.

In the second temptation in Matthew (the third in Luke), Satan removed Jesus from the wilderness to the pinnacle of the temple and challenged him to throw himself down. The tempter doubted Jesus's role as the "beloved of God" and asked him to prove himself by having God save him. Had Jesus consented he would have been tempting God far more severely than the children of Israel did by complaining. In Matthew 4:7, Jesus again answered from Deuteronomy ("You shall not tempt the Lord your God") and refused to test God by demanding mighty acts and proofs of his power.

For the third temptation in Matthew (and second in Luke), Satan took Jesus to a "very high mountain" (Matt. 4:8) and offered him all the kingdoms of the world if Jesus would only fall down and worship him. The Omnipotent God appeared on a mountain a number of times in the Old Testament. The children of Israel went to the Mountain of God to worship and Elijah contested Baal's priests on Carmel. In Matthew 4:10, Christ answered again from Deuteronomy ("You shall worship the Lord your God and him only shall you serve") and avoided the apostasy of the children of Israel at the golden calf.[14]

In the wilderness, Satan challenged the core of Christ's relationship to God and the foundations of his coming ministry. This was not an accident. The spirit "drove" or "led" Jesus into the desert, suggesting the period of forty days in the wilderness was necessary and divinely ordained. The time in the wilderness became an internal or personal test of Christ's relationship with the Father and therefore of his role as the Son of God. Jesus's "sonship" would be subject to tremendous attack from the religious authorities when Jesus entered public ministry.

The outcome of the temptation becomes an interesting contrast to the history of Israel. At the end of the Exodus, the Hebrews were not a completely new people. The younger generation had replaced the older, but the people were still failing whenever they faced stress. They had stumbled at every test and had angered the Lord multiple times, yet into the Promised Land they went. When Christ was baptized by John, the Baptist himself acknowledged Christ's freedom from the need for repentance. Unlike the children of Israel who had been living among the idols of Egypt, there was no reason to think Christ "needed" to cleanse himself. Yet he went into the desert to confront Satan and unflinchingly tossed aside the same temptations that had conquered and bound the children of Israel for centuries.

The temptations also demonstrate a key relationship between Christ and the nation of Israel. When the Hebrew people entered the desert, they were to become a "nation of priests." The Old Testament is a long chronicle of their successes and failings in this regard. When Satan tempts Christ to act as an earthly ruler would act, Christ does not succumb (a very important matter considering Christ's ultimate sacrifice on the cross and role as Savior of the world). Israel went into the desert to emerge as an "earthly" nation. Christ went into the wilderness so he would *not* become an earthly ruler. Much of the wilderness imagery early in the Gospel narratives concerns the differences in calling between the Kingdom of Heaven and the kingdoms of the earth. [15]

He Went up on the Mountain to Pray

There can be little doubt the emphasis on the desert and the Jordan Valley at the beginning of Christ's ministry was symbolic of "a new Exodus" and tied John and Jesus to the prophets. The desert fast was not the end of wilderness or isolated sojourns for Jesus, however. Like Elijah, Jesus returned from the desert to confront the social and political powers of his time, and like Elijah, he never completely removed himself from the rural countryside. In the Gospel of Mark, we find Christ withdrawing from the towns and the crowds to pray by himself or with his disciples. After teaching in Capernaum, and spending time healing and casting out demons, "in the morning, a great while before day, he rose and went out to a lonely place, and there he prayed" (Mark 1:35). The Greek for "lonely place," *eremos topos,* is related to the Greek for "desert," *eremia,* and might also be translated "desert place." Mark 6:30–32 reports that after the beheading of John the Baptist: "The apostles returned to Jesus, and told him all that they had done and taught. And he said to them, 'Come away by yourselves to a lonely place, and rest a while.' For many were coming and going, and they had no leisure even to eat. And they went away in the boat to a lonely place by themselves." And

Mark 6:45–47 reports that just prior to a storm on the Sea of Galilee: "Immediately he made his disciples get into the boat and go before him to the other side, to Bethsaida, while he dismissed the crowd. And after he had taken leave of them, he went up on the mountain to pray. And when evening came, the boat was out on the sea, and he was alone on the land."

The same pattern is found in the other Gospels. In Luke 5:16, after healing a man of leprosy, great multitudes gathered to hear Jesus teach and to be healed, "but he withdrew to the wilderness and prayed." Before Christ selected the twelve apostles from among his disciples, Luke 6:12 remarks: "In these days he went out to the mountain to pray; and all night he continued in prayer to God." In Luke 9:28, Christ "took with him Peter and John and James, and went up on the mountain to pray" prior to the transfiguration. John does not mention prayer in the desert or on the mountain, but John 6:15 does record that when the people wanted to make Jesus king, "Jesus withdrew again to the mountain by himself."

In these texts we find repeated a withdrawal to isolated sites for prayer—and, secondarily, rest—in the midst of Christ's active ministry. There is no evidence in the passages of continuing confrontations with Satan. The texts imply Christ and his disciples needed to remove themselves from the stresses of ministry and to have some time to themselves. The Gospel's provide enough detail for us to infer that most of these sojourns were relatively short—a morning or an afternoon and evening, or perhaps, at most, two or three days. The spiritual exercise varied from very intensive (praying all night) to more restful (going away in the boat to rest awhile). Mark and Luke give the impression it was necessary for Christ to withdraw to these isolated locations just to escape the crowds. Considering the number of times it is mentioned, it must have been a regular practice.

The texts also specifically mention Christ going to these wild or isolated sites alone or with a small group of disciples. Unlike John, Jesus did not call the people into the wilderness and most often goes by himself into the desert or on to the mountain. If we look at the total number of times the Gospels mention Christ or his disciples praying, wild nature and the garden of Gethsemane are the two most important locations for this spiritual activity.

It should be recognized both John and Jesus initiated ministry to the people in uncultivated and wild sites. Teaching and preaching took place at the river, by the seas, in the desert or isolated places, and on mountains. Prophecy and healing were more urban occupations, as was dealing with human antagonists, including the scribes and Pharisees. There are, in fact, confrontations between human antagonists and Christ and his disciples in cultivated settings and gardens—but not at wild sites. When John called the Pharisees and Sadduccees a brood of vipers, he apparently sent them scurrying back to town for good. Most of the threats of violence and verbal confrontations in the Gospels are in Jerusalem, in the synagogues, or in houses.[16] In the Gospels, the sojourns in wild nature are marked by a freedom from adversity of human origin.

Blessed on the Mountains

Like the sojourns in wild areas, the wilderness theophanies do not end with Christ's forty-day fast. The Gospels record more heavenly voices and appearances of angels, and they describe several events where Christ himself could be considered a form of theophany, including all the postresurrection appearances. Christ's encounters with demons should also be counted as contact with the otherworldly and the evil counterpart of an angelic meeting.

The theophany most in keeping with the Old Testament concept of divine appearances was the transfiguration. According to Matthew 17:1–2, Jesus took Peter, James, and John "up on a high mountain apart" where "he was transfigured before them, and his face shone like the sun, and his garments became white as light." According to Luke 9:29, "as he was praying, the appearance of his countenance was altered, and his rainment became dazzling white." Moses and Elijah appeared in glory, a bright cloud overshadowed them, and a voice from heaven said, "This is my beloved Son, with whom I am well pleased; listen to him!" (Matt. 17:5). (Or "This is my Son, my Chosen, listen to him!" [Luke 9:35].)

This event was similar to the coming of the Holy Spirit at the Jordan and to the theophanies in Sinai. The transfiguration revealed the true nature of Jesus as the Son of God to the disciples. Christ appeared in his divine aspect. God, in the Old Testament, was associated with fire, clouds, and shining light. Christ assumed his radiance. The coming of Moses and Elijah, both of whom had encountered Yahweh at Mount Horeb, served as further evidence of the divine nature of the event. Like the theophany at the baptism, the voice discussed Christ's relation to the Father, but, in this case, also tied Christ to his disciples by the command "Listen to him!" Just as Yahweh provided the Law on Sinai, Yahweh gave Christ's teachings validity at the transfiguration.

We so often think of the risen Christ meeting Mary Magdalene in the garden that we neglect the other postresurrection appearances. A majority are outdoors. John 21 records the very critical encounter of Peter with the risen Christ on the beach. After a breakfast of bread and fish cooked over charcoal, Christ asked Peter three times if he loved him. Peter answered in the affirmative each time, and Christ instructed Peter to feed or tend his sheep. In the discourse, Jesus sent Peter on one of the most influential missions of all history, the establishment of the church.

In Matthew 28, Christ directed the disciples to a mountain in Galilee and met them there. When they saw him "they worshiped him; but some doubted" (Matt. 28:17). On the mountain, Jesus gave the disciples the "great commission" to "make disciples of all nations, baptizing them in the name of the Father and of the Son and of the Holy Spirit, teaching them to observe all I have commanded you . . ." (Matt. 28:19–20). The mountain again became an important site for revelation and a place where a new ministry was initiated. The instruction to take the teachings of Christ to all the nations was as important as the giving of the Law to Israel.

A number of authors have suggested the Gospels associate the demonic with wild nature.[17] Christ does encounter evil spirits outdoors, such as the Gadarene demoniacs at the tombs. Luke 8:29 mentions a man who was driven into the desert by a demon. Several other texts concerning the presence of demons obviously refer to populated country, however, and, in Mark 1, Christ drove out an unclean spirit from a man in a synagogue. Another passage—sometimes quoted to support the concept of the wilderness as the home of demons—is the healing of the dumb demoniac in Luke 11:14–26. Jesus commented: "When the unclean spirit has gone out of a man, he passes through waterless places seeking rest. . . ." This implies an affinity between the unclean spirit and the desert. However, Christ continued: ". . . and finding none, he [the unclean spirit] says 'I will return to my house from which I came.' " The Greek for house in this case is *oikos,* implying a demonic preference for the comforts of the human.

Ulrich Mauser, in *Christ in the Wilderness,* suggested that Christ's stilling of the storm also represented a contest with the demonic in wild nature. In the Old Testament, there is no interaction between a prophet and nature with the same structure as Jesus' rebuke to "the wind and raging waves" (Luke 8:24), followed by a calm. The story of Jonah includes a storm, but Jonah did not start it or still it—God did. The Gospels do not mention a demon or spirit steering the wind, so there is no reason to assume demons were present. The actions of Christ in quieting the forces of wild nature may be interpreted as exactly that—a demonstration of power over the creation, proving his role as the Holy One of God. As we have already seen, the Hebrews considered Yahweh to be in charge of storms. Although it is a reversal of the Sinai theophany, in which the display of natural force accompanied the presence of God, Christ's simple rebuke implies the same prerogatives, a matter well understood by the disciples, who wondered at the miracle. Christ's humanness caused him to look at the storm from the bottom up, unlike Yahweh at Sinai who looked from the top down.

The Spirit in the Wilderness

Before extracting any New Testament models of wilderness experience, we should ask: was John the Baptist really the last of the Old Testament prophets and, therefore, a representative of the prophetic model? The major arguments against placing John with Elijah are that (1) John prepared for his ministry in the wilderness, but there is no evidence of personal transformation via divine encounter, and (2) John's wilderness existence was more completely ascetic. John's experience relates as much to that of Christ as it does to that of pre-Hellenistic prophets.

Both John's extended and Christ's limited asceticism were purposeful and related to preparation. John may have been a Nazarite, since he drank no wine, but his dietary restrictions seem to have exceeded both the requirements of Nazarite vows and the practices of Qumran (where eating bread was allowed). In the Old Testament, asceticism in preparation for ministry was not a well-formed concept. The prophets certainly fasted, but this was a general cultural practice. Denial of physical sustenance was an important theme of the Exodus, but the people were not purposefully choosing the life-style. Deuteronomy 8 suggests the hunger was a form of testing, but there is nothing to indicate it was a conscious form of preparation. God very quickly provided manna, quails, and water. The Exodus does not report continuing self-denial, but rather immediate provision, long before the need was severe. The children's ungrateful attitude rather than an inability to survive an ordained fast became the root of the difficulties.

Other Old Testament wilderness models also seem to favor divine provision over controlling hunger and thirst through self-will. Ravens cared for Elijah at wadi Kerith, and angelic provision allowed Elijah to travel "in the strength of that food forty days and forty nights to Horeb" (1 Kings 19:8). Fasting was certainly important to the Hebrews, but, in the case of wilderness sojourns, the repeated theme was Yahweh's ability to provide sustenance.

The Gospels indicate that John and Christ went into the wilderness in response to God's will. In the case of John, the long desert sojourn functioned as a conscious preparation for ministry. In the case of Jesus, the Spirit "drove" or "led" him into the wilderness. We see this type of leading in Abraham's call to Moriah, the children of Israel following the pillar of cloud, and Elijah at wadi Kerith. Fleeing into the wilderness does not appear as a major theme in the Gospels. Christ went into the isolated places or on to the mountain to escape the crowds, but this is not of the same type of flight as Elijah's escape from Jezebel or David's escapes from Saul.

In the Old Testament, the children of Israel traveled to Horeb specifically to meet with God. Many of the other Old Testament theophanies and divine encounters in the wilderness were "coincidental," however, at least in terms of human motives. Hagar was not looking for angels, and Jacob went to sleep on the rock because he was tired. Elijah fled to the Negeb out of fear and was not purposefully trying to find the "still small voice." In the Gospels, the pursuit of the divine in wild nature was most frequently purposeful. John baptized at the Jordan in preparation for the coming kingdom. Christ went in the desert before starting his ministry and purposefully selected isolated sites for prayer. Christ led the disciples up on the mountain of the transfiguration and told them to meet him on a mountain in Galilee. The exceptions are Peter meeting Christ on the beach and other similar incidents, such as the risen Christ meeting Mary in the garden or the two disciples on the road to Emmaus. The clarity of "wilderness objectives" in the Gospels was a function of the spiritual vision and prophetic insight of Christ and John. They knew what they were doing, who they were in relation to God, and what was coming. Like the wandering Hagar, the very human and wayward Peter had to be taken by surprise.

In the Gospels, we find a new form of wilderness sojourn—a withdrawal of limited duration for prayer and rest. Christ and his disciples were the primary practitioners, and they retreated to mountains, isolated places, far lakeshores, or the desert after very intense periods of ministry, or when finding time for prayer had become difficult. The Gospel writers suggest that, when Christ would leave, the crowds would follow desiring teaching and healing. This implies the withdrawal was not during lulls in the ministry.

The Gospels thus present three types of wilderness sojourns:

1. John's long preparation for ministry, accompanied by self-denial. This type of wilderness use may have continued into John's active preaching period and may have been imitated by John's disciples who fasted and participated in other ascetic exercises. Although it originates in more ancient Hebrew traditions, it extends and intensifies previous prophetic practice.

2. Christ's forty-day complete withdrawal for fasting and prayer, including a confrontation with Satan over his relationship to the Father and over his role as the Christ. Christ went into the wilderness alone prior to beginning ministry.

3. Short periods of withdrawal, extending from a few hours to two to three days (the maximum time is not clear from the Scriptures), for rest and prayer. Christ might go alone or might take a small group of disciples. These short sojourns were in the midst of active ministry and appear to have been free from demonic attack.

Another new element in the wilderness experience of the Gospels was John's baptism at the Jordan; it called the people to symbolic participation in an exodus. Elijah had crossed the Jordan, but he had not taken the nation with him. John's call was the only occasion in the Scriptures when the people as a whole were drawn to the wilderness and there was not simultaneously a military or governmental concern.

A last new element is the appearance of the demonic and satanic among the wilderness theophanies. This may be the result of a greater emphasis on demonology in Palestinian Judaism,[18] but it is also certainly related to some major changes in theology. The Gospels partially remove the question of righteousness before God from the types of environmental response seen in the Exodus and cast the problem in terms of human and demonic reaction to Jesus. In the Synoptic Gospels (Matthew, Mark, and Luke), the "heart condition" of all the personages, human and otherworldly, in the narratives is displayed in their relationship to Christ. The early confrontation with Satan not only proves Jesus' role and abilities, it also sets the stage for the challenges of the scribes and Pharisees. The final temptations come in the questioning of the high priest and Pilate, after the arrest, and in the comments of passersby and of religious leaders at the cross:

> "You who would destroy the temple and build it in three days, save yourself! If you are the Son of God come down from the cross. . . . He saved others; he cannot save himself. He is the King of Israel; let him come down now from the cross, and we will believe in him. He trusts in God; let God deliver him now, if he desires him; for he said, 'I am the Son of God.' "   (Matt. 27:40–43)

These remarks tossed at a man dying in pain were a bitter echo of the offers Satan made to a man hungering in the desert.

In the Gospels, unlike during the Exodus, God did not discipline the fallen nation. Creation, however, responded to the tragedy. When Jesus "yielded up his spirit, . . . the curtain of the temple was torn in two, from top to bottom; and the earth shook, and the rocks were split; and tombs also were opened" (Matt. 27:51–52), and ". . . there was a darkness over the whole land until the ninth hour, while the sun's light failed" (Luke 23:44–45). God did not send a plague among the people nor cause the ground to open and devour the Chief Priest. Resting in the hands of God, creation, like the women at the cross, was torn by grief.

In the Gospels, people who did not repent faced a final judgment in a divine court, while people pleasing to God entered a kingdom not of this world. During the Exodus, the wayward nation encountered environmental troubles and finally entered a real geographic location, the Promised Land. The New Testament removed the issue of the relationship of the nation Israel with the land and replaced it with a spiritual kingdom. The New Testament wilderness remained a place of spiritual seeing and hearing—a place where God spoke and Christ was observed transfigured. The role of the wilderness environment itself—both in expressing God's providence and in executing God's wrath—had, however, been much reduced. Christ, as the center of the gospel narratives, directly mediated both blessings and wrath. Ravens no longer brought meat, but Christ, sitting on the shore in a lonely place, provided bread and fish for the faithful. When Peter ate with the risen Christ by the lake, the conversation with Christ became the key to the encounter.

In the ministries of Christ and John, the Old Testament pattern of the wilderness as a place for dealing with one's personal relationship with God was extended. The wilderness was for the contemplative and the internal, as well as for vision of the divine. Christ's continued withdrawal for prayer, conducted away from his public ministry, filled his need to speak to the Father. His battle with Satan was an examination of his internal spiritual state and faithfulness to his calling. Christ returned to isolated places repeatedly for the prayer and rest which were a necessary complement to his intensive outreach to the people.

We have in the Gospels two related models of wilderness experience: the *ascetic model* of John the Baptist and the *contemplative model of* Christ. In the *ascetic model,* we find:

1. Wilderness served as a site for extended preparation for a difficult ministry.
2. The strenuous environment of the wilderness provided freedom from social temptations and allowed a perspective on the culture to develop. Asceticism was consciously practiced to set an example for the culture.
3. Wilderness was associated with repentance and freedom from sin.
4. Wilderness was associated with both personal and cultural transformation.

The *contemplative model* contains the above elements and:

1. The Old Testament pattern of wilderness as a place to see and hear God clearly was reaffirmed. Wilderness remained a location for theophanies, but also became a site where one can wrestle with one's internal spiritual state or with Satan.
2. Wilderness and isolated places became preferred sites for rest and prayer. The withdrawal became a counterpoint to Christ's pattern of ministry.
3. The duration of withdrawal to the wilderness was determined less by the physical length of the journey and more by the relative need to

recover from the stresses of ministry, to speak to God, or to deal with internal spiritual matters. Short stays with little environmental stress were important enough to be mentioned multiple times in the Gospels.

4. The environment continued to be an indicator of God's will (as in the cloud at the transfiguration, Jesus calming the storm, or the darkness at the crucifixion), but interactions with Christ take precedence (the loaves and the fishes and Peter meeting the risen Christ on the shore).

In both the ascetic and contemplative modes, retreat to the wilderness was voluntary or spirit-led. Although wilderness provided relief from human antagonists, the theme of flight was much reduced, and replaced by conscious withdrawal for spiritual purposes.

Across the Lake to Rest

From a contemporary perspective, we can perhaps more easily identify with Christ's comings and goings from the isolated places than with those of any other biblical figure we have yet encountered. The short withdrawal for rest and prayer is similar to most modern wilderness sojourns. Christ's modification of the Old Testament patterns of wilderness use was, in fact, primarily a by-product of his style of ministry, which was more similar to contemporary practices than to that of the ancient "forthsayers." Christ's full-time commitment to reaching the people, which required an intense schedule of preaching and healing, kept the crowds constantly around him—a rare concern for an Old Testament prophet. Unless he made an effort to withdraw, Jesus had difficulty finding time to himself to rest or pray.

In the rush of our daily schedules, we may also find extended times for prayer or solitude can be few and fleeting. Both people in full-time ministry and those who spend as much of their spare time as possible in church activities often end up burned out and frustrated with their spiritual lives. In Christ, we see three years of extremely stressful ministry fueled by short intervals of prayer and supported by rests away from the crowds. Christ's choice of quiet locations included the desert wilderness, the mountains, isolated sites away from the towns, and the garden of Gethsemane. If we intended to preach or to heal, we need to spend at least as much time with God as Christ spent—if not more. Wild nature can provide refuges, free from human distraction, for contemplative prayer and for rest from our day to day responsibilities, and places where we, like Christ, can seek deep contact with the divine and relocate ourselves on our spiritual road.

With their emphasis on purposeful wilderness withdrawal, the Gospels raise an important question for contemporary wilderness travelers—how long should one stay in the wilderness? The wilderness retreats taken by Christ and John ranged from a few hours to several years. The length of a wilderness stay in the Gospels does appear to be related to purpose. The longer sojourns were for "preparation," the shorter ones were for prayer, rest, and renewed contact with God in the midst of ministry. Our contemporary vacation of one to two weeks (a relatively recent innovation in the industrial West) was not in vogue. The shorter stays in the Gospels ranged from a few hours to the equivalent of a long weekend. The longer stays were at least forty days, which is about six weeks.

The lack of one week vacations in the Bible does not mean that one should not take such a period of rest. Most of us would find it difficult to pray for a week in any case and would probably only spend part of a vacation in serious pursuit of God. The Gospels do verify the value of the short period of withdrawal and the potential spiritual benefit of spending an isolated afternoon alone in prayer. A two- or three-day retreat, taken every now and then, makes sense and is a potentially powerful source of renewal. The pattern of short retreats is supportive of "routine" prayer, relief from daily stresses, and spiritual exercise such as fasting or worship.

The major problem is the role of the longer sojourns. Are such initiatives important for leaders and others entering missions or new ministries? By contemporary standards, forty days in isolation is a long time. Yet the longer withdrawal may offer psychological advantages. Several days may be required for someone to free themselves of fears, preoccupations, and personal concerns and begin to open themselves to deeper issues. Cultural attachments weaken as one adapts to a nonhuman setting. Elijah offers a valuable case study. First there was the short interval under the broom tree, during which God spoke to him about his fear of Jezebel, and an angel provided food. Then there was the journey to Horeb, giving Elijah time to consider his circumstances. Then, in the cave, Elijah saw all of raging nature thunder before him just before he heard God in the gentle little breeze. Finally Elijah had to grasp his new mission. All this took time—even for the perceptive prophet.

The story of Elijah makes another point—the knowledge of God is not cheap. Television evangelism often gives the impression one can drive up to the call box and order holiness and spirituality—the way one orders a hamburger and a milkshake—and then drive around to the window and pick them up, wrapped in grace—all for a small contribution. The biblical wilderness experience declares this is not the case—and a great deal more than forty days may be required.

Ironically, we often prescribe three years or more of seminary education for Christian ministers, but rarely send them out into the desert for a few days to wrestle with their calling. Against hundreds of hours of classes, term paper writing, and practice ministry, forty days seems like a small investment to get to know God.

Before rejecting the forty days as too lengthy to be practical, we must also examine the frequency of wilderness sojourns. How often should these occur? Forty days is an exceptional sojourn, much longer than the regular rest of Christ and his disciples, and there is no evidence it was repeated more than once or twice in a person's life, or that it was something everyone did. The long sojourn belongs to a deep search for God, a major transition, a struggle with sin or evil, or the preparation for formal ministry. It should also be recognized that more is not necessarily better (as American popular culture so often declares). A major insight may spring from an afternoon of disciplined prayer. Two or three full days of seeking God is adequate for many purposes and can offer a substantial supplement to daily devotions. Most of us will go into the wilderness only to be quickly called back to service.

At this point it should be noted that some other models of wilderness spiritual experience, besides the ascetic and contemplative, may require longer stays. This is particularly true of the leadership model which produces physical, social, and spiritual fitness. Outward Bound courses usually last twenty-one days, and secular wilderness leadership schools can exceed two months. David certainly spent extended periods—of several months and more—in the biblical wilderness. For training leaders or for building character in adolescents, programs of several weeks to several months are justified. (My own experience indicates it takes about four to ten weeks for someone to adapt to working comfortably in eastern forest wilderness—and this does not include slaying lions and bears!)

What of longer stays of a year to a lifetime? The wilderness vocation, or complete withdrawal for a lifetime, appeared with monasticism. Although the Jewish Essenes practiced an unending retreat to the desert, the emergence of that life-style among Christians is really post-biblical and a phenomenon of the early church period. The only major New Testament figure who remained in the wilderness for more than a year was John the Baptist, who did not have a complete wilderness vocation and returned from the wilds to challenge the social and political authorities of his day. Some biblical critics deprecate John's life-style as "the last of the Old Testament prophetic ministries," and thus dismiss his life as a model to be followed by Christians who are citizens in the New Kingdom. Yet, John reached beyond Old Testament standards in the way he prepared for his mission. John's approach was more than a proof of his calling by God (the major reason he would have been an Old Testament imitator), it was an answer to a series of cultural problems common in Hellenistic Judaism and a means of speaking to the scribes and Pharisees. John repudiated the desire of the "religious" people of his time for personal possessions and social status by abandoning both and by casting godly spiritual practice in a mold so extreme it was impossible to replicate by half measures. If we consider John the last of the old order, to be replaced by Christ, then John's wilderness spirituality is indeed archaic. If we consider John an essential predecessor to Christ and a prophet who helped to prepare the people for their Savior, then John's ministry may still have an important place in the New Kingdom. Worth noting is John's role within a believing but apostate culture. Although he certainly attracted the attention of the Romans, John was calling out to the people of Israel. Unlike Peter or Paul, John never walked the streets of Antioch or set foot on the beaches of Greece, and he was not part of the New Testament wave of cross-cultural evangelization. John can, in fact, seem like a lost cause as minister to the people—a man whose only real role was to announce the Messiah. Yet there can be little doubt many of those who first answered John's call for repentance went on to follow Jesus.

The question of wilderness sojourns lasting a year or more revolves around the role of John and the prophets and any modern equivalents they might foster in the church. Does some portion of the church need to voluntarily undergo cultural deprivation to provide clear vision for the remainder of the body? Are the voices of the Baptist or of Elijah critical—not to the unbelieving scattered around the world, but to "the people of God" within supposedly Christian cultures? Wilderness spirituality in its more strenuous formats—the prophetic model and the ascetic model—could potentially help contemporary Christianity in its battles with the cults of personality, prosperity, national superiority, and middle-class values. Long wilderness sojourns would thus be justified for those who need to free themselves of cultural baggage thoroughly enough to be able to call the people of God to repent.

It should be noted in closing this chapter that the gospel writers perceived no conflict between "wilderness vocation" and ministry. Time alone with God was necessary to the implementation of service. The Gospels place a greater emphasis on personal purification prior to initiation of ministry than is common in the twentieth-century church and also assume the prophet must receive a deep understanding of the person of God before beginning to speak. The Gospels also suggest that contact with the divine must be renewed through prayer and solitude, and that even the ministry of Christ was refreshed by periods of rest.

9
The Desert Road
(Acts and Revelation)

Down to the Water

The coming of Christianity brought a new meaning to the old Hebrew concept of the "converted heart." In his resurrection appearances, Christ instructed the disciples to carry the good news to all the nations. The Jews had always accepted "god-fearers"—those who believed in the God of Israel—into the congregation. An outsider could not, however, become part of a tribe or of the Levitical priesthood, and, thus, was never a full member of the Hebrew religious community. With the establishment of the New Testament church, the chosen of God were no longer limited to the biological seed of Abraham—anyone could become a Christian.

The Gospels present more wilderness imagery than the other New Testament books, perhaps because the Gospels are in a more completely Jewish setting and because they concern the revelation of the person of God in Christ. Discussions of Hebrew experience in the wilderness do, however, appear as parts of other New Testament teachings. In the book of Acts, for example, the deacon Stephen retells the trials of Moses in the wilderness. The purpose of the lecture was to remind the Jews of their continued apostasy before God and to associate the mission of the Christian apostle with that of the prophet Moses. Stephen declared to the council in Jerusalem that, even after God called Moses on Holy Ground and made him "mediator between the angel who spoke on Mount Sinai"[1] and their forebearers, the people ignored him and "brought sacrifice to an idol and rejoiced in the work of their hands. "[2] Stephen implied that the Hebrew people were just about to relive the mistakes of the desert wanderings and reject Stephen's preaching about Christ, just as they had rejected the leadership of Moses and the vision of Yahweh in the wilderness. Sorrowfully, although appropriately, Stephen's speech ended in a vision of "the glory of God, and Jesus standing on the right hand of God,"[3] which his accusers could not see. The people at the hearing rose up in a rage, cast the saint out of the city, and stoned him.

The New Testament, aside from the Gospels, not only utilizes the wilderness passages from the Hebrew scriptures, it also presents new wilderness experiences in two fresh forms. The first new expression of wilderness is found in a series of historic events in Acts. These begin with a desert adventure in the best Hebrew tradition. An angel commanded the apostle Philip, who has just completed a very successful evangelizing campaign in Samaria, to: " 'Rise and go toward the south to the road that goes down from Jerusalem to Gaza. 'This is a desert road' " (Acts 8:26). Philip found an Ethiopian eunuch, who was also a high governmental official, reading the book of Isaiah aloud while riding in his chariot. Joining the man in his conveyance, Philip convinced the eunuch that "the sheep led to slaughter" in Isaiah is Christ. Philip then expounded "the good news of Jesusc to the eager Bible student (Acts 8:35). When they came to some water along the road, the Ethiopian requested baptism. Afterwards, "the Spirit of the Lord caught up Philip; the eunuch saw him no more, and went on his way rejoicing" (Acts 8:39).

This passage is followed by the call of the dedicated Jew, Saul, who had been present at the stoning of Stephen, and was actively attacking the young Christian church. Saul, accompanied by other men, was traveling to Damascus to find and arrest Christians, when "suddenly a great light from heaven flashed about him" (Acts 9:3). Saul "fell to the ground and heard a voice saying to him, 'Saul, Saul, why do you persecute me?' " (Acts 9:4). Saul, not unlike Moses at the burning bush, asked, "Who are you Lord?" The voice answered, " I am Jesus, whom you are persecuting" (Acts 9:5). The men who were with Saul heard the voice but did not see anyone. When the theophany departed, Saul stood up and found he was blind. His companions led him into the city where a man named Ananias interpreted the incident and laid hands on the repentant persecutor in order to fill him with the Holy Spirit. Saul, of course, became Paul, the great apostle to the Gentiles.

If we assume that Luke and Acts were originally distributed together and not separated by the book of John, these two events were actually part of a series of three theophanies, conversions, and calls "on the road." In Luke 24, two disciples of Christ were on their way to Emmanus, a village several miles from Jerusalem. Jesus joined them and walked with them to the town. The two did not recognize the resurrected Christ until he broke bread with them and then disappeared. This incident parallels both the eunuch's and Paul's encounters with Christ. The two men going to Emmaus initially did not see because they thought Jesus was dead; the eunuch did not see because he could not understand the passage in Isaiah and had not heard about Jesus; and Paul did not see because he was blinded by his religious zeal. All three encounters on the road ended with holy vision. The two men recognize Jesus and go tell the others they have seen him alive; the eunuch drives away rejoicing in salvation; and Paul, temporarily blinded, regains his sight and assumes a new mission. The roads were a very important feature of the Roman world and were to facilitate greatly the spread of Christianity throughout the Mediterranean region. It is hardly surprising that, in the Lucian descriptions of the rise of the early church, the theophanies or miraculous appearances of Christ are often tied either to Jerusalem or to travel between towns.

The baptism of the eunuch is an important biblical event that sent the Word of Christ outside the Levant in the person of a convert who had not personally seen Jesus. The incident tempts comparison to the desert wanderings of Hagar and other narratives of the foundational model of wilderness spiritual experience because it concerns personal revelation. On the desert road, however, Philip, rather than the angel who sent Philip, meets the eunuch. In contrast, angels minister directly to Hagar. The story of the eunuch describes evangelism as a means of conveying information about the person of God. The Lord spoke through Philip, rather than via an otherworldly messenger. In Acts, God is more likely to move a human to minister to someone else than to provide an insight through strictly divine means. Philip and the eunuch find water, just as Hagar did, but it is less an example of a mighty act of God and more an example of the perfection of divine timing. The provision of water in the wilderness facilitates baptism and allows the eunuch to proceed joyfully on his way.

In view of the nasty mood in Jerusalem at the time, the meeting on the Jerusalem road may have offered protection from social interference. If Philip had approached the eunuch, who was returning from a pilgrimage, in the temple, it might have been difficult to interpret Christ to him, much less to publicly baptize him. The desert road offered Philip a chance to peacefully discuss Isaiah with a man of willing heart and to spark a conversion, just as the desert offered Hagar a chance to find her future without interference from Abraham's household.

The theophany appearing to Saul is obviously a parallel event, but it concerns a man with different attitudes. Saul had violence on his mind and was unavailable for gentle evangelization. In this case, the Divine did act directly. The theophany was not in wilderness *per se,* but in isolation nonetheless—outside the city and away from the multitude. Unlike many of the theophanies of the Old Testament, which tend to concern individuals (or the nation Israel), the voice was audible to Saul's companions—people who weren't "involved" in his theophany. The circumstances of the theophany appearing to Paul are more similar to John's baptism of Jesus or the Transfiguration than to most of the theophanies in the Old

Testament. Like Jacob at Bethel, Paul was in transit. Further, as was the case with Jacob at wadi Jabbock, a name change resulted, and Paul received a new understanding of himself relative to God. Both the patriarch and the apostle became "progenitors" of a holy people, Jacob in a physical and Paul in a spiritual sense.

After meeting Christ, Paul spent most of his ministry moving from city to city. When faced with wild nature, however, Paul responded within the expectations of the prophetic tradition. Under arrest and on his way to Rome by ship, Paul suffered through a stormy "northeaster," during which the ship ran aground and broke apart in the surf. Paul with all the crew and passengers swam or rode planks to the shore of Malta. A viper, driven out of its hole by the heat of a warming fire, bit Paul before he was completely recovered from his desperate swim. The evangelist, much to the surprise of the natives of Malta, shook the poisonous snake off his hand and remained unaffected by the venom (Acts 27 and 28).

The narratives provide an interesting comparison to prophetic interactions with storms and Christ's stilling the tempest. The ship had left late in the season, and Paul had advised them not to start. The unpleasant weather described in Acts was not unusual for the time of year, and there is no implication of divine wrath as in the story of Jonah. Once caught in a gale, Paul did not attempt to still the storm, but rather advised the centurion how to save the soldiers and passengers. Without direct divine intervention (e.g., walking on water), the entire group safely reached shore.

The Old Testament almost invariably credits events such as severe storms as acts of God. The implication that the storm was normal winter weather and part of a "natural" process, places the emphasis in the story on the personal reactions of Paul to the dangerous situation. No angels arrived to carry Paul to safety; the evangelist had to swim for himself Yet, no one was lost to the sea. God appeared to be favoring Paul and perhaps even more important, verifying his words and his advice.

The viper managed to bite Paul (unlike the lions in Daniel who never got a taste of the prophet), but Paul survived. Acts does not say Satan inspired the viper to nip the evangelist, nor is there any evidence of demonic action. Again, the heat of the fire driving the snake from hiding was a natural event, and Paul was, inopportunely, picking up firewood at the time. The miracle in Acts was not that Paul took control of nature (he never really tried to still the storm or to stop the snake), but that the evangelist got where he was going. The message is twofold: first, if you listen to the evangelist, you will be saved from the storm, and, second, if you travel for God, God will help you through difficulties.

The Woman with Eagle's Wings

The second new form of wilderness imagery in the New Testament is the use of wilderness locations in apocalyptic writing. Much like the writings by or about the later prophets, the book of Revelation places some of its visions and key events in the desert or on mountains. (The authorship of Revelation is tradionally attributed to the apostle John, but many modern scholars consider the author unknown.) In Revelation 12:1, a "woman clothed in sun," usually interpreted as Mary or the nation Israel, is about to give birth. A dragon, interpreted as Satan or an evil force, appears in the sky and threatens to consume her son. God takes the child up to the holy throne. In the tradition of Hagar and the children of Israel, the woman flees into the wilderness (literally desert) to "a place prepared by God, in which to be nourished for one thousand two hundred and sixty days" (Rev. 12:6). The woman becomes symbolically the early Christian church. The wilderness setting evokes not only thoughts of Elijah sitting by wadi Kerith and waiting on the Lord, but also the flight of the Christians in Jerusalem across the Jordan into a town called Pella before the fall of Jerusalem in 70 C.E. The early church had to "flee into the wilderness"[4] just as the ancient Hebrews and their prophets had done so many times before.

In Revelation 12:14, the woman is "given the two wings of a great eagle that she might fly from the serpent into the wilderness, to the place where she is to be nourished." The serpent, not giving up the chase, "poured water like a river out of his mouth after the woman, to sweep her away from the flood. But the earth came to the help of the woman, and the earth opened its mouth and swallowed the river which the dragon had poured from its mouth" (Rev. 12:15–17). Here a dozen motifs from the Hebrew prophets converge in this description of the fate of the Christian community. The wings of the eagle recall Exodus 19:4 when God spoke out of the mountain to Moses and said, "You have seen what I did to the Egyptians, and how I bore you on eagles' wings and brought you to myself," or Isaiah 40:31, which declares, "But they who wait for the Lord shall renew their strength, they shall mount up with wings like eagles." The eagles' wings are the arms of God which carry the church to safety. When the serpent breaks forth the waters, we see God's order and boundaries—so well described in Genesis and in the psalms—violated. The serpent unleashes chaos against his fleeing adversary. Just as the bed of the Reed Sea helped to mire Pharaoh's chariots and the wind had pushed back the waters allowing the children of Israel to safely pass, the earth comes to the aid of the blessed woman and swallows the flood.[5] The Greek word used in this passage for "earth" is not *kosmos,* but rather *gi,* which can mean literally ground or soil. As the wilderness protected the anointed king of the ancient land of Israel, the wilderness and the earth itself will protect the chosen church of the Beloved Lord.

In a superb literary counterpoint to the woman clothed in sun, Revelation 17:1 introduces "the great harlot who is seated on many waters." (The waters are the peoples of the earth.) The author of Revelation reports that one of the seven angels [of the seven bowls] "carried me away in the Spirit into a wilderness [desert], and I saw a woman sitting on a scarlet beast which was full of blasphemous names, and it had seven heads and ten horns" (Rev. 17:3). The woman, sporting gold, pearls, and jewels, is clothed in purple and scarlet. The portrait is that of a wayward empress or of an expensive whore. Revelation calls her Babylon, but symbolically she is the city of Rome. The angel carries the author of the Apocalypse to the wilderness, not because Rome is situated in the desert, but because the wilderness offers an appropriate visionary landscape. As Conn notes in his commentary on Revelation: "John is taken to the wilderness to get a better perspective on the idolatrous city. In the Exodus, the desert is where the Jews went to receive a truer perspective on God. "[6] In the wilderness of the Exodus the Jews also learned how deadly idolatry could be and freed themselves from the bonds of an unholy culture, just as the author of Revelation is implying the young Christian church should do. The beast, Revelation reports, will destroy the harlot.

In Revelation 21:1–2, the author sees "a new heaven and a new earth" and observes "the holy city, new Jerusalem, coming down out of heaven from God." Then one of seven angels carries him away in the spirit to "a great, high mountain," where the vision again is of "the holy city Jerusalem coming down out of the heaven from God" (Rev. 21:10). This passage parallels Ezekiel 40:2 in which the Spirit brought Ezekiel "into the land of Israel, and set [him] down on a very high mountain . . ." in order that Ezekiel could see the design for the new temple. The choice of the mountain as the visionary landscape also reminds the reader of the magnificent theophanies on Mount Sinai and the Transfiguration of Christ. If John saw "horrid Rome," alias Babylon, from the desert, it was only appropriate that he should see the beautiful New Jerusalem from the mountain. To end the revelation, an angel shows the author "the river of the water of life, bright as crystal, flowing from the throne of God and of the Lamb through the middle of the street of the city; also, on either side of the river, the tree of life with its twelve kinds of fruit, yielding its fruit each month; and the leaves of the tree were for the healing of the nations" (Rev. 22:1–2). Wicked Babylon is thus replaced by holy Jerusalem. In this apocalyptic work, the Christian visionary, via the mountain, returns to Eden to "see" the throne of God and the tree of life. The "seeing" and "hearing" and closeness to God lost in the Fall are regained.

To See the Crystal River

The resurrection of Christ hardly terminated wilderness spiritual experience or banished wilderness as a visionary landscape. Desert or isolated settings play a role in Christian conversions and in the very important divine call of the apostle Paul. The story of the young church is largely one of urban evangelism as well as conflict and imprisonment. The mission of Peter, Paul, and the other apostles was to introduce people to the new faith. Following the Spirit's leading, they went where potential converts could be found. The theme of wilderness vision evolves towards a theme of "highway vision" as theophanies and spiritual transition occur in isolated roadside sites.

His survival of the shipwreck and the viper served to identify Paul as a godly man and to verify the truth in his testimony, just as remaining unharmed by the wild had marked the true prophet. Paul displayed a prophetic knowledge of the outcome of the voyage, and, like Elijah sitting in the wadi while waiting for the drought to end, Paul waited for God to still the waters and send him on to Rome. Christ's control over the storm had been replaced by missionary tolerance.

Jewish apocalyptic literature probably greatly influenced Revelation, the book of the New Testament most like the work of the Old Testament prophets. This mystical volume happily accepts motifs from the ancient "forthsayers," including their "spiritual" or visionary landscapes. The author of Revelation also assumes the earth will assist in defending the church as the new holy people of God and displays the interaction between the divine and creation in a favorable light.

In the New Testament as a whole, wilderness remains associated with spiritual transition, in the form of call and conversion and in the arrival of a new kingdom. Wilderness and solitary experience also retain their association with very spectacular spiritual vision and with deep revelation about the person and intent of the Savior. The holy conveyers of God's word continue to be released from the ravages of the wild (to be injured or killed by erring humans instead). The wilderness protects the true people of God, and, in Revelation, the earth even attempts to thwart a dragon.

These passages bring up two concerns for today's readers. First, is the vision of John locked in a book, lacking color and drama on a black and white page, or is it available for the contemporary Christian to "see?" The canon of Holy Scripture is now closed for Christians. Is the walk (or flight) to the edge of the New Kingdom also closed? Throughout the New Testament there are visions of angels and of Christ. Are these also gone? The answer requires discussion within the Christian community. If one takes a dispensationalist position, then the visions belong to an earlier age, and nothing of the sort is available today. If one believes the New Testament church remains the model for the modern church, then some amount of visionary experience should be expected. Such visions have always been rare and, throughout the history of the Hebrews, were far more evident in some eras than others. Our attitudes towards vision will, however, influence our attitudes towards wilderness—if deep wisdom and exceptional vision is still available, the wilderness is a potential setting for the experience.

Second, is the "desert" sometimes the best place for a conversion or call? Are there times when a trip alone or with a small group can aid spiritual understanding or reduce interference? One could not plan an experience such as Paul's. One could, however, respond to a divine call to meet the Ethiopian eunuch, or stumble into someone else struggling with the book of Isaiah. Despite Christendom's usual reliance on group meetings, there may be intellectual or psychological problerms best dealt with in isolation. Wilderness may not be the majority solution, but, in cases where negative social pressures are likely to arise, it could offer a "safe location" for religious activity. To evangelize in the wilderness, presumably, one must follow Philip's example, by responding to the call and making an effort to reach the desert road.

10

The Original Desert Solitaire
(Desert Monasticism)

In Wadi Qilt

I first saw Wadi Qilt from an overlook below the old Roman road from Jericho to Jerusalem. Traveling with a geography class from the Institute for Holy Land Studies, I had dutifully followed my instructor down the goat trails from Cyprus, a ruined Herodian fortress, and settled on a cliff edge to peer into a canyon several hundred feet deep. While the professor talked on about the demise of the Herods, I scanned the limestone walls across the gorge. The stream valley, lying at the bottom of several rock ledges, was wide enough for someone on foot to traverse. What looked like a donkey trail wound along a terrace in the middle of the far cliff faces. "It must be possible to hike up the wadi," I thought to myself, while shuffling notes on archaeology.

After we returned to our bus, our driver started up a route originally constructed for horses and chariots and proved that ancient thoroughfares can handle modern group transport. With rolling chalk hills on one side and the depths of the gorge on the other, we left Cyprus and climbed cautiously out of the Rift Valley. This was the same road Jesus walked, traveling from Jericho to Jerusalem before the triumphal entry. The Scriptures provide no physical details of the setting, but it is wilder and more beautiful than I had imagined. The stark skyline, the white ridges, the jagged red rock of the wadi . . . I wondered if Christ, on his long uphill trudge, had tarried to appreciate the colors of the landscape or had admired the vistas of the distant cities.

We stopped on a high stony knoll marked by a large dark cross and poured out on the slope to search for the best view. Below us, Monastery St. Georges rose from a palm covered spot in the depths of the canyon. Unfazed by the challenges

147

of topography, the monks had built their haven into the sheer cliff walls above the desert stream. The monastery stood as a sophisticated construction in a land of scattered bedouin tents—but it belonged in Wadi Qilt. The monks had taken the desert, combined it with the need for a place to worship God, and produced the pale towers of Monastery St. Georges.

We sat on bare, stony ground and sang a few hymns and songs from the Scriptures. The evening sun ignited the gray haze of the valley above the Dead Sea and turned the plains around Jericho a fiery red. The burning twilight swept up the slope, converting the skeletal white of the wilderness to a deep, glowing pink. The monastery rested quietly in the low light and then was covered by the shadow of the surrounding heights.

I moved over and sat down beside the instructor. "There must be a way to the monastery. How do you get down in the wadi?" I asked.

"Oh, there's a trail all the way through the canyon . . . and there's a path just below us, leading off from the Roman road."

I looked back towards Jericho and the Jordan, then smiled to myself. I was planning a trip into "the wilderness."

* * *

The most difficult part of visiting Wadi Qilt was getting out of Jerusalem. A number of my classmates at the Institute for Holy Land Studies weren't in shape for such a trip, and most of the remainder weren't interested in "wasting" one of our precious free days by revisiting a place we had already "seen." They decided to go shopping in Bethlehem instead. A college student who had worked at a wilderness camp in Colorado was game, however, and he rounded up two more adventurers. A bomb scare at the bus station (there had been a real bomb on a bus the day before) delayed our start. Confusion over finding the right bus—in the midst of both armed soldiers and hundreds of people hurrying to work or school—slowed us further. When the bus driver finally opened the doors by an unsigned dirt road, we fled the crowded vehicle. Each pale hill said "peace," and every rocky pinnacle said "isolation" as we headed down a jeep track towards a desert spring.

The major part of our route turned out not to be a footpath, but the side of a Roman aqueduct. Still efficiently carrying water after twenty centuries, the aqueduct followed the middle of the slope, sometimes crossing dry washes, sometimes skirting cliff edges. Nature offered alternatives to this marvel of ancient engineering, however, and, at one point, we traversed a small natural arch bridging the bed of an ephemeral stream.

Despite the unwanted slopes and the stern topography, evidence of human activity, past and present, was everywhere. On a steep ridge above a natural seep, we saw a herder gathering his flock and riding his donkey up the dusty slope. At one wide spot in the valley we observed a small garden, earth mounded to conserve water, and borders fenced to ward off voracious livestock. Goat droppings littered every ledge, and, when we came upon a few scraggly trees, goats occupied the branches.

As we headed downstream, the canyon grew deeper and the walls grew steeper. Finally, the aqueduct ended abruptly, releasing its torrent of water to race down the cliff. The Romans had captured the water again as it rushed into the natural stream course and had channeled it into another aqueduct on the other side of the canyon. The change of the aqueducts also marked our passage into the territory of the monastery. We walked up to one of the large dark crosses the monks had constructed to mark the corners of "their" land. Uncertain of how to enter the monastery buildings that extended several stories above the bottom of the gorge, we stopped and examined the potential routes. One of my companions took a path from the base of the cross and managed to pull himself out onto the roof of the upper story of the cliff dwelling. Despite the presence of a door, this entrance was not open to strangers.

Undaunted, we decided, if the high road wasn't open, we would take the low road. Scrambling down two steep, ill-defined trails, we reached the bank of the Qilt, a stream still waiting for winter rains. The quickest route to the main door of Monastery St. Georges was a civilized path shaded by palms. There was no evidence of a truck or jeep road, but the monks had built a foot bridge across the gravel scoured stream bottom. Donkey trails led out to Jericho and to the ridge intersecting the Roman road.

We hesitated at the metal door. Would we be welcome? Some Israeli hikers were approaching down the opposite side of the canyon. Reasoning they were probably also planning to tour the monastery, we opened the door and went in.

It is daunting, after walking for a half day through biblical wilderness, to arrive at someone's residence and, with no one to greet you, enter to see what's inside. The fear of adverse social encounter is especially strong if you suspect the people who live there belong to neither your time nor your culture, are ignoring the rest of the world, and may be purposeful in trying to get away from people just like you. We poked around carefully, hoping we wouldn't irritate anyone in authority.

The monks, as it turned out, expected visitors. A bearded young man in black orthodox garb found us and politely guided us into the chapels. Decorated from floor to ceiling with icons and holy ornaments, these sanctuaries were of the same genre as the churches and shrines in Jerusalem. The desert life might be simple, but its centers of worship had acquired all the complexity impassioned artists could create.

Entering a small room, one of my companions stepped back in shock. Human skulls and bones lay in neat stacks in boxes on the floor. "Martyrs," said the young monk, "killed by the Persians." We looked at them with more curiosity than reverence. "Why would anyone keep all these bones here?" asked my young friend. A Lutheran, she had never seen such relics in a church in Washington state.

A monk, bent with age, motioned to us to come outside. Uncertain, we followed him up another flight of stairs to a high balcony beneath the bell tower. He took us into a badly lit room and pointed towards the paintings on the walls. He spoke no English and kept repeating his message until he thought we understood. "It's John the Baptist," I suddenly realized, "and there is Elijah being fed by the ravens." The old monk smiled and nodded. I looked at the uneven surfaces and the soot blackened ceiling. The chamber was a cave, probably a natural formation in living rock. Prophets and hermits had sought this sheltered spot long before the monastery edifice covered its mouth.

The old monk held out a plate for money. I made a donation and wandered outside. To my left on the cliff face were additional stories of the building, stacked one on top of the other—the monks' quarters, where we couldn't enter. To the right was a wooden scaffolding jutting out of the cliff face below a rough opening in the rock. No ladders or stairs were visible, but a basket suspended by a rope and pulley hung from a wooden beam overhead. I pointed it out to my friends and suggested it was a cave reserved for a monk seeking complete solitude. He could draw food up in the basket without having to leave his cell.

"Why would he want to do that?" asked the Lutheran.

My hiking companions returned to the chapels to purchase postcards, but I lingered on the balcony and absorbed the setting. Elijah's sojourn here was probably legend, but he may well have known the site or traveled through the valley on the way to Jerusalem. John the Baptist might also have found the wadi in his wanderings. Christ had certainly seen the gorge from above. The traditional Mount of the Temptation was not too far distant. Had Jesus visited this wadi or a similar canyon during his forty days with the wild beasts?

Monks had first occupied the caves and, like John, resided in the wilderness while realizing their calling. The austerity, intensity, and continuity of this spirituality were difficult to contemplate. Wearing hiking shoes and blue jeans, I was a "day tripper," here until I had to catch the last bus back. The prophets stayed until God spoke, Christ stayed until he defeated Satan, and the monks stayed for a lifetime. Were we all part of the same story?

We cheery, well-fed Protestants had an easy walk down the gorge. The "fathers" who first came to Wadi Qilt would have thought us "spiritual wimps." They were disciplined, striving beyond anything any of my troop would contemplate. Is that why they came here—to prove themselves in a great spiritual contest? What did they gain? Was their asceticism pride, or a move of the Spirit? Why sit in a cave by yourself? Why not evangelize thousands?

I leaned off the balcony . . . a beautiful vista of a green oasis among the precipices, graced by the cross on the far peak. Did the monks see this as I did? Was the *kosmos* a concern or a demonic hazard? I would enjoy staying here a week, but I would soon start to wander up and down the wadi. I looked up at the cell with the scaffold and basket . . . stationary solitude.

My friends were getting restless and worrying about the bus schedule. We snapped a few tourist photos and, rather than walk back up to the Roman road, took the trail towards Jericho.

Uncertain of the time, we began to rush. I wanted to stop and look under the rock shelters and investigate the small crevices. I found a cave fronted with crumbled masonry, but was pulled onward without discovering if it was an hermitage or part of a *lavra,* a cluster of separate monastic cells.

Suddenly the path grew narrow, and it was necessary to place one foot directly in front of the other to progress safely. My friend from the Colorado wilderness encouraged me onward. Ahead, our companions had lost the trail completely and were grasping rock ledges to lift themselves above an awesome drop straight to the bottom of the canyon. A thorn bush jutting out of a crevice caught my shirt, but I had no room to pull away.

"We can't go on like this all the way to Jericho," I complained. "This is impossible."

I stopped. Something wasn't right. Overhead I saw human handiwork, a row of rough stones set at the top of the terrace.

"There's the trail up there."

Kicking loose rock into the air below, I scrambled upwards, over scree. In our hurry, we had missed a turn and walked out on a dangerously narrow ledge.

As we rounded another bend in the trail, the sound of a flute drifted up from the far slope. A goatherd was entertaining himself while his charges grazed around the aqueduct. The music was simple and sweet. I stopped to listen for a moment, then bolted off again. The bus schedule. . . .

And finally Cyprus, the place where we'd first looked into the wadi. What had started as a backcountry trek had become an encounter with other people from other times. As I turned to admire the cliffs and crags for a few final seconds before dashing to the nearest paved road, I thought again about the monks and wondered what they were like before they found a ministry coping with tourists. Were we very different or very similar, those monks and I? What was it that drew them into the wilderness?

The Wilderness Vocation

Monasticism, more than any other movement in the history of Christendom, has been associated with wilderness. Following the analysis of the major biblical models of wilderness spirituality, the next logical step is to investigate wilderness use in the early church. Monasticism is a complex phenomenon with many orders and schools of thought. For simplicity's sake we will consider monasticism in three different settings in three different eras. This chapter is concerned with the desert fathers who began monasticism's long wilderness sojourn. The following chapters consider Celtic monasticism and the Franciscans, respectively. In each case, we will investigate the monks' attitudes towards their wilderness environment and toward wild nature, as well as the importance of wilderness to the monks' spiritual practice.

Before the close of the Age of Martyrs, the first Christian hermits fled to the deserts of Egypt and Palestine. In the beginning, these ascetics were mostly solitary, residing alone in caves, abandoned buildings, simple shelters, or even tombs. They then began to form loose groups, called *lavra,* and eventually organized the communities, or *cenobia,* that have housed the majority of monks from late Roman times to the present.

For the modern student, both the motivation and the practices of the earliest monks may be puzzling. The first monks were usually not priests and sometimes were not even literate. In the beginning there were neither orders nor organization, just individuals who felt called to leave the world, trek into solitary locations, and live by themselves. Unlike Christ and John the Baptist, they felt no strong call to minister to people in nearby cities. They considered not only sexual desires, but also food and sleep to be sources of temptation. Self-denial was the ultimate virtue. Despite their great "fear" of the world, they proceeded, with unquenchable enthusiasm, to battle demons by day and by night. The desert fathers were convinced that, from their hidden rock shelters, they were engaging in spiritual warfare which would help to free the universe of the powers of evil.

These monks were, on one hand, vigorous in their pursuit of fasting and prayer and absolute in their dedication to God. On the the other hand, they were prone to extremes of asceticism, such as eating nothing but five figs a day, or sitting on top of a column for thirty-seven years.[1] Early monastic practice may appear to have been more of a spiritual sideshow than a serious search for God.

We need to make an effort, however, not to let the great distance in time and culture—or some of the monks' less disciplined pursuits—discourage us from investigating their motives for withdrawing into the wilderness. We can ask legitimately if their asceticism was a continuation of New Testament teachings. We can also ask if their wilderness practices have anything to offer us today.

Roderick Nash, in *Wilderness and the American Mind,* repeats a widespread idea that the early church saw the wilderness as "the earthly realm of the powers that the Church had to overcome."[2] A number of writers have concluded the first monks went into the desert because the desert was the home of demons.[3] They suggest, therefore, that the monks had no appreciation for the desert itself and considered it a very negative and unpleasant environment. If they had been able to get into hell, the desert fathers might well have gone there instead.

Nash also proposes that the monks paid little attention to their natural environment and had few, if any, aesthetic concerns. Nash distinguishes St. Basil's fourth-century written description of "the forested mountain on which he lived" as unusual for the time in that it seemed to recognize natural beauty in wilderness. "[4] Nash concludes: "On the whole monks regarded wilderness as having value only for escaping corrupt society. It was the place in which they hoped to ignite the flame that would eventually transform all wilderness into a godly paradise."[5]

In his brief analysis, Nash implies that monastic appreciation of wilderness as a thing in itself was very rare. Yet a great many monastic writers other than St. Basil mention the desert, wilderness, or wild nature. Did the monks have a sense of place or an appreciation for their environment? Why did many monks go into the wilderness rather than just begin separate "Christian" villages or farms? Is there something about wilderness that is beneficial to monastic spirituality?[6]

The Inner Mountain

To do justice to what the desert fathers and early hermits thought about their environment, we should look at their own sayings and biographies. This material is different from modern historical sources in several respects. First, although the monks produced a number of "histories," these documents were intended primarily to instruct others in Christian virtues and to extol monastic ideals, rather than record events exactly as they occurred. The histories and sayings contain heroic tales and folklore as well as theological material. Most of the people in the documents are historic figures, but some may not be. Material may have been borrowed from one source and added to another. Some motifs are repeated from manuscript to manuscript. Other than archaeological artifacts and their own writings, however, we have few sources of information about these early wilderness residents. The writings, however, give us an opportunity to determine attitudes towards nature since they teach values and were written by the monks themselves. Determining the literal facts of history can prove more difficult.

One of the earliest and most influential of the monastic biographies (saints' lives are also called hagiographies) is *The Life of Saint Antony* by Athanasius, who was also a monk. Antony, born in 251 C.E. and entering the ascetic life in 271 C.E., was not the first Christian in the desert, but he became so well known for his holiness that many young men followed his example and retreated from the towns of the Nile Valley into the dry lands beyond. Leaving his wealth behind him, Antony went first to visit an old ascetic outside his home village; then he moved to a tomb and had himself locked in. After a series of conflicts with Satan and his demons, he relocated to an abandoned fort in the desert.[6] Finally, as his reputation spread, he found himself "beset by many and . . . not permitted to withdraw as he had proposed to himself." He thus headed toward Thebaid and a "people among whom he was unknown."[7] As he waited to cross the Nile, a voice from above asked him where he was going. Antony reported his flight from the annoying crowd. The voice then said, "Whether you go up to Thebaid or, as you have been considering, down to the pastures, you will have more—yes, twice as much trouble to put up with. But if you really wish to be by yourself, then go to the inner desert." Antony followed some Saracens for "three days and three nights and came to a very high mountain."[8]

The available descriptions of Antony's new and final home not only contain some interesting natural detail, but also Athanasius's direct statement: "Antony, as though inspired by God, fell in love with the place, for this is what He meant who spoke to him at the riverbank." The Greek verb used for "fell in love" in this sentence is a form of *agapao,* implying spiritual or godly love.[9] The passage clearly indicates Antony not only loved his haven in the wilderness, but that this love was divinely ordained. Athanasius describes the mountain as having "water, crystal-clear, sweet, and very cold" and "a few scraggy date-palms."[10] Antony, to avoid troubling the Saracens who were going out of their way to bring him bread, found "a small patch that was suitable, and with a generous supply of water available from the spring, he tilled and sowed it. This he did every year and it furnished him bread. He was happy that he should not have to trouble anyone for this and that in all things he kept himself from being a burden. But later seeing that people were coming to him again, he began to raise a few vegetables too, that the visitor might have a little something to restore him after the weariness of that hard road."[11]

St. Jerome, a monk who translated the Greek and Hebrew scriptures into Latin, described Antony's mountain in *The Life of St. Hilarion:*

> There was a high rocky mountain about a mile in length. Water sprang forth from the crevices and flowed down to its base, where some of it was absorbed by the sand and the rest, falling deeper, gradually formed a river shaded by numerous palm trees overhanging from both shores, making the location very pleasant and comfortable. . . . On the very top of the mountain there were two . . . cells, accessible only after a difficult spiral climb. Here, Anthony had sought refuge from the multitudes who came to him and from the companionship of his disciples.[12]

The monks were sensitive to the site and its qualities. These descriptions were— for ancient writings intended as spiritual instruction—both detailed and affirming in terms of natural features.

Although wild animals were potentially a threat to solitary desert dwellers, Antony's approach was considerate of the other creatures. Athanasius reports that Antony found that "wild animals in the desert coming for water would damage the beds of his garden. He caught one of the animals, held it gently and said to them all: 'Why do you do harm to me when I harm none of you? Go away, and in the Lord's name do not come near these things again!' And ever afterwards, as though awed by his orders, they did not come near the place."[13] Jerome, in a version of the same story, reports the animals, wild asses, returned after the rebuke from Antony, but only drank water and never molested the garden again.[14]

Antony's theology of creation was simple and profound. Once, when a philosopher asked Antony, "How . . . dost thou content thyself, Father, who art denied the comfort of books?" Antony answered, "My book, philosopher, is the nature of created things, and as often as I have a mind to read the words of God, it is at my hand."[15]

It should also be noted, Antony's life-style, which was widely imitated, was based on subsistence gardening and encouraged wildlife to continue to share the resources of the Inner Desert with him. Although monks did alter desert water sources and farmed where they could, their harvest of natural resources was very restricted and limited to necessary items, such as palm fronds for baskets. This life-style was the result of self-denial, rather than of environmental concern; however, their simplicity and restraint present an interesting contrast to more materialistic movements in Christendom.

Holiness and Wild Beasts

Several important themes in the writings of the desert fathers concern wilderness or wild nature, and natural examples are often used to make spiritual points. Among the most interesting are the numerous stories about wild beasts. Although some of these stories portray the monks as having special power over wild animals (they could command the animals to do or not do things), others portray the monks as living in a special relationship with animals. The monks not only could communicate with the animals, but they could also coexist with them. The monks had thus regained the position Adam had before the Fall.[16]

The History of the Monks of Egypt, for example, describes the life of a hermit named Theon: "They said of him that at night he would go out to the desert, and for company a great troop of the beasts of the desert would go with him. And he would draw water from his well and offer them cups of it, in return for their kindness in attending him. One evidence of this was plain to see, for the tracks of gazelle and goat and the wild ass were thick about his cell."[17] Theon appears in this passage not as master, but more as friend to his fellow creatures. The description of his relationship to the animals, in fact, completes a long list of his other noble attributes. Theon was described as wise in discourse, and many sought him for his healing abilities. He was so visibly holy "that he seemed an Angel among men, so joyous were his eyes and so full of grace did he appear."[18] The implication in the text is his friendship with the animals was the final proof of his blessedness.

An even more dramatic example of the wild beasts responding to a Christian holy man is found in The Life of St. Paul the First Hermit, by St. Jerome. According to this work, St. Antony had heard that there was a man who had been in the desert longer than he had and was much holier. Antony, seeking the hermit, wandered the desert unsure of his goal. When he saw a she-wolf "panting in a frenzy of thirst"[19] enter a cave at the bottom of a mountain, Antony guessed that Paul was providing the wolf with water, and he entered the cave himself. After Paul received Antony, a crow "softly flying down"[20] deposited an entire loaf of bread in front of them. Paul was already near death when Antony arrived, and after Antony departed, the older hermit expired. On his way home, Antony had a vision of Paul climbing the steps of heaven among angels, prophets and disciples. Antony hurried back to the cave to find the corpse "on its bent knees, the head erect and the hands stretched out to heaven."[21]

Antony wanted to bury the body, but he had no tool available for digging the ground. He was torn by indecision because the return trip to his own monastery involved a three-day journey and that would mean leaving the body exposed; but, if he stayed at Paul's cell, there was really nothing he could do.

But even as he pondered, behold two lions came coursing, their manes flying, from the inner desert, and made towards him. At the sight of them, he was first in dread; then, turning his mind to God, he waited undismayed, as though he looked on doves. They came straight to the body of the holy dead, and halted by it, wagging their tails, then couched themselves at his feet roaring mightily; and Antony well knew they were lamenting him, as best they could. Then going a little way off, they began to scratch up the ground with their paws, vying with one another in throwing up the sand, till they had dug a grave roomy enough for a man; and thereupon, as though to ask for reward of their work, they came up to Antony, with drooping ears and downbent heads, licking his hands and feet. He saw that they were begging for his blessing; and pouring out his soul in praise to Christ for that even the dumb beasts feel there is a God. "Lord," he said, "without whom no leaf lights from the tree, nor a single sparrow falls upon the ground, give unto these even as Thou knowest."[22]

Again the theme is a relationship between a human and animals, not raw power over nature. Reciprocity marked Paul's dealings with his animal neighbors. The wolf arrived seeking water, and the crow came bringing bread. When Antony, with his limited human means, could not provide for the deceased saint, God acted through Paul's beloved wilderness. Even the mighty lions requested an exchange: in return for digging the grave, they desired the blessing of the servant of God. Antony did not think it beneath him to speak to God on their behalf.

Some monastic histories do include instances of power being employed to subdue nature, but it is always in response to need. A monk named Benus, for example, drove away a hippopotamus that was "laying waste to the neighboring countryside"[23] (presumably damaging crops) and did the same to a troublesome crocodile. Rarely in the histories is violence done to an animal, although Hilarion does burn a boa that has been devouring not only livestock, but also farmers,[24] and Macarius of Alexandria, when bitten by an asp, tears the animal apart and says "If God did not send you, how dare you to come?"[25]

In a reversal on the theme of divine power, a young monk saw old men walking along the road and ordered wild asses to transport them to St. Antony. When Antony heard of the miracle he began to tear his hair and lament because he felt the young monk had fallen from grace. The young monk wept for his sin, but within five days he was dead.[26] In this case, although a miracle occurred, it was rejected, presumably since it was viewed as an act of pride. Another story relates how a monk who met a leopard on a cliff path at night commanded the animal in the name of Jesus to make way. The leopard jumped down the cliff twelve feet, but the monk "showing no fear of God, threw great stones at the leopard."[27] The leopard ran up another way and, very angry, attacked the monk, swatting him with his paw. The monk was injured, but survived. Among the desert fathers, the mere demonstration of power over nature was not, by itself, holy.

The first monks often found the desert a harsh environment, but, again, the use of power was mitigated by the offerings of the desert itself. When Macarius of Alexandria was traveling in the arid lands and ran out of water, he found a herd of antelope and took milk from the udder of one with a calf. She followed him back to his cell and continued to nurse him.[28] Antony, lost without water, knelt and prayed, and "the Lord made a spring come forth where he was praying."[29]

Some of the stories describe monks managing with only what was naturally available. Two monks going into the "utter desert" could find only bitter squills (a plant) to eat, so one of them prayed, and they were able to subsist on the squills for four years.[30] Another monk, living in a very arid site, collected dew off the rocks during the months of December and January by using a sponge; he stored it in pots for use the rest of the year.[31] Another monk dug a well where the water was bitter. He accepted this "so that he might show the power of his endurance."[32] The same monk, while assisting others, prayed, and a cistern was filled. In all these accounts the difficult conditions proved the spiritual worth of the brother, either through a miracle or through ascetic endurance.

One of the interesting features of the writings of the desert fathers is the number of times large carnivores, particularly lions, appear in the histories. Antony's adventures with the lions were only the first of many. Poemen was an old man "of so great virtue . . . [that he] would welcome the lions into his cave with him, and offer them food in his lap; so full of grace was the man of God."[33] A visitor to the solitary spent an extremely cold night in Poemen's cave. The hermit, however, had been warm enough–a lion had come and slept beside him.[34] A monk called John was living away from his *lavra* as a hermit at a time when desert marauders were terrorizing the neighborhood. The fathers of the *lavra* called John back, but he did not want to leave his solitude. He entrusted himself to God who sent a great lion for his protection. Although at first he was uncomfortable to find the animal lying beside him, the lion stayed with him and warded off the raiders.[35] The lion, which is thoroughly representative of wild nature, was almost always portrayed as a friend of the monks. The large carnivores not only got along well with them, they also offered protection. The monks, who were spiritual warriors for God, may have identified with the lions. The relationships also demonstrated the monks' courage.

In some of the histories, predators do damage people or, more frequently, livestock. The two stories cited above, in which monks destroy a boa and an asp, are not, however, completely typical either of the monks' treatment of reptiles or of their dealings with harmful animals. Crocodiles carried St. Pachomius across a river on their backs, and St. Simeon the Stylite healed a blind dragon.[36] The latter tale clearly indicates that the desert fathers found no creature beneath God's mercy.

A majority of the stories dealing with troublesome animals portray the animal as repentant. A typical example is the story of a she-wolf who regularly had dinner with a hermit who offered her whatever scraps were left. Once when he was late returning, she came at the usual time and finding some loaves of bread unattended, devoured one. The hermit returned and noticed the loss.

> Then as the days went by and the creature did not come—too conscious of her bold act to come to him she had wronged, and affect innocence—the hermit took it sorely to heart that he had lost the company of his pet. At last when the seventh day had gone by, his prayers were answered: there she was, as he sat at his meal, as of old. But it was easy to perceive the embarrassment of the penitent: she stood, not daring to come near, her eyes fixed in profound shame upon the ground, and plainly entreating pardon. Pitying her confusion, the hermit called her to come near, and with a caressing hand he stroked her sad head: and finally, refreshed the penitent with two loaves for one.[37]

The wolf felt grief and guilt, and the hermit forgave her. This story teaches Christian values through the actions of a wild animal.

A hyena brought a blind pup to Macarius of Alexandria—he who tore apart an asp and nursed from an antelope. Macarius put spittle on its eyes, as Christ had done with the blind man, and "immediately it saw the light."[38] The mother nursed it, picked it up, and left. She returned the next day with "the hide of a large sheep" as a gift. Macarius was distressed and made the hyena promise not to harm the poor by eating their sheep and never again to kill another creature, but only to eat what was already dead. He offered to feed her if she could not find carrion. The hyena did as the hermit asked and continued returning to the old man for food if she could not find any herself.[39] The tale of the hyena features reciprocity and portrays the animal in need as receiving a biblical style miracle. When she tried to show her gratitude inappropriately, Macarius instructed her, and she repented.

In some of these tales, the animal isn't the one in the wrong. The Abbott Gerasimus found a lion with an infected paw, removed the reed causing the injury, and cleaned the wound. The lion refused to leave him, so the abbot gave him charge of the monastery's donkey, which was used to draw water. When a camel driver stole the donkey, the lion returned downcast to Gerasimus. The abbot assumed the lion had eaten the donkey and commanded him to do the donkey's work. The lion, thus, began to haul water on his back. The lion was vindicated, however, when the camel driver returned and, seeing the lion, fled, leaving his camels. The lion picked up the donkey's halter rope in his mouth and led him, with three camels, back to the monastery. At the death of the abbot, the lion was so grieved he lay down, roaring, on the old man's grave and eventually died there.[40] Again, the theme is Christian virtue, including honesty and fidelity. The predator, in this case, never committed a sin, but, when incorrectly judged, takes the donkey's service without complaint. The desert fathers saw their own values operating in nature and perceived wild nature as directly responsive to the Holy and to the Will of God. Penance and withdrawal from sin were extremely important goals for the early Christian ascetic, and animals were portrayed as participating in these. The monk not only thought the lion and the wolf capable of joining him in service to God, he found that they could express the foundational virtues of faith, hope, and love.[41]

It is tempting to dismiss these stories as legends or inventions and, therefore, as poor representations of monastic attitudes towards the wilderness and wild nature. Although some of them are fiction—or highly decorated versions of fact—this does not make them less valid as indicators of environmental perceptions or values. The purpose of these histories was spiritual instruction—to teach the values of the desert fathers. The consistency with which certain themes appear in the writings of a number of different authors demonstrate the cultural rather than the individual basis of these environmental views.

Heuristic value aside, some of the stories may have been based on real events. There have been numerous verified cases of hermits or other people who have lived in wild areas for a long time befriending animals. If left unmolested, animals lose their fear of humans and, when provided with food or water, may learn to trust individuals or safe locations. There could have been monks such as Theon who provided water and soon had animals gathering around their cells. Although some of the stories, such as that of Antony and the lions burying Paul the Hermit, depend on the miraculous, many do not. (Note that in this case it was also a miracle that Antony knew Paul had died.) In the *Life of Malchus, The Captive Monk* by St. Jerome, for example, the hero and his wife (both celibate) were escaping from their captors. Desperate, they fled into a cave. Their master sent a slave in to drag them out. The slave entered and, searching for them in the dark, was attacked and killed by a lioness. The master, missing the slave, made the same mistake and was also killed. Malchus and his wife, afraid to move, spent the night in the cave. In the morning, the lioness, sensing the presence of humans, took her cub in her teeth and left.[42] The point of the story was not only the role of providence, but also the suggestion that the chastity of Malchus and his wife served "as a wall of protection."[43] If their chastity alone protected them, it would indeed be a miracle, but the lion may also have overlooked or ignored their presence. The story could certainly be based on actual events.

Another example—which may be better natural history—is the historic account of Sabas establishing a monastery. He had built a tower but water was still a problem. "But as he prayed in the little oratory one night of a full moon, he heard the beating of a wild ass's hoof in the valley below, and leant out to see the animal digging deep into the gravel, then bending down and drinking. He went down and found, indeed, living water at the foot of the cliff—an even supply that never fails."[44] Wild horses and asses have been known to excavate for water, so it is possible a wild equid led Sabas to the site of the spring. The monk may have known enough about the behavior of wild equids to suspect the animal might be looking for the aquifer. Providence again is seen as arising from or compatible with wild nature.

Desert Darkness, Desert Light

The monks got along so well with the wildlife that it is difficult to imagine them battling demons at the same locations. Yet battle they did, and the demons sometimes took animal form. St. Antony during his life in the tombs experienced "phantoms of lions, bears, leopards, bulls and of serpents, asps, and scorpions, and of wolves . . ." breaking through the walls of his chamber. "The lion roared ready to spring upon him, the bull appeared about to gore him through, the serpent writhed without quite reaching him, the wolf was rushing straight at him; and the noises emitted simultaneously by all the apparitions were frightful and the fury shown was fierce."[45]

The evidence indicates that the monks purposefully withdrew to the desert to encounter the satanic, but our modern supposition that they thought the desert itself was evil may not be correct. Jeffrey Burton Russell, in his book *Satan: The Early Christian Tradition,* notes that the early Christians believed that "the prayers of the communities in the increasingly Christianized empire were driving the demons out of the cities and that they were now congregating in the desert instead."[46] The assumption was not that the desert somehow produced demons or that they lived there naturally, but that they had been forced to dwell there by the faithful. The monks, in fact, thought that demons lived in the air and could move long distances at will.[47] According to Athanasius, the Enemy was quite perturbed when Antony decided to become a hermit—because "he feared he would fill the desert too with his asceticism."[48] A tall demon appears to Antony at one point and complains: "Now I have no place, no weapon, no city. Everywhere there are Christians, and even the desert is full of monks."[49]

In the many monastic tales of the demonic, a direct connection between the demons and the desert landscape is absent. There were no specifically demonic animals, places, or landscape features. Going to town was usually held to be a far greater danger. The stories also portray the monks as attracting the demons owing to the monks' holiness—something which the demons could not stand.

Early monasticism was, if anything, antiurban. St. Jerome wrote in one of his letters: " O desert enameled with the flowers of Christ. O solitude where those stones are born of which in the Apocalypse is built the city of the Great King! . . . How long (he asks a friend) will you remain in the shadow of roofs, in the smoky dungeon of the cities? Believe me, I see here more of the light."[50] On another occasion Jerome advised:

> If you wish to take duty as a presbyter, then live in the cities and walled towns. . . . But what has the monk to do with cities, which are the homes, not of solitaries, but of crowds? . . . We have our masters in Elijah and Elisha, and our leaders in the sons of the prophets, who lived in the fields and the solitary places, and made themselves tents by the waters of Jordan. After the freedom of their lonely life they found confinement in the city as bad as imprisonment.[51]

The antiurban feeling is also found in the numerous stories of monks who went to cities and abandoned their calling or fell into sin. A monk named Heron, for example, as a result of self-exaltation for his asceticism, was "driven as though by fire" to Alexandria. In the metropolis, he "fell willfully into a dissolute life" and visited the theater, horse races and tavern. He met an actress—his ultimate downfall. After an illness, he repented and returned to the desert fathers.[52]

Many of the passages mentioning the desert suggest the monks had a longing to be there, a strong feeling it was *their place*. Malchus, the captive monk, enjoyed tending sheep in the desert because he was away from his masters, and he thought himself to be like Jacob or Moses. He stated: " I was delighted with my captivity and thanked God for His judgment, for the monk whom I had nearly lost in my own country I had found again in the desert."[53]

The feelings described in some of the monastic writings even have the implication of pleasure in the isolated places. A story about St. Antony demonstrates the association:

> A hunter in the desert saw Abba Anthony enjoying himself with the brethren and he was shocked. Wanting to show him that it was necessary to sometimes meet the needs of the brethren, the old man said to him, "Put an arrow in your bow and shoot it." So he did. The old man then said, "Shoot another," and he did so. Then the old man said, "Shoot yet again," and the hunter replied, "If I bend my bow so much I will break it." Then the old man said to him, "It is the same with the work of God. If we stretch the brethren beyond measure they will soon break. Sometimes it is necessary to come down to their needs." When he heard these words the hunter was pierced by compunction and, greatly edified by the old man, he went away. As for the brethren, they went home strengthened.[54]

Antony described the monk who is away from the desert as a fish out of water. The monk belonged in the wilderness.

Russell suggests the desert had a dual meaning:

> It was the place of refuge from the temptations of society, but it was also a place where temptations came directly from the Devil. In the desert one could get away from petty distractions, from small vices and small virtues, and take part directly in the cosmic struggle between Christ and Satan. And the monks were quite right.
>
> Whether one interprets demons as external beings or as internal psychological forces, no doubt exists that the monks felt themselves under the most incessant attacks from the powers of evil. . . . The demons attacked the hermits more than the cenobites, because the higher one rose in the spiritual life the more impressive the attacks of the enemy on one became. The monks replaced the martyrs as the "athletes of God."[55]

Dualism was so predominant in early monastic thought that it became characteristic of it. The strong monastic rejection of the human as evil is often attributed to Platonism, however, rather than to Hebrew theology. The hellenized Alexandrian Christianity of Clement and Origen applied philosophy to the structure of the spiritual world and produced a hierarchical cosmos with good and evil at odds eternally with one another. The thought of the desert fathers often reflects Greek ideals; yet it presents an interesting contrast in its treatment of the human and the natural. On one hand, the monks were constantly fighting personal sin and attempting to control their very humanness. On the other hand, the early monastic writers assume the natural world is composed of lower beings, but they do not portray sin as an overwhelming force in nature. Nature is sometimes represented as harsh, but, rarely as demonic. Satan's attacks on the monks were direct, rather than operating through nature. Despite their rejection of the worldly, the monks do not appear to have been at war with the world itself.

Reciprocity in the relationships between humans and animals hardly seems like classical Platonism, which pays very little attention to wild nature as such. Platonic thought idolized the rational, sometimes depreciating nature as less than the ideal of Plato's pure "forms." The concern of the desert fathers for wildlife could, however, have been influenced by several schools of thought circulating in the Roman world.

Jewish and Christian traditions provided models for some of the stories. The story of the crows feeding Paul the hermit is similar to the account of Elijah at wadi Kerith, where ravens brought him bread and meat (1 Kings 17:1–6). The stories about the lions reflect the Old Testament idea that a true prophet would not be injured by wild animals. The tale of Macarius and the hyena replicates a healing performed by Christ. The monks were certainly highly aware of Christ's forty days in the wilderness with the wild beasts and of other Biblical wilderness traditions. Another important set of passages, which could have justified the proliferation of animal stories, are the prophecies in Isaiah, including Isaiah 11:6, "The wolf shall dwell with the lamb" and Isaiah 43:20, "The wild beasts will honor me." Unlike the Essenes, the desert fathers no longer had any reason to expect the Messiah to appear in the wilderness. They may, however, have expected the prophecies of Isaiah to be fulfilled literally in the New Kingdom, and they saw themselves as representative of the new Adam returning to Eden.

The stories about Old Testament prophets escaping from, or, if unfaithful, being killed by lions, are not an exact parallel to the desert fathers literature on large predators. In the story of Daniel, for example, the accounts do not personify or even deal with the actions of the lions, but rather describe the interaction between the king and Daniel. Daniel told the king an angel shut the lions' jaws (Dan. 6:17–24). The lions did not make a decision concerning Daniel's fate nor did they show any virtues. God prevented them from acting like wild lions. Some tales of the fathers, such as *The Life of Malchus,* are similar to this, but others, such as Gerasimus and his lion, provide the lion with much more character. Gerasimus and the lion with the injured paw is far more similar to Androcles and the lion and to Greek fables than to any passage of Hebrew scripture. The lion made decisions, such as returning the missing donkeys, and mourned the death of his master, just as the Androcles's lion recognized him and decided not to eat him.

The histories of the fathers composed of short passages about a number of different monks are very similar to Greek fables in structure. A fable usually presents a setting and a small group of characters, portrays one incident, and then ends with a statement of moral instruction. In Greek fable, animals speak and are capable of ethical thought. The desert fathers do not report animals as speaking like humans (nor does the Bible, except in the case of Balaam's ass [Numbers 22:21–35]) but they do use animals as moral models. Biblical animals are usually either passive agents, directed by the Spirit of God, or active cooperators with the divine, rather than independent souls acting in response to ethical dilemmas. Animals, plants, and landscapes are sometimes personified in the Bible, but this is usually in poetic or prophetic passages and not in records of historic events.

In addition to Greek fables, the lives of Greek philosophers may have influenced the desert monks. A number of philosophers of the Pythagorean school of thought, who were an important source for Neo-Platonists, were supposed to have had special relationships with animals. Pythagoras "was said to have called down an eagle from the sky and tamed it, and to have made friends with a bear."[56]

Apollonius of Tyana (late second to early third century C.E.) was supposed to have known the language of the animals. The accounts of Apollonius's life document not only his vegetarianism and aversion to animal sacrifice, but also his natural history observations on the character, intelligence, and affection for their young of animals, such as elephants, wolves, and bears.[57] The desert fathers may not have read Pythagoras or Philostratus' *Life of Apollonius of Tyana,* but the values expressed by these works were presumably widespread in Hellenistic culture.

As previously suggested, some of the stories may be based in fact. Animals gathering around a monk's cell at a spring could have reminded the monks of the writings of Isaiah. We cannot assume all the animal stories are mere inventions to prove that a prophecy had been fulfilled. The literature of the desert fathers probably grew from a mixture of actual desert experience, interpretations of the Hebrew and Christian scriptures, local folklore, and Greco-Roman myth and philosophy. The intrusion of Greek literary models was not necessarily in conflict with Christian values. Gerasimus's lion, for example, does not present a major theological threat. A Greek interest in using animals as moral examples seems to have been comfortably incorporated into a monastic attempt to renew human friendship with creation and, thus, fulfill the prophecy of Isaiah about the coming Kingdom of God.

Perhaps the deepest theological problem in monastic attitudes towards creation lies in the body/soul dualism, so central to Neo-Platonism. In their treatment of animals and of their desert environment, however, the monks also incorporated Hebrew thought, which holds creation to be a great work of God and, therefore, not intrinsically evil.[58] The monks clearly thought providence was operating in their lives and presented survival in the desert as an affirmation of their own holiness. These stances are much closer to the biblical concept of the wilderness's providing for the anointed and the righteous, than they are to Greek perception of nature. The monks were striving for the time when all would be made new. The Neo-Platonic dualism was, therefore, muted by the addition of Hebrew creation theology and apocalyptic vision.

The Original Desert Solitaire

For the early Christian ascetics, the desert was neither a random choice of residence, nor, as with the Essenes (the hiders of the Dead Sea scrolls), a location selected to fulfill a prophecy or vision. The desert offered an environment conducive to spiritual exercise. It provided, above all, solitude and freedom from external sources of temptation. The wilderness surrounded them with quiet and simplicity.

The monk could contemplate God, or struggle with evil, without distraction. The monk could fight the ultimate internal battles. The desert hermit lived in a cell within a greater cell—the wilderness around him.

The wilderness values of the monks were not tied to material or aesthetic values, but to spiritual and personal values. Silence, for example, is a key monastic virtue. *The History of the Monks of Egypt* first describes Nitria, an important monastic community, and then goes on to describe the territory of the hermits:

> Beyond this (Mount Nitria) there is another place in the inner desert, about nine miles distant: and this place, by reason of the multitude of cells dispersed through the desert, they called Cellia, The Cells. To this place those who have had their first initiation and who desire to live a remoter life, stripped of all its trappings, withdraw themselves: for the desert is vast, and the cells are sundered from one another by so wide a space, that none is in sight of his neighbor, nor can any voice be heard.
>
> One by one they abide in their cells, a mighty silence and a great quiet among them: only on the Saturday and the Sunday do they come together to church, and there they see each other face to face as folk restored in heaven. If by chance any one is missing in that gathering, straightway they understand that he has been detained by some unevenness of his body and they all go visit him, not indeed all of them together but at different times, and each carrying with him whatever he may have by him at home that might seem grateful to the sick. But for no other cause dare any disturb the silence of his neighbor, unless perchance to strengthen by a good word, or as it might be to anoint with the comfort of counsel the athletes for the struggle. Many of them go three and four miles to church, and the distance dividing them one cell from another is no less great: but so great is the love that is in them and by so strong affection are they bound towards one another and towards all brethren that they be an example and a wonder to all.[59]

Solitude, and therefore, the wilderness, were necessary to purification.

Early monastic ideals set the contemplative life above the most noble of Christian services to others. Among the traditions of the fathers there is the story of three friends, one of whom went to be a peacemaker, one of whom went to care for the sick, and one of whom withdrew to the desert. The first two found their vocations too difficult, and, "failing in spirit," they traveled to the desert to see the third.

> And he was silent for awhile, and then poured water into a vessel and said, "Look upon the water." And it was murky. And after a little while he said again, "Look now, how clear the water has become." And as they looked into the water they saw their own faces, as in a mirror. And he said to them, "So is he who abides in

the midst of men: because of the turbulence, he sees not his sins: but when he hath been quiet, above all in solitude, then does he recognize his own default."[60]

An even stronger recommendation for solitude was made by the abbot Moses who instructed: "Go and sit in thy cell, and thy cell shall teach thee all things."[61]

Although the sayings of the fathers indicate that wandering is counter-productive ("Even as a tree cannot bear fruit if it be often transplanted, no more can a monk that is often removing from one place to another"[62]), as the cenobia developed, the need for solitude moved individual monks to depart to the far reaches of the wilderness for a season. Euthymius, in 411, C.E., about a hundred years after Anthony emerged from his fort, initiated the practice of leaving the monastic community and traveling into the Inner Desert in Palestine. He would depart—following the example of Christ—on the Feast of Our Lord's Baptism (the 14th of January) and return in time for Palm Sunday. Until his death at age 90, he went annually with his friends, subsisting on herbs and what water they could find, to such sites as Masada and the wild gorges above the Dead Sea. On their return, they would either go back to the monasteries, or, "armed with those lovely little flowers of many colours that show themselves in the spring in the most barren places of the wilderness," travel to Olivet or Jerusalem.[63] Euthymius not only initiated new communities deeper in the wilderness, he also renewed the importance of solitude as the *cenobia* grew in size and became embroiled in a series of theological debates. Solitude was so important to monks such as Euthymius and Sabas that, having left the "world" for the monastery, they felt the need to withdraw further into the wilderness for at least part of the year.

Monastic standards on the duration of such solitude varied from maintenance of a lifelong hermitage to shorter periods. By modern standards, however, the isolation was lengthy, and the competitive nature of monastic ascetic practice tended to extend it. The abbot Sisois, while residing by himself on St. Antony's mountain, saw no one for ten months and then encountered a herder (who was not a monk) with his cattle. On discovering the herder had seen no one for eleven months, the abbot "went to his cell and smote himself saying, 'Lo, Sisois, thou didst think thou hadst done somewhat, and thou hast not done as much as this man of the world.'"[64] Euthymius traveled in the most arid parts of the Dead Sea valley for about three months a year. Monastic values called for complete commitment. If one were going to be quiet, an hour or two—or even a week—were not enough.

The Sayings of the Fathers lists a number of monastic virtues or concerns, including quiet, self-restraint, avoidance of fornication, lack of material possessions, patience, sober living, continual prayer, and contemplation.[65] The desert life, by its isolation and austerity, encouraged or assisted in development of these virtues. In the wilderness there was no need for money. Wanton company was a rarity. Without the pressures of continual social intercourse, prayer and contemplation could be pursued without interruption.

The monks absorbed the qualities of the desert and developed a strong sense of place. Their treatment of the desert and its creatures was never far from their own spiritual goals, yet they seem to have adjusted well to the environment and to have held both the desert and its animal inhabitants in high regard. The fathers wrote, for example, that Macarius of Alexandria "was a lover beyond all other men of the desert, and had explored its ultimate inaccessible wastes."[66] To love the desert or to explore the desert was to become holy. The desert was austere, and the desert was silent. The monks were as the desert, and the desert was as the monks.

How Long a Voice in the Wilderness?

The *early monastic model* of wilderness experience has obvious similarities to the *ascetic* and *contemplative models* of the New Testament. First, freedom from sin was a major concern, both for the desert monks and for John the Baptist. The monks imitated Christ in his battle with the demonic, but, rather than finishing the war in one trip, they continued the struggle for a lifetime. Second, like John, the monks could attract others to them and distribute their message from the wilderness. Third, both Christ and the monks had a concern for prayer, contemplation, and communion with God. The wilderness became a quiet place for spiritual exercise.

The monks also modified some New Testament practices. The most important change was the extension of the wilderness retreat from a few years or forty days to the edge of eternity. Christ and John withdrew and then returned to the people. The desert fathers withdrew and never came back. In the lives of Christ and John, there was little evidence that asceticism itself was ever a goal. The competitive religious disciplines of the first monks emphasized less food, longer withdrawal, and more prayer. John limited what he consumed much more strictly than the Nazarites, and probably the Essenes, but he seems to have avoided severe fasts and long stationary periods, sitting in a wilderness cell. Christ participated in fasting and contemplative prayer for limited periods. From his appearance at a marriage feast and the accusation of the Pharisees that he was a drunkard and a glutton,[66] we may assume Christ participated in fasting and other obvious forms of self-denial periodically rather than continually. Christ's withdrawal to the isolated places was in response to specific social or spiritual needs, rather than for its own sake.

Since there were others, such as Bannus and the Essenes, in the desert before John, it seems likely the association of extreme asceticism with holiness
proceeded the desert fathers. In their zeal, and, like the Essenes, in a reaction to
Hellenistic culture, the first monks took asceticism to its human limits. During
the lives of Christ and John, ministry always remained central, whereas personal
ascetic practice served as preparation for ministry and as a means of contact with
God. The desert fathers changed this balance, placing self-denial before personal
ministry. The first monks certainly drove out demons and healed the sick. They
did not, however, travel to preach or seek out people to teach. They also preferred
to engage the demonic on a cosmic level and to upset the evil principalities and
powers. Both Christ and John the Baptist attacked the dark forces through ministry to the crowds or to individuals.

Ironically, in the Greco-Roman world, isolation in the desert did not
suppress the monks' message. The faithful traveled hundreds of miles in order to
visit them, and their writings were widely circulated. They set an example concerning resistance to the unholy aspects of Eurasian culture that was to influence
both the western and eastern branches of the church for centuries. Although they
appear to be introverted, primarily pursuing their own spiritual purity, we have
to acknowledge they conceived their task as necessary to the purification of the
entire church and their battle with the demonic as critical to defeat of the forces
of evil. And the communities from which they came thought that what they were
pursuing was a spiritually critical goal. Some of the infusion of Platonism into
Christian thinking may have been unfortunate, but these "athletes for God" attracted many who valued the classical ideal of the heroic. The monks were products of Hellenistic culture speaking to a Hellenistic world.

Theophanies graced the lives of the monks, but this direct contact with the
divine was usually interpreted as the outcome of a life of holiness or as a personal
test, rather than as an unexpected revelation of the Person of God. Since the
monks had the Holy Scriptures to tell them who God was and what God was like,
the need for the type of wilderness theophany found throughout the Bible may
have been diminished for them. In early monastic writings, angelic messengers
tend to guide the monks' actions, assist with temptations, or verify righteous
actions, and thus serve as affirmations of ascetic ideals, rather than as sources of
promises or of deep insights into the being of God. The monks engaged demons
more frequently than angels and tended to take the warfare with the demonic,
which began in the Gospels, to an extreme.

The monastic associations with wild beasts were similar to those of the
prophets, but received far more attention in the monastic literature than in the Holy

Scriptures. The types of beasts were variable, but the principles of contact are similar to the *prophetic* and *new Exodus models.* The holy person got along well with the wild, while the unholy man faced the claws of the lion. The personification of the animals, however, was new and probably a result of Greek influence. In the Hebrew scriptures, wild nature aligned with the will of the God. The anointed of the Lord did as God wished, and nature responded. In the tales of the desert fathers, the peaceful interaction with wild nature was in direct response to the holiness of the monks. The monks considered themselves part of the New Kingdom and expected the prophecies of Isaiah to be fulfilled—not just in a positive response from the wildlife, but also in the support of the desert itself. In the monastic literature, the animals become "junior monks" acquiring the values of their mentors. In the Old Testament, wild nature is more a creation of God, continuing to respond to divine will, and excused from any call to penance or social reprogramming.

In summary, the *early monastic model* incorporated:

1. A reciprocal relationship between the monks and wildlife, particularly with large mammals. The monks lived in peace with large predators.
2. A love for the desert, which in return provided *a modus vivendi* for the monks.
3. An appreciation of God as Creator and for God's actions in creating a New Kingdom.
4. Use of the desert for isolation and removal of temptation. The desert became a huge disturbancefree "cell," surrounding the monk's own cave or hut.
5. Extreme asceticism, with fasting and isolation taken to their human limits. They did not, however, disparage the natural world because of this—only their own consumptive desires.
6. Complete withdrawal from the world, with no return for preaching or evangelism. Communication is indirect. The monks told the world about Christ as much through their actions and life-styles as through their speaking and writing.
7. A preference for battling the demonic over meeting the angelic. Much of the early monastic literature emphasizes defeat of sin rather than encountering the person of God directly.

In comparison to the austere life-style of the desert fathers, the modern wilderness ethic has a much greater stress on adventure and experience as well as on aesthetics. Development of individual character is a goal of some contemporary wilderness sojourners, but unrelenting pursuit of virtue and complete attention to God are not twentieth-century priorities. The modern wilderness user is more likely to go touring and seek out varying landscapes and scenes than to sit in the same place for months. The number of miles traveled or the number of places visited, rather than the length of stay, are used to evaluate the value of the trip. The back-packer usually wishes to be free of the distractions of the urban environment, but it is a temporary freedom. The unwanted intrusion in the modern wilderness is human construction, not the demonic or the ungodly. Some moderns make a profession out of wilderness, but few make a life out of it.

We may find, however, some of the monks' attitudes instructional in pursuing our own spiritual exercise in the wilderness. The monastic practice of staying in the same location for a period of time fosters contemplation. The monks chose environments with low levels of sensory input for prayer and fasting. They valued silence. They recognized the wilderness offered a tremendous reduction in the cultural temptations of cosmopolitan Greco-Roman society. Although we may have little interest in replicating their ascetic feats, we may find the wilderness is still an advantageous site for meditation and exercises in self-denial.

Monastic attitudes towards wildlife could also stimulate us to review our relationship with wild nature. The monks didn't just travel through the wilderness, they made friends with it and, through divine inspiration, came to love it. We have a tendency to "use" wildlife, and when we feed animals, such as bears or squirrels, it is usually for our viewing pleasure. We do not see other species as having spiritual interests—much less spiritual value as potential teachers or participants in spiritual events. The desert fathers' relationship with the wild was characterized by either *philia* (brotherly love) as demonstrated by reciprocity, or *agape* (divine love) as demonstrated by giving without any demand for return. They strove to live in peace with other creatures, including the potentially dangerous. Today, we view the book of Isaiah as symbolism or allegory, or, if we dare to accept the prophecies as literal, we think of them as simultaneously being fulfilled at some future time. The monks believed that Christ was the shoot of Jesse and that part of their earthly task was to help establish the New Kingdom in preparation for his reign. Perhaps if we took the words of the prophets more literally and thought of God's will for creation as a current issue, we too could sleep with lions.

11

Oaks, Wolves, and Love
(Celtic Monasticism)

To the Western Isles

We sometimes mistakenly think that an appreciation of natural diversity developed as a result of the age of science, or, perhaps, with the rise of the conservative movement in the nineteenth century. In Western religious traditions, appreciation of the form and variety of nature is far older, winding back to the time of Psalms and before. Our present interest can, in fact, hardly exceed that of our predecessors. As the monastic movement spread out of the Levant, it produced new schools of Christian spirituality in new settings. One of these offspring of St. Antony, Celtic monasticism, is now rarely discussed outside scholarly circles, but, in terms of western wilderness heritage, it is among the most important sources of nature-oriented art. Centuries before the birth of St. Francis of Assisi, Celtic scholars were producing rich religious manuscripts, full of natural imagery and wilderness narratives. Modern European literature owes a great debt to these European saints who enjoyed the song of the thrush in the oak and the stride of the red deer through the heath and who recorded these divine graces in their religious poetry. The Celts took the best of their ethnic heritage as nature lovers and combined it with an adoration for the Savior from the East to produce a magnificent "wilderness literature," now too oft neglected.

The monastic movement had barely matured in the arid Middle East when it began to spread through Europe and into the northern forests. About 360 C.E., Martin of Tours founded the first monastery in Gaul, and others soon followed his example, including John Cassian, who arrived from Palestine to live in what is now southern France in the early fifth century. In 432 C.E., less than eighty years after the death of Antony, a cleric named Patrick (who may have actually been either one person or two) set out to convert the people of Ireland, some of

173

whom were already Christian. Contrary to legend, Patrick did not live to see the entire island converted, but he did establish an Irish church with an episcopal system of government which incorporated monastic elements.[1] Monasticism continued to grow and, by the sixth century, had become a major force in Irish Christianity. In its early days, Irish monasticism was probably influenced by Christians in Britain. As the monastic movement gained strength and numbers, missionaries were sent to Scotland and Britain (where there had been Christian communities since the Caesars' legions imported the new religion from Rome and Asia Minor) and eventually to the Continent. The "golden age" of Celtic monasticism extended from 550 to 660 C.E., but the monasteries continued to be a center of Irish religious life until the Tudors attempted to dissolve them during the sixteenth century. One of the first great flowerings of monasticism away from the deserts and from Greek and Hebrew culture, thus, grew out of the conversion of the Celts.[2]

The Celts were a people of forest and sea coast and, therefore, had substantially different attitudes towards wild nature than those of the desert fathers. As one of the most important groups in the development of Christian arts and literature in the Middle Ages, the Celts provided ample materials for evaluating the relationship of early Christianity to wilderness. Both Roderick Nash, in *Wilderness and the American Mind,* and George Williams, in *Wilderness and Paradise in Christian Thought,* seem to have ignored the Celts. This is especially surprising in the case of Williams who reviewed numerous literary sources. Celtic monks not only produced hagiographies (saints' lives) full of natural imagery, but also decorated some of the finest Christian poetry ever written with descriptions of wooded landscapes and native wildlife. The oversight of modern commentators may be due partially to the Celts' lack of distinction between wilderness and other types of settings and to their distance from theological discussions in Rome. Although ancient Celtic writers did mention "desert" places, the common practice was move characters from wild to cultivated landscapes with little emphasis on the transitions. Withdrawal into uninhabited regions was a repeated motif in Celtic monastic literature, but the wild sites, usually forests, lake shores, and islands, are only occasionally referred to as "wilderness." The Celt was already a resident of pasture, woodland, and moor, and had no great cities, such as Alexandria, to abandon. There is thus little concept in Celtic writing of reverting to the wild or withdrawing to the desert. Nor is there much anti-urban discourse, the unholy cities of the Roman Empire being very distant both physically and culturally. The Celtic monk sought a yet more isolated spot within an already rural landscape and entered into spiritual warfare, not with the debauched Hellenists, but with the druids and the old religion.

The purposes and literature of the Eastern church influenced the Celts from the beginning. *The Life of St. Antony,* with its desert ideals, reached Gaul within fewer than twenty years after it was written.[3] Celtic scholars studied the classical languages and founded great libraries. The Irish saint, Columban, writing in 600 C.E. to Pope Gregory the Great, not only composed the letter in passable Latin, but also discussed the writings of St. Jerome, another of the desert monks.[4] The eastern influence can be seen clearly in Celtic liturgical and hagiographical efforts.[5] The Celts, with their indigenous and loosely structured monasticism, became the true wilderness heirs of the desert fathers.

In looking at Celtic monastic literature, we should note, first, that all types of Celtic religious literature had secular parallels, and, in many cases, incorporated pre-Christian themes or mythology.[6] The reworking of older sagas and the inclusion of druidic motifs and magic is far more conspicuous in the hagiography than in the shorter devotional poetry (some saint's lives were actually long poems or contained lays, etc.).[7] The Celtic saints' lives were spiritual adventure stories. Unlike the works of the desert fathers, they did not emphasize wise sayings or examples of ascetic endurance. The Celtic saint, rather than gaining holiness through training and self-discipline, was born to be holy, was recognizable as exceptional even as a child, and could have been involved in miraculous acts before assuming any public religious role. Celtic hagiography constructed long lists of miraculous and mystical events. Often, the saint's power obscured the saint's personality. In ancient Celtic culture, inclination to learning, foretelling and using of magic were considered to be inherited traits, belonging to certain social classes, such as the druids. Celtic hagiography did not teach righteous action as directly as the literature of the desert fathers did, although examples of righteous acts were included. It tended, in contrast, to teach transfer of spiritual power by association with the elect of God—or by association with their families or the lands they occupied. Celtic hagiography frequently asserted, for example, that people buried in the cemeteries established by certain saints could not go to hell.[8]

In the ancient Celtic world, land ownership was linked to families, and a person's identity was, in turn, very closely tied to family membership. Land and a sense of place were thus continuing themes in Celtic religious writing. Celtic monks were not as completely withdrawn from the remainder of their society as were the desert fathers. They involved themselves in politics and clan business, as well as in establishing churches. We can thus assume some of the geographic description in the hagiography reflects territorial interests and family spheres of influence.

Celtic literature and other early Christian materials are quite different from modern historic sources and offer numerous problems in interpretation. The hagiographies, although they supposedly recorded the lives of saints, were not intended to be detailed historic commentaries or even accurate records of land exchanges or family relationships, although they served as the latter to some extent. The hagiographies were intended to verify the spiritual and territorial claims of a saint's followers. They, therefore, contain heroic tales, folklore, information on property acquisition, religious teachings, and theological material, as well as biography. Some of the people in the documents probably never existed, and many of the events are exaggerations. Material may have been borrowed between manuscripts, and it may be difficult to tell the origin of some segments of text. Most of the hagiographies now available to us are from manuscripts copied or composed long after the deaths of the saints whose lives were considered worthy of a biographer's attention.

Despite their shortcomings as history, these sorts of writings lend themselves to an attempt to determine attitudes towards nature because one of their purposes was to teach values. It should be remembered, however, that the examples of the saints and the religious communities may have been only partially reflected in the actions of the people around them. Religious values state what one ought to do, not necessarily what one does.

The literature presented in this chapter extends from about the seventh century to the twelfth century and is primarily Irish although some examples from related British materials are included. Many older writings were incorporated in later documents, thus providing numerous difficulties to accurate dating. In most cases, it is difficult to discern which portions of a text represent those
attitudes or events contemporary with the saint and which are later additions. It is clear, however, that the interest in nature was present from the beginning of the Christian period and continued undiminished through the height of monastic development and the Middle Ages until the dissolution of the monasteries. (The reader should recognize the English translations used in this paper may have modified the tone or changed the emphasis of the originals. The translations of Lady Gregory and Helen Waddell are less literal than those of Charles Plummer, for example, and may reflect more modern "nature romanticism.")

Compassion for Wolves

The Celts had a love for animals. Many stories about Celtic saints concern domestic animals, such as cattle, oxen, and horses, and there are also numerous tales of wolves, deer, wild boar, otters, whales, and other wild creatures. The Celtics used a greater diversity of natural imagery than the desert monks, who concentrated on mammals and a few reptiles. The Irish monks frequently mentioned birds, fish, plants, and other types of organisms. The difference between the literatures is partially a reflection of the differences in the natural diversity between the British Isles, with their forests, heaths, and rocky coasts, and the deserts of Palestine and Egypt. The Celts also had a greater interest in landscapes (or territorial claims), and their hagiography includes more travelogues that describe the places visited by saints.

The Celtic monks showed little animosity toward predatory beasts, and, as several authors have noted, wolves were very popular with the saints.[9] A repeated motif was the caring saint who fed starving wolves. Saint Maedoc of Ferns, for example, fed wolves more than once. When, as a boy, Maedoc was minding the sheep of his foster mother, eight wolves came fawning, "poor, weak and starving." Maedoc, "seized with pity," took eight wethers from the flock and fed the wolves. When his foster mother heard of his charity, she was livid. Before she could accuse him of losing the sheep, however, eight wethers appeared miraculously to replace those given to the wolves.[10] Years later, at a church Maedoc had built:

> Maedoc was one day alone there indoors in his cell. He saw some wolves coming to him, and they went round him gently and fawning. Maedoc understood they were asking for food. He was moved to compassion for them; he gave the calf (his brother's) to them, and bade them eat. When the woman came in the afternoon (at milking time), she looked for the calf to let it in to them. Maedoc said to her: "Do not look for it, for I have given it to the wolves." One of the brothers said: "How can the cows be milked without their calf?" Maedoc said to the brother: "Bend thy head towards me," said he, "that I might bless it; for when the cows see it, they will give their milk humbly and obediently." And so it was whenever the cows saw the head of the brother, they would suddenly lick it and so give their milk.[11]

This tale not only shows compassion for wild animals as a virtue, but, with Irish humor, teaches humility as well.

The care of the saints was also extended to animals fleeing from hunters. Saint Coemgen (St. Kevin) once had a wild boar rush to him for protection. The dogs chasing the animal found their feet bound to the ground and were unable to pursue the boar while it was under the spiritual power of the saint.[12] There are several stories of saints rescuing hunted deer. St. Godric (of Britain, from a literature related to Irish hagiography) found a stag being pursued by hounds. The animal, "shivering and exhausted," seemed "by its plaintive cries to beseech his (the saint's) help." Godric let the stag into his hermitage where it dropped at his feet. When the hunters arrived, they questioned Godric about the stag, but "he would not be a betrayer of his guest and he made the prudent answer, 'God knows where it may be.'" The hunters, recognizing the saint as holy, asked his pardon and departed. The stag remained with Godric until the evening, when the saint released it. The stag "for years thereafter, . . . would turn from its way to visit (Godric) and lie at his feet, to show what gratitude it could for its deliverance."[13] Maedoc, who cared for the wolves, was "praying in the recesses of a wood, when he saw a stag pursued by hounds, and the stag stopped by him. Moeog (Maedoc) threw the corner of his plaid over the horns, to protect it from the hounds; and when these came up, they could not find trace or sight of it; and it afterwards betook itself to the wood in safety."[14] In another version of this story, Maedoc puts his rosary on the stag's horns and the hounds perceive the stag as a man and cease to pursue it.'[15] In the first version it is the saint's compassion that provides the protection, but, in the second, it is the power of the rosary extended to wild nature.

In some stories the protection is not from hunters, but from the monk's own party. St. Patrick, for example, was climbing a hill in Ard-Macha to look at a grant of land the king had given him. On the hill "they found a roe deer and a little fawn lying on the very spot where the altar of the great northern church of Ard-Macha now stands." Patrick prevented his companions from killing the fawn. "Tenderly he took up the fawn and carried it on his shoulders, the doe following close at his heels, until he laid it down on another height at the north side of Armagh." Patrick, in the image of the good shepherd of the Hibernian mountains, carried the fawn the way he carried the Irish church.[16]

There is an interesting passage in "The Life of Maedoc of Ferns" in which the saint commands the family of Ragallach not to trespass on his land (the land of the church), "and of any living creature not to kill so much as a hare or an angled trout within the territory of his church or sanctuary." The punishment for disobedience was "a short life and hell, and disease and famine."[16] Although this passage may be merely an assertion of property rights to the fullest, it also implies that animals should be left alone on holy ground.

Celtic saints used spiritual power directly for their own protection and had no qualms about cursing an enemy—or his land. An interesting case of an animal helping to execute a saint's judgment concerns St. Berach and Diarmait, a poet and "chief master of druidism." Diarmait was occupying land that Patrick had bequeathed to Berach, and Diarmait did not want to leave. After a long series of magic contests and tests of power, Berach was awarded the land and said that Diarmait would die on a specific day at the end of the year. On that day, Diarmait began to revile Berach, since he was still alive. A bishop rebuked Diarmait and suggested that he shut himself up in a church until the day was over. A stag, pursued by hunters, stopped by the east window of the church. On hearing shouting, Diarmait went to the window and looked out. One of the hunters threw a spear, which hit Diarmait in the throat. Dairmait died on the floor of the church, while the stag escaped unharmed.[18] In an even more direct example of the use of magic, St. Brendan exercised vengeance on a man called Dobarchu who had killed the saint's oxen. Brendan prayed that Dobarchu would be turned into an otter, and indeed it happened when Dobarchu, after seeing a salmon leap and catching a trout, fell into a lake.[19] Both these tales suggest "shape-chang-ing" and pre-Christian mythology.

Animals and other elements of wild nature frequently offer assistance to Celtic monks. As with the desert fathers, a major theme is the working of the hand of God through providence. Animals sometimes offered themselves up to be eaten, and the sea frequently provided sustenance. An anonymous life of St. Cuthbert (of Lindisfarne in Britain) reports the holy man, when visiting a monastery, kept his nightly prayer vigil. After soaking in the sea up to his arm pits, he withdrew to the shore to pray, and "immediately there followed in his footsteps two little sea animals (otters), humbly prostrating themselves on the earth; and licking his feet, they rolled upon them, wiping them with their skins and warming them with their breath. After this service and ministry had been fulfilled and his blessing had been received, they departed to their haunts in the waves of the sea."[20] Here the response of the animals identified the man as holy.

On another occasion Cuthbert needed a beam for his chamber, and the sea de-livered a timber exactly twelve feet in length.[21] While on an evangelizing trip, the saint not only had an eagle provide food for him, but also incorporated the eagle as a part of his teaching:

> On a certain day, he (Cuthbert) was going along the river Teviot and making his way southward, teaching the country people among the mountains and baptiz-ing them. Having a boy walking with him in his company he said to him: "Do you think that someone has prepared you your midday meal to-day?" He an-swered that he knew of none of their kindred along the way and he did not hope for any sort of kindness from unknowing strangers. The servant of God said again to him: "My son, be of good cheer; the Lord will provide food for those who hope in him, for he said, 'Seek ye first the Kingdom of God and his righ-teousness and all these things shall be added unto you' in order that the saying of the prophet may be fulfilled: 'I have been young and now am old, yet have I not the righteous forsaken', and so forth. 'For the laborer is worthy of his hire.' " After some such words he looked up to heaven and saw an eagle flying in the sky and said to the boy: "This is the eagle which the Lord has instructed to provide us with food to-day." After a short time, as they went on their way, they saw the eagle settling on the bank of the river. The boy ran towards the eagle in accor-dance with the command of the servant of God, and stopping, he found a large fish. The boy brought the whole of it to him, whereupon Cuthbert said: "Why did you not give our fisherman (the eagle) a part of it to eat since he was fast-ing?" Then the boy, in accordance with the commands of the man of God, gave half the fish to the eagle while they took the other half with them, and broiling it in the company of some men, they ate it, and gave some to the others and were satisfied, worshiping the Lord and giving thanks. Then they set out according to God's will to the mountains, as we have said, teaching and baptizing the people in the name of the Father and of the Son and of the Holy Ghost.[22]

Here wild nature provided for the saint and several others, but Cuthbert did not neglect the needs of the eagle. The story is reminiscent of Christ's loaves and fishes and of St. Peter's fishing adventures.

Animals volunteer for other services, including finding missing items and carrying things. Coemgen accidentally dropped his psalter into a lake, and an otter returned it to him, dry and readable.[23] Ciaran had a stag that came to him daily, lay down before the monk, and allowed him to use its antlers as a reading stand.[24] When Columba once forgot his books, a stag returned them to him on his back.[25]

Although the Celtic hagiographies display some of the reciprocity with animals found in the writings of the desert monks, the passages where the animals do something for the monks and those where the monks do something for the animals are more often separated from each other in the texts. In addition, the Celtic accounts contain numerous stories of saints ordering animals to provide services or otherwise exercising control over them. These passages were intended to prove the spiritual power and influence of the saint. Abban, for example, ordered wolves to tend his sheep,[26] and Coemgen, needing milk for a foster child, commanded a doe to provide it. When a wolf came and killed the doe's fawn, Coemgen then ordered the wolf to take the place of the fawn in order that the doe would continue to provide milk.[27] Mochuda provided two deer to pull a poor man's plow.[28] Coemgen put a stag to pulling a chariot,[29] as did St. Declan when his horse became lame.[30]

Commanding the Wind

The saints' power extended beyond animals to plants, the tides, and even the winds. Berach, in his contests with Diarmait, prayed and caused a thorn tree to rise into the air.[31] Celtic saints were frequently reported to have crossed rivers on dry ground or withstood storm tides. Maedoc made some horses cross the sea without wetting their feet,[32] and Brendan prayed and calmed a sea full of whirlpools.[33] Broichan, a druid, declared he would interfere with a voyage of St. Columba and stirred up the elements against him. Columba put to sea anyway and caused the adverse winds to change direction, pushing his boat to the desired port.[34]

The saints also proved their power by making their way through forests and bogs, or over mountains. Maedoc, forced by an enemy to fetch firewood, drove a team of oxen and a cart over bog where "God made a smooth and easy road, and a firm level path through the soft and yielding surface of the bog."[35] When some admirers of St. Coemgen attempted to carry him on a litter through a thick wood, the trees bowed to allow him to pass and then stood up again.[36]

In a few cases the saints' exercise of power over nature is destructive. St. Columba, for example, killed a wild boar by praying that it would die.[37] More typical, however, are St. Brigit and other holy personages who tamed boars, lions, or wolves.[38] Cuthbert was gentle in driving the birds away from his crops, when he asked: "Why . . . do you touch the crops that you did not sow? Or is it perchance, that you have greater need of them than I?" He then challenged them by saying: "If, however, you have received permission from God, do what He has allowed you; but if not, depart and do not injure any more the possessions of another." The birds left and did no further damage.[39] This story is very reminiscent of Antony's protection of his garden in the desert.

The Celtic saints had a passion for destroying or driving out two types of organisms—monsters and venomous reptiles. Perhaps a majority of the saints had a victorious encounter with either a monster or snake of some kind, and some saints, like Abban and Brendan, were credited with disarming, exiling, or killing several. A few of the monsters lived on land, like Abban's lionlike monster,[40] but most were water monsters living in lakes or in the ocean.

The relationship of this theme to the Celts' attitudes towards wilderness and wild nature is difficult to determine for several reasons. First, most of these creatures are clearly mythological. Unlike wolves, there were neither lake monsters nor snakes in Ireland in late pagan or early Christian times. The monsters seem to be evil personified through nature, and again the purpose of the stories is to prove the saints' power. The author who described Abban repelling a monster on Loch Garman stated, "It was the devil who caused the monster to come to them in that form to destroy the saints."[41]

The association of monsters with the ocean would have stemmed from sailors' encounters with real sea creatures. Celtic seafarers, however, would have had little reason to attack sharks, whales, or other ocean giants in self-defense. If it is heroic to fight maritime giants, then why not terrestrial wolves? Perhaps the sheer size of many marine creatures put them beyond the reach of the fisherman or warrior. Wolves, on the other hand, could be killed easily with a spear. The ocean monsters may have been symbolic of the unknown and unconquerable. The hero needed a worthy adversary, and his points of iron had long ago defeated the native wildlife.

The association of monsters with water is even more difficult to explain in terms of inland lakes. Ironically, the lakes were often sacred sites in pre-Christian times, and the saints frequently chose their shores for monasteries. These stories may reflect an intrusion of pre-Christian Celtic myth. Battling with monsters may also have been symbolic of the defeat of the ancient religion or a concession to its remaining power. The saints forced the monsters from the lakes, making their shores "spiritually" safe, and reducing the fear of dark things hidden in the murky depths.

Wild Boar as Monks

Although the animals in the Celtic stories often lack personality, the same may be said of some of the saints. There are some beautiful exceptions to this, however, and, if the histories of the desert fathers implied that animals can show Christian virtue or become "monks," the Celtic hagiographers stated this directly. One of the best examples is from the life of the hermit Ciaran:

> And Ciaran settled himself (at the Well of Uaran), and he alone, and a great woods all around the place; and he began to make a little cell for himself, that was weak enough. And one time as he was sitting under the shadow of a tree a wild boar rose up on the other side of it; but when it saw Ciaran it ran from him, and then it turned back again as a quiet servant to him, being made gentle by God. And that boar was the first scholar and the first monk Ciaran had; and it used to be going into the wood and to be plucking rods and thatch between its teeth, as if to help towards the building.[42]

Ciaran did not stop with one animal; he added several to his flock. He found the natural characteristics of some of these creatures troublesome, but he proved the monastic life can overcome even inherited evil, an important matter to the Celts:

> And there came wild creatures to Ciaran out of the places where they were, a fox and a badger and a wolf and a doe; and they were tame with him and humbled themselves to his teaching the same as brothers, and they did all he bade them to do. But one day the fox, that was greedy and cunning and full of malice, met with Ciaran's brogues and he stole them and went shunning the rest of the company to his own den, for he had a mind to eat the brogues. But that was showed to Ciaran, and he sent another monk of the monks of his family, that was the badger, to bring the fox back to the place where they all were. So the badger went to the cave where the fox was and he found him, and he after eating the thongs and the ears of the brogues. And the badger would not let him off coming back to Ciaran, and they came to him in the evening bringing the brogues with them. And Ciaran said to the fox, "O brother" he said "why did you do this robbery that was not right for a monk to do? And there was no need for you to do it" he said "for we all have food and water in common, and there is no harm in. But if your nature told you it was better for you to use flesh, God would have made it for you from the bark of those trees that are about you." Then the fox asked Ciaran to forgive him and to put penance on him; and Ciaran did that, and the fox used no food till such time as he got leave from Ciaran; and from that out he was as honest as the rest.[43]

In most of the cases in which an animal committed a misdeed, the saint commanded the animal to cease or to make reparation. St. Cainnic, for example, was staying on an island, but the birds were constantly noisy and bothered the saint. He rebuked them and they "got together and set their breasts against the ground, and held their peace, and until the hour of matins on Monday morning they stayed without movement and without a sound, until the Saint released them by his word."[44] In a number of stories a wolf killed a calf or fawn, and a saint ordered the wolf either to bring another, or it take the place of the lost juvenile in comforting or suckling the mother.[45]

Animals did sometimes voluntarily seek forgiveness or do penance, thus indicating Christian virtue. Cuthbert found ravens tearing thatch off his guest house. "He checked them with a slight motion of his right hand, and bade them cease from injuring the brethren. When they ignored his command, he said; 'In the name of Jesus Christ, go away forthwith, and do not presume to remain any longer in the place that you are damaging.'" The ravens left at once, but in three days one returned. "With its feathers sadly ruffled and its head drooping to its feet, and with humble cries it prayed for pardon, using such signs as it could; and the venerable father, understanding what it meant gave it permission to return." The raven came back with its mate and brought "a worthy gift, namely a portion of hog's lard" to the saint.[46]

Flight of the Crane

The Celts had a special love for birds. From the diminutive wren to the tall, graceful crane, the saints showed compassion for these delicate creatures. The birds, furthermore, were associated with angels and the souls of the departed. Several saints were thought to have birds light on them or fly around them. When asked why birds were not afraid of him, Columba of Terryglass answered "that his thoughts ceased not to fly like birds to the sky."[47]

St. Coemgen (Kevin) was praying one Lent, and, as he knelt in a little solitary hut, with hands outstretched towards heaven, "a blackbird settled on it, and busying herself as in her nest, laid in it an egg. And so moved was the saint that in all patience and gentleness he remained, neither closing nor withdrawing his hand: but until the young ones were fully hatched he held it out unwearied, shaping it for the purpose."[48] Coemgen was also said to pray with flocks of birds circling around his head.[49]

St. Columba showed the same care for a crane. Since Columba was in exile on Iona, off the coast of Scotland, the crane symbolized his wish to return home to Ireland. Columba instructed a brother to go and sit above the shore and look "for a stranger guest, a crane, wind tossed and driven far from her course in the high air: tired out and weary . . ." she would fall on the beach. Columba told the brother to "tenderly lift her and carry her to the standing near by; make her welcome there and cherish her with all care for three days and nights . . . ," and then to release her. The brother found the crane and did as Columba had instructed him; ". . . for when her three days housing was ended, and her host stood by, she rose in the first flight from earth into high heaven, and after a while at gaze to spy out her aerial way, took her straight flight above the quiet sea, and so to Ireland through the tranquil weather."[50]

An ultimate sign of saintliness was never to have molested a bird. St. Molua was said to have "never killed a bird or any other living thing." When a cleric saw a bird lamenting at Molua's death, he wondered about it. An angel spoke to him and said: "Molua MacOcha has died, and therefore all living creatures bewail him, for never has he killed any animal, little or big; so not more do men bewail him than the other animals, and the little bird thou beholdest." When St. Cellach was dying, all the birds of the forest waited in hushed appreciation.[51]

Birds appear in Celtic hagiography as messengers and as spiritual beings. In *Fis Adamnain,* an eighth century account of a vision of heaven and hell: "Three stately birds are perched upon that chair in front of the King (God), their minds intent upon the Creator throughout all ages, for that is their vocation. The birds celebrate the eight (canonical) hours, praising and adoring the Lord, and the Archangels accompany them. For the birds and the Archangels lead the music, and then the Heavenly Host, with the Saints and Virgins, make response." Over the head of God "six thousand thousands, in the guise of horses and birds, surround the fiery chair."[52]

From very early in the history of the Christian church, figures such as St. Ambrose have interpreted the songs and activities of birds as offering praise to God. In the voyage of Bran, the hero discovered a blessed island where birds sing the praises of God as canonical hours.[53] Brendan similarly discovered an island, Paradise of Birds, where not only do the birds sing joyously all day long, but the human occupants only speak to praise God.[54]

Under Northern Oaks

Unlike the desert fathers, the Celtic monks did not have a large consolidated "wilderness" available for withdrawal. They certainly sought isolation, however, and showed strong preferences for certain types of landscapes or natural features including oak forests, lakes, and islands. Although they sometimes ascended mountains for prayer, this motif is surprisingly infrequent considering the availability of open mountain tops in the British Isles. They do not seem to have associated God with the heights, as did the prophets of the Middle East.

In investigating the relationship of Celtic Christianity to wild landscapes, we must first look at the sanctuaries of pre-Christian Celtic religion, which "were found frequently in sacred woods and near lakes, including it would seem, what are now bogs and swamps." Nora Chadwick suggests the Celts did not care for life in towns and they had "a ritual preoccupation with the natural environment."[55] Classical authors report sacred oak groves, untouched by the woodsman's ax, where "every tree was sprinkled with human gore."[56] Archaeological evidence supports the use of such sites, and excavators have recovered many Celtic votive offerings tossed into the water during ancient times.[57]

Little is known of the Celts' pre-Christian beliefs; the written descriptions of these practices were authored largely by Romans who observed the Celtic sanctuaries but were not privy to the rites, or by Christian monks who were not contemporary with the druids. The sacred groves and lakes may have had spirits associated with them. Lucan, a Roman, reported about a wood near Marseilles where "subterranean hollows quaked and bellowed, that yew trees fell down and rose again, that the glare of conflagration came from trees that were not on fire, and that serpents twined and glided around the stems." Lucan suggested that even the priest would not enter the grove at night and that the people feared "the lord of the grove."[58]

The Celtic monastic literature suggests that the monks found the forest peaceful, except for an occasional confrontation with a wild boar or disruption by hunts. Nash, however, accuses Christians of purposefully removing the natural sanctuaries and cutting groves.[59] Sacred groves, certainly, were cleared. The practice began, however, with the Romans, well before the time of Christ. Caesar's army cut the grove Lucan described near Marseilles and probably destroyed others like it in subduing their "uncivilized" adversaries. Christians did not have the political power to do anything similar on a large scale until the time of Constantine (after 300 C. E.).

Indigenous Celtic monastic clerics were more interested in occupying oak forests than in destroying them. The saints did have trees cut to build churches and to provide fire wood,[60] but their writings indicate they often left the forest standing at the sites of their sanctuaries and hermitages. Aedh, king of Ireland, gave a piece of land in Doire to Columba (Columcille):

> And he [Columba] had so great a love for Doire, and the cutting of the oak trees went so greatly against him, that he could not find a place for his church the time

he was building it that would let the front of it be to the east, and it is its side turned
to the east. And he left it upon those that came after him not to cut a tree that fell of itself
or was blown down by the wind in that place to the end of nine days, and then to share
it between the people of the townland, bad and good, a third of it to the great house
and a tenth to be given to the poor. And he put a verse in a hymn after he was gone
away to Scotland that shows there was nothing worse to him than the cutting of
that oakwood:
"Though there is fear in me of death and of hell, I will not hide it that I have more
fear of the sound of an axe over in Doire."[61]

A hymn attributed to Columcille (Columba) praises Doire for its quiet and its
oaks:

It is the reason I love Doire, for its quietness for its purity; it is quite full of white angels
from the one end to the other.

It is the reason I love Doire, for its quietness for its purity; quite full of white angels is
every leaf of the oaks of Doire.

My Doire my little oakwood, my dwelling and my white cell; O living God in Heav-
en, it is a pity for him that harms it![62]

Columba was so opposed to unnecessary felling of trees that he did not place his
sanctuary in the traditional position, facing east.

Columba's wish for protection of his woods was taken one step further by
Coemgen (Kevin). This saint, with an authority modern preservationists should
envy, "promised hell and short life to any one who should burn either green wood
or dry from this wood (where the trees bowed down for him) till doom."[63] The
manuscripts frequently mention saints praying or conducting devotions in the forest. Mochuda walked through the woods singing psalms,[64] and Maedoc prayed "in
the recesses of a wood."[65]

The choice of sites for monasteries was similar to the choice of devotional
sites. Patrick, in selecting a site for Berach, "ordained that it should be in the
meadow on the brink of the lake. . . . And he ordered that the sanctuary ground
should be all that lies between the bog and the lake, that is the plain with its
wooded meadows and boggy oak-groves."[66] "The Life of Berach" called this location "the sacred glen."[67] St. Patrick predicts, concerning Maedoc's coming to
Druim
Lethan (Drumlane), that "though marvelous in your eyes be the number of trees
on the ridge on which ye are, Druim Lethan, not more numerous are they than
the prayers and hymns, the psalms and genuflexions, the alms and Masses which
will be performed on it in time of the noble angelic saint (Maedoc)."[68]

Most interesting of all is Cocmgcn's (Kevin's) choice of Glen da Loch (Glen-dalough, or the glen of the two lakes). Patrick prophesied thirty years before Coemgen's birth that Coemgen would build a monastery on the site. The glen of the cliffs is supposed to have pleased Patrick.[69]

Coemgen had to overcome a monster in the lower lake before he could occupy his divinely chosen home. He is also supposed to have spent a long period in solitude there and prayed standing in the lake or lying on the stones at its edge. A "Life of Coemgen" relates:

> Seven years in tangled deserts
> Wert thou (Coemgen) in gentle sort,
> Dwelling beside thy people,
> Without food, except (the fruits of) Cael Faithe.
>
> Coemgen (was) for length of years
> Among deserts in the woods,
> And he saw no man,
> Nor did any man see him there.
>
> Far from his friends was Coemgen
> Steadfastly among the crags;
> Nobly and alone he saw the order
> Which was brought to the brink of the fair lough.
>
> At night he would rise without fear
> To perform his devotion in his fort;
> There he would early recite his hours
> (Standing) habitually in the lough up to his girdle.[70]

Coemgen's love and appreciation for the landscape of Glen da Loch expressed themselves most fully in his negotiations with an angel. The angel came to Coemgen to take him to the east end of the lesser lake in order to show him his place of resurrection (the place where the saint would be buried, to rise when Christ returned—the Irish took resurrection very literally). Coemgen, in Irish style, objected to the angel's suggestion on the grounds that the valley was too hemmed in by mountains and the monks would not feel comfortable there. The angel promised that if Coemgen would bring fifty monks to the site, as each one of these died he would be replaced by a new monk until the Day of Judgment. Coemgen objected. Fifty was too few. The Angel then offered to make the number of monks many thousands and to raise a great city there. The Angel closed by proposing: "And verily if thou shouldst will that these four mountains which close this valley in should be leveled into rich and gentle meadow lands, beyond question thy God will do it for thee." Coemgen, despite the offer of a great

following, declined the city in favor of leaving the mountains as they were created by God. He replied: "I have no wish that the creatures of God should be
moved because of me: my God can help that place in some other fashion. And
moreover, all the wild creatures on these mountains are my house mates, gentle
and familiar with me, and they would be sad of this that thou (the angel) hast
said."[71] To turn down such a religious honor in favor of preserving the mountains,
makes Coemgen truly a saint of the wilderness.

Mass on the Back of a Whale

Coemgen was not alone in his prayers in water. Cuthbert, as already mentioned,
prayed standing in the sea, and other saints sought out shores or rocky islands.
Ciaran would go with his foster mother to pray "on a flood-surrounded rock,
which was in the sea amid the waves to the south of Ross Banagher" (it was not
clear how they crossed the water to the site).[72] Declan built "a great monastery
by the south side of the stream which flows through the island (Ait-mBreasail)
into the seas."[73] Columba established his mission on Iona, and Cuthbert chose the
solitary life on an island, "shut in on the landward side by very deep water and on
the seaward side by the boundless ocean."[74] The choice of islands as monastic sites
may be explained best by their isolation. There were pre-Christian sanctuaries on
some islands, however, and pre-Christian legends associated islands with *Tir na
nOc*—the land of the blessed or the land of the forever young, which also plays a
role in the heroic voyage sagas of the Christian saints.

Unlike the desert fathers, who stayed at one location for years on end and
moved only to establish new monasteries or to withdraw deeper into the desert,
the Celtic saints were travelers. The Celts idealized the exploratory voyages to
the outer islands in the tales of Brendan and Bran. The narratives of the voyage of
Brendan, for example, depict paradise as an island. Brendan is searching for *Tir
Tairngiri,* the promised land,[75] which owed as much to Celtic myth and landscape
preference as it did to imagery from the Scriptures. The old Celtic yearning for
the other world combined with Christian concern for praise and worship produced some descriptions of ideal monastic retreats.

Brendan finds first the Paradise of the Birds (which is not *Tir Tairngiri*):

> And when they reached the island they landed there. And the island was extraordinary
> in appearance, for there were many excellent fruits there, and marvelous birds dis-
> coursing joyously from the tops of their trees, and little bees gathering and collecting
> their harvest and household store for their dwellings, and strangely beautiful streams
> flowing there, full of wondrous jewels of every hue. And there were many churches
> there, and a monastery in the middle of the island full of an excellent variety of things
> of every hue; and a venerable wise decorous and devout order in it.[76]

The narrative describes a beautiful natural setting, and then provides a parallel discussion of the religious institutions of the site. Even the singing of the birds and the activities of the bees provide a subtle comparison to the monks who do not speak except to praise God and who busy themselves with the affairs of the church. Although idealized, the Paradise of Birds was probably not dissimilar to some actual offshore monasteries, such as Skellig Michael, whose birds (likely raucous seabirds) flew about constantly.[77]

Symbolic of Brendan's own relationship to God and God's provision for the voyage was the annual Easter celebration. As Easter drew near the first year, Brendan's company began to encourage him to land so they could celebrate the holy day. Brendan replied: " 'God is able . . . to find a land for us in any place He pleases.' When Easter came, a great whale raised its shoulders high above the surface of the waves, so that it formed dry land. And then they landed and celebrated Easter on it. And they were one day and two nights. When they had entered their boats, the whale dived into the sea at once."[78] This continued for seven years. Appropriate as the sanctuary of an heroic wanderer, the whale's back represents the extreme in Celtic use of natural areas for devotions.

Brendan's eventual arrival at Paradise is anticlimactic. The hagiographer did not describe the land in detail, but rather told the story of an elder who has not aged since he arrived at the island. Having waited sixty years for Brendan's arrival, the elder promptly sent Brendan and his party home to "instruct the men of Erin," and then departed for heaven himself. At first glance, Brendan seems to have found the place all men would wish to reside:

> An island rich, everlasting, undivided,
> Abounding in salmon, fair and beauteous.[79]

Brendan, however, dutifully went home and established a monastery at his place of resurrection, Clonfert. The disjunction here may result from a mixture of the pre-Christian ideal of the Land of the Forever Young as an actual place and the Christian concepts of heaven and resurrection. "The life of Brendan of Clonfert" does not indicate Brendan will ever return to *Tir Tairngiri.*

Poets of the Solitary Wood

The Irish religious poets concerned themselves with nature even more than the hagiographers had. Their products, more descriptive and freer from proofs of spiritual prowess, served as hymns of praise or testaments to God's goodness. Throughout, wild nature operates as a beauty-filled backdrop, if not the actual focus of the works.

The use of natural imagery in Celtic poetry is almost certainly pre-Christian. Although no manuscripts remain from before the time of Patrick, some examples of earlier poetry may appear in later works. A piece of alliterative chain verse, actually an invocation, ascribed to Aimigrin, who lived hundreds of years before Christ, can be found in the *Leabhar Gabhala* (The Book of Invasions). The poem is of unknown date, and it is not certain to what extent it represents pre-Christian religious writing:

> THE MYSTERY
> I am the wind which breathes upon the sea,
> I am the wave of the ocean,
> I am the murmur of the billows,
> I am the ox of the seven combats,
> I am the vulture upon the rocks,
> I am a beam of the sun,
> I am the fairest of plants,
> I am a wild boar in valor,
> I am a salmon in the water,
> I am a lake in the plain,
> I am a word of science,
> I am the point of a lance of battle,
> I am God who created in the head the fire.
> Who is it who throws light into the meeting on the
>   mountains?
> Who announces the ages of the moon?
> Who teaches the place where couches the sun?
>   [If not I][80]

The natural imagery of "The Mystery" is found in (presumably) later Celtic religious poetry, but the pantheism is replaced by a trinitarian view.

The "Deer's Cry," an invocation of St. Patrick attributed to the seventh century, utilizes creation imagery in one of eight verses intended to summon divine assistance. Patrick is supposed to have written the poem after he and his monks escaped an ambush by his enemies by appearing to them as wild deer with a fawn.

> DEER'S CRY
> I arise to-day
>   Through a mighty strength the invocation of the Trinity,
>   Through belief in the threeness,
>   Through confession of the oneness
>   Of the Creator of Creation.

I arise today
> Through the strength of Christ's birth with his baptism,
> Through the strength of His crucifixion with his burial,
> Through the strength of His resurrection with
>   His ascension,
> Through the strength of His descent for the
>   Judgment of Doom.

I arise today
> Through the strength of the love of Cherubim,
> . . . .
> [a list of angelic helps follows]

I arise today
> Through the strength of heaven:
> Light of sun,
> Radiance of the moon,
> Splendor of fire,
> Speed of lightning,
> Swiftness of wind,
> Depth of sea
> Stability of earth,
> Firmness of rock.

I arise to-day
> Through God's strength to pilot me:
> . . . .
> [a list of God's characteristics follows][81]

Here orthodox Christian theology frames the power of nature and brings it into a Christian context. The use of natural imagery in Celtic poetry extends to the person of God, as in this poem about the crucifixion:

> At the cry of the first bird
> They began to crucify thee, O cheek like a swan !
> It were not right ever to cease lamenting-
> It was like the parting of day from night.[82]

The poetry, much of it monastic, provides insights into the lives and values of the Celtic clerics. The monks extolled their forest sanctuaries. A poem attributed to St. Gall, who founded a monastery on Lake Constance in Switzerland and died in 635 C.E.,[83] described a scribe who is working outdoors. Note that the European blackbird is a member of the thrush family and has a sweet, flutelike voice:

WRITING IN THE WOOD
A hedge of a wood-thicket looks down on me;
a blackbird's song sings to me
    [a message not concealed]
above my little book, the lined one,
the twittering of birds sings to me.

The clear-voiced cuckoo calls to me, a lovely speech
in a gray mantle from bushy dwellngs
God's Judgment! The Lord befriends me!
I write fair under the great wood of the forest.[84]

One of the best known of the hermitage poems is: "The Hermit's Song." The verses describe the pleasant environment of a monk's forest retreat; he refers to his wood as "beautiful." He has placed his hut to face south, near a stream. He ends by thanking God for His wonderful provision.

THE HERMIT'S SONG
I wish, O Son of the living God,
O ancient, eternal King,
For a hidden hut in the wilderness
That it may be my dwelling.

An all grey lithe little lark
To be by its side,
A clear pool to wash away sins
Through the grace of the Holy Spirit.

Quite near, a beautiful wood,
Around it on every side,
To nurse many-voiced birds,
Hiding it with its shelter.

And facing the south for warmth;
A little brook across its floor,
A choice land with many gracious gifts
Such be good for every plant.
. . . .
[five verses on a church and other monks]

This is the husbandry I would take,
I would chose, and will not hide it:
Fragrant leek,
Hens, salmon, trout, bees.

Rainment and food enough for me
From the King of fair fame,
And I to be sitting for a while
Praying God in every place.[85]

Another well known hermitage poem is "Guaire and Marban," also known
as "King (Guaire) and Hermit." The work describes a discussion between two
brothers, one of whom has chosen the religious life, and the other of whom,
although also a saint, has chosen to remain in the royal court. The hermit describes his simply woodland life, and then his brother expresses the wilingness to
trade all his kingdom for it.

GUAIRE AND MARBAN

GUAIRE:

O Marban, O hermit,
why do you not sleep on a bed?
More frequent for you was the passing of the
    night outside,
The end of your tonsure on the ground
    among the pines

. . . .

MARBAN:

The size of my bothy [hut]—small, not small—
a place of familiar paths;
a woman in a cloak of blackbird color
sings a sweet song from its peak.

Stags of Rolach Ridge spring
out of its clear stream of the plain;
from it red Roigne is visible,
mighty Mucruime and Moenmag.

Little lonely dwelling,
it has an estate of forest paths.
To see it, will you go with me?
I found it calm indeed.

Mane with twists
of yew of gray trunk
[famous omen],
beautiful the place,

the great green oak,
besides that augury.

. . . .

Choice wells,
falls of water
excellent for drinking—
they gush forth in abundance;
berries of yew,
bird-cherry, privet.

Lairs around it
of tame swine,
goats, pigs,
wild swine,
tall deer, does,
badger's brood.

. . . .

Produce of rowan,
black sloes
of dusky blackthorn,
food of acorns,
bare fruits of bare slopes.

A clutch of eggs,
honey, acorns, heath-pease
[it is God who sent it]
sweet apples,
red bog berries,
bilberries.

. . . .

Bees, beetles,
humming of the world,
gentle crooning:
barnacle geese, brants,
shortly before the beginning of winter,
music of a dark torrent.

A lively songster,
fussy brown wren,
on hazel bough; in a piebald hood
the woodpecker on an oak
in a great crowd.

Fair white ones come,
cranes, gulls;

the coast sings to them;
no mournful music
of the grey-brown hen
in the russet heather.

. . . .

Voice of the wind
toward the branching wood,
a very gray cloud;
falls in the river,
trumpeting swans:
beautiful music.

A beautiful pine
makes music for me
not having been bought;
to Christ each fault
is no worse for me
than for you.

Though you are pleased
with what you enjoy
more than any treasure,
I am grateful
for what is given me
by my fair Christ.

Without a time of combat,
without the din of battle
that assails me,
I am grateful to the Prince
Who gives each good
to me in my hut.

GUAIRE:

I will give my splendid kingdom
with my share of Colman's patrimony
in undisturbed pos
session to the hour of my death
for being in your company, O Marban.[86]

    This monk was very aware of his environment. He observed the landscape
and the seasons. He valued nature not only for sustenance, but also for its beauty.
The glories of the forest were not just visual, but also auditory. The poetry is

characterized by its great diversity of imagery; everything in the environment is of interest. The political power of Guaire and his honor in battle did not equal the calm surrounding Marban. Guaire envied his brother's "estate of forest paths" and the peace of the solitary life.

The joys of the hermitage were not limited to the forest, but also found their way to seacoasts and islands. A poem, attributed to St. Columba, but probably of a later date, expresses the feelings of a monk residing on an island:

> Delightful to me on an island hill, on the crest of a rock, that I might often watch the quiet sea;
>
> That I might watch the heavy waves above the bright water, as they chant music to their Father everlastingly;
>
> That I might watch its smooth, bright-bordered shore, no gloomy pastime, that I might hear the cry of strange birds, a pleasing sound;
>
> That I might hear the murmur of the long waves against the rocks, that I might hear the sound of the sea, like mourning beside a grave;
>
> That I might watch the splendid flocks of birds over the well watered sea, that I might see its mighty whales, the greatest wonder.
>
> That I might watch its ebb and flood in their course, that my name should be—it is a secret that I tell—"he who turned his back upon Ireland" [referring to Columba's exile to Iona];
>
> That I have a contrite heart as I watch, that I might repent my many sins, hard to tell;
>
> That I might bless the Lord who rules all things, heaven with its splendid host, earth, ebb and flood;
>
> That I might scan one of the books to raise up my soul, now kneeling to dear heaven, now chanting psalms;
>
> Now gathering seaweed from the rocks, now catching fish, now feeding the poor, now in my cell;
>
> Now contemplating heaven, a holy purchase, now a little labor, it would be delightful.[87]

The poem has a balance of concerns. The monk appreciated the waves praising God, and then admires God's artistry. The monk used the setting for gaining humility and seeking repentance. After declaring how well creation worships, the monk wished to be able to bless the Lord himself. He desired to study for edification, to pray, and to sing psalms. His time was divided among maintenance tasks (gathering seaweed), charity (feeding the poor), and contemplation (residing in his cell). For him the life was "delightful." The driving pace and social fragmentation of the modern world have all but extinguished this mode of Christian existence.

From This Wood to the Next World

The Irish monks are the first group we have investigated who were physically removed from the religious traditions of the Middle East. They had no direct contact with Judaism and relatively limited intercourse with Greco-Roman culture. The first Celtic monks in Gaul met monks who had been in the Levant and had access to copies of the Scriptures and the works of the church fathers. Their reliance on the written word to convey previous Hebrew and Christian spiritual experience in wilderness was greater than that of the desert fathers. The Celts were also living in a landscape full of old pagan worship sites and in a culture honoring the intellectual and religious traditions of their ancestors. As we summarize Celtic monastic attitudes towards wild nature, we must also ask which of their practices appear to be Christian in origin and which appear to relicts of druidism. Environmental ethicist Jay Vest has suggested Celtic Christianity was an interference with more ancient ideals, such as the maintenance of sacred groves, and actually degraded Celtic relationship with the wild.[88] Other authors, such as Roderick Nash, portray Christianity as attacking the more environmentally oriented northern Europe religions. Is there evidence that Christian practice restricted rather than encouraged Celtic interaction with wild nature?

As astute a Celtic scholar as Charles Plummer has attributed much of the natural imagery in the Irish saints' lives to old solar worship.[89] It is, however, impossible to separate an historic continuum into its numerous threads and declare one to be all important. The Celtic and British hagiographers quote the Holy Scriptures and discuss numerous Christian virtues. The histories and sayings of the desert fathers clearly influenced these writings. Bede's "Life of Cuthbert," for example, mentions the similarity between Cuthbert driving the birds from his garden and St. Antony exhorting wild asses.[90] Many of the events in the Celtic saints' lives actually happened, and many of the places described are real. Even the fantastic sites of Brendan's voyage probably represent actual geographic locations. Brendan's fiery island may well have been an Icelandic volcano, and his crystal column may have been an iceberg.[91] One or more of the Saint Ciarans could have had animals roaming around his cell. A fox may really have attempted to eat his shoes.

A number of themes concerning wild nature are identifiable in Celtic literature. First the location of sites for prayer favored wild lands. To attribute the use of woodlands and lakes solely to pre-Christian traditions, however, overlooks some other considerations. The precedent for the use of isolated sites was found in the Scriptures and in the writings of the eastern church. The search for isolation was part of the monastic movement spreading through all Europe. The Celts lacked deserts, but had other possible uninhabited locations. The most available were deep oak forests, the shores of mountain lakes, the mountains themselves, caves, swamps, bogs, and rocky seacoasts. Of these, swamps and bogs are very difficult to occupy and poor places to construct churches (although Berach placed his sanctuary between a bog and a lake). Of the other five types, all were used for devotions or establishing chapels or monasteries. The Irish hagiographies mention caves and mountains less frequently than woods. A. P. Forbes has noted, however: "The custom of retiring for a time to a cave was very common among the British and Scottish saints."[92] St. Patrick is said to have fasted for forty days and forty nights on the mountain Cruachan Aigle, and St. Brendan is said to have been the founder of a small beehive chapel on the top of Mount Brandon on the Dingle peninsula (Ireland). In the stories of Brendan's voyages, he saw the "promised land" from a mountain (as did Moses). Mount Brandon, 3217 feet high, can offer a view up to a hundred miles—as far as the peaks of Connemara.[93] Celtic monasteries and churches were sometimes built on hills or promontories, but usually not on mountain peaks.

The limited use of mountains may not have been due to a preference for druidic oak groves, but rather a response to the climate of the British Isles. Even though ridges afforded expansive vistas, the summer season was short, and the mountains were difficult places to grow vegetables or maintain cattle through the year. Unlike the peaks of the arid Middle East, where the vistas are interrupted primarily by awe-inspiring thunder storms, the peaks of Ireland are shrouded in mist and sleet a good deal of the time.

When establishing a residence or monastic community, the potential of a site to supply basic needs became critical. Among the isolated mountain ranges, the valleys surrounding the lakes were more sheltered than the peaks and provided easy access to water. The oak groves were even more protected from the elements, often had richer soil, and provided wood. Living on the islands was probably difficult in early Christian times, but these offshore sites had several ready sources of sustenance, including fish, birds, and seaweed. Island climates were moderated by the low elevations and the surrounding ocean.

In Ireland, pines were extinct or nearly so by 200 C.E.[94] About 300 C.E., just prior to the Christian era in Ireland, there was "devastating clearing away of the elm and the ash."[95] Rowans (mountain ash) are usually smaller trees growing at higher elevations, and other species, such as birches and alders tend to be on ridges or in swamps. Thus, by the time monasticism caught on, most of the remaining mature low elevation forests on dry ground in the kingdoms of the British Isles were oak forests. If the monks wished to occupy nonagricultural sites, they would have chosen oak groves because of their availability, rather than because of their association with the old religion.

We can conclude that the Celtic monks used what isolated habitats they had. Some of these were certainly pre-Christian sanctuaries (reuse of more ancient holy sites was a tradition started by Abraham), but the monastic literature rarely suggests that the monks were purposefully displacing the old religion with their activities. The extent texts suggest, rather, that certain topographic locations and certain natural features may have been preferred. The site selected for a monastery was very important to the Celt and received considerable attention from hagiographers. Some of these wild sites were places of resurrection, thereby deemed suitable residences for eternity. (This undoubtedly encouraged selection of very aesthetic locations since the monks thought they would rise from the dead on site and live there forever.) When Maedoc and his party went to his place of resurrection at Rossinver, they "drew near to the fair shining Cuillin [meaning wood] and the beautiful wooded forest which was near the mighty lough, [and] they heard the sweet harmonious singing, and the melodious words of chanting, and the loud musical voice and heavenly shouts of fair wondrous angels above the Cuillin."[96] One has to admire Coemgen's place of resurrection at Glen da Loch, with its lakes and wild crags. One also has to admire his concern that it stay mountainous and filled with wildlife.

The Celtic monks followed the practices of monastic establishments elsewhere by praying in their cells and keeping the canonical hours. As we have already seen, however, the hagiographies make numerous references to praying outdoors, and some of these incorporate angelic or demonic visions. Ryan notes in *Irish Monasticism:* "Visions of angels were so frequent that they almost appear commonplace. . . . Whatever be the truth of many of these stories they show at least the conviction that those who had been raised to the higher forms of prayer should receive at times external rewards for their intimacy with God."[97] The writings, cited earlier in this chapter, about both the wood at Rossinver and Columba's oak wood associate the woods themselves with angels. Columba is also supposed to have gone to a hill top on the far side of the island where he was living and to have had a whole troop of angels descend to speak to him.[98] Most of these outdoor theophanies are not of the Old Testament variety where there is information conveyed about the character or will of God. Most appear to be "personal favors" or confirmations of the saint's holiness.

The Celts were less interested in the demonic than the desert fathers and turned instead to battle the remaining druids. Their first spiritual conflicts were

less a matter of personal holiness and more a matter of converting a pagan culture. There are, however, a few references to demons in the wild, although these
are greatly outnumbered by references to the angelic. Before Cuthbert could establish a monastery on the island of Lindisfarne it necessary to drive the demons
out.[99] A deacon of St. Galls, staying with the Celtic saint in the Swiss Alps,

> . . . went to fish in [a] torrent, and, as he threw his net, two demons appeared to
> him under the form of two naked women about to bathe, who threw stones at him,
> and accused him of having led into the desert [the wilderness] the cruel man who
> always overcame them. Gall, when he came, exorcised these phantoms; they fled,
> ascending the course of the torrent, and could be heard on the mountain, weep-
> ing and crying as with the voices of women: "Where shall we go? this stranger
> hunts us from the midst of men, and even from the depths of the desert;" while
> other voices asked, "whether the Christian was still there, and if he would not
> soon depart."[100]

These stories have strong similarities to passages in the *Life of Antony* and were
almost certainly influenced by works in Greek.

The Celtic view of the other world was related to their concept of the control
of nature through spiritual power. The saint did not employ the spiritual power
routinely, but used it to provide in emergencies (i.e., a stag replacing an ox, or a
horse pulling a plow or chariot), to aid the poor (Columba giving a poor man a
special stake for hunting),[101] to defeat druids (a thorn bush lifted into the air, or
the winds changed), to expedite passage (a road appears over a bog), or to prove
the power of God (mass celebrated on the whale, or Columba's eagle delivering
a fish). The druids in the stories used power in a very similar way, although they
are not as benevolent. St. Berach dealt with Diarmait the druid in terms of miracles
a druid would understand. Berach, for example, at Diarmait's challenge turns
two mounds of snow into fires.[102] This confrontation is reminiscent of Elijah's
igniting the wood in front of the prophets of Baal on Mount Carmel.[103]

The Celtic monks preferred to do their spiritual battling in this world, using
otherworldly power as a weapon. The desert monks, in contrast, employed human stamina to attack the demonic. The Celts actually may be closer to ancient
Hebrew traditions when they confront unbelieving elements in their own culture.
Their understanding of spiritual power, however, lacks the theological sophistication of the Hebrew prophets. There is nothing in the Irish saints' lives which ap-
proximates Elijah's flight into the wilderness and his reception of the "still small
voice."

The "lives" do occasionally show the saints abusing power. St. Brendan did
penance for turning Dorbarchu into an otter.[104] An angel rebuked Maedoc for at-

tempting to go without a ship from Ireland to Britain to seek a confessor. The angel said: "Presumptuous is thy deed." Maedoc replied: "Not out of presumption was I minded to do it . . . but through the power of God." The angel told him he needed no confessor, "but the God of the elements." If, however, he wanted one, Molua MacOcha would do.[105] These narratives demonstrate more knowledge of right versus wrong, however, than of the omnipotence of God. In the Celtic hagiographies, God rarely works directly—the Divine is almost always routed through the saint.

The implementation of spiritual power by Celtic saints was, in general, respectful of the environment. When a saint commanded an animal to provide a service, the animal was usually released afterwards or left unharmed (except when feeding the saints). In some cases, where natural objects or landscape features were involved, the changes could be more permanent. Declan, for example, caused the sea to recede for a mile around the island on which he established a monastery.[106] Ciaran of Clonmacnois is supposed to have removed a lake because the rowdy occupants of an island in its center were disturbing his prayers. Ciaran, though, did not destroy the lake, he merely had the Lord shift it "far away to another place."[107] The saints did not usually use spiritual power to alter the wild character of sites—except for removing monsters. The wildlife were not driven forth, they were encouraged to stay.

The Celtic concept of spiritual power usually does not portray the environment as an independent agent responding to the will of God. As already cited, there are numerous cases in the Celtic literature of divine provision, but spiritual battles are usually fought directly by the saint. It is also notable in the Hebrew scriptures that either God works through the environment to provide for an individual, or, if an individual is going to mediate the divine action, other people almost always receive the results of the wrath or of providence—that is, the prophet does not cause the rock to open to provide water solely for himself. Celtic hagiography, on the other hand, includes numerous instances of miraculous manipulation of wild nature for personal provision.

Jay Vest has suggested the pre-Christian Celts worshiped power in nature. He calls the "nature awe."[108] If the Celtic hagiographies incorporate pre-Christian magic to any extent, it would seem power over nature was more of an issue than power in nature. The tales of the druids also indicate this. MacNickle has pointed out that many of the stories concerning wild animals pulling plows and wolves tending flocks suggest a lingering interest in domestication.[109] Some of the stories, therefore, may reflect folk concern for the development of agriculture and the slow conversion from nature controlling the food supplies of hunter/gatherers to humans beginning to plant and harvest. Tales of saints causing trees to bloom or of saints planting ordinary forest trees but reapng a spectacular fruit harvest may also suggest attempted domestication.

Despite numerous "natural miracles," the hagiographies do not teach nature worship. The saint was almost always using the power of the creator God (mentioned frequently in this context) and operating through nature. If the stories encourage idolatry, it would be worship of the saints rather than worship of natural objects.

In regard to the person of God, the Celtic literature makes repeated reference to God the Creator. This would be in contrast to naming God savior, judge, help, deliverance, righteousness, etc. The saints' lives found in Plummer's collection call God, "the Creator," "King of the elements," "dear Creator," "God, the Creator of all things," and "King of the stars."[110] The early Celtic church—at a distance from the violent theological arguments engulfing the church fathers of the Mediterranean region—may not have wrestled as strenuously with questions concerning the person of God and the means of salvation as the ecclesiastical bodies of the dying Roman empire. The Celts emphasized the aspect of God most interesting to them, the Lord as the wellspring of the universe. The Celtic poet, unlike the Psalmist, tended to describe natural features for their own sake and as a setting for religious activity. The poems vary in content, but God's love and providence as expressed through nature are repeated themes.

Stopping the Hunt

A very important motif in the Celtic hagiographies—and one which we have not encountered in the literature we have already reviewed—is the protection of wild animals and plants from human abuse. In Leviticus, the Hebrew Law does require conservation of nesting wild birds, and there are references in the ancient Rabbinical literature to protection of forests and prevention of over-grazing. We do not, however, see major biblical characters portrayed as heroes for the preservation of wild beasts or forests. The desert fathers befriended animals and sometimes rescued them from natural difficulties. Gerasimus helped a lion with a reed in his paw, and Marcarius healed the blind hyena pup. Protection from human antagonists—in contrast—was rarely a concern.

Many of the stories in the Celtic hagiographies concern the feeding of starving animals or protection from hunting. Again the question of pre-Christian influence arises. The hagiographers favored wolves and stags. Is this a reflection of the sacred nature of these animals or the influence of pre-Christian cults, such as that of the horned god? The answer is difficult to determine because the druids left no written documentation of their practices.

Plummer argues that frequent mention of wolves in the lives of the saints implies a pre-Christian cult; thus many of the "preservationist" stories could be carryovers from the old religion. Plummer's hypothesis, however, has several weaknesses—at least as the sole explanation for these motifs. First, the literature of the desert fathers emphasized lions. If the Celtic monk were influenced by St. Antony, St. Jerome, and others, they may have preferentially told stories about wolves, the only large predators in the western isles. Second, the lives of the Celtic saints most frequently discussed large conspicuous animals, such as wolves, deer, and whales, as did the lives of the desert fathers. In heroic sagas, the repeated use of the more spectacular beasts is not surprising. MacNickle may be posing a better question when she writes: "One can understand the inclusion of the deer and wolf, but why not such common animals so familiarly known as the hare, rabbit, squirrel, even if not the hedgehog, weasel, stoat, vole, polecat, bat?"[111]

The natural histories of the animals themselves may be part of the key. The animals included are not only the largest, but are also the best known in terms of behavior or the most likely to provide certain services, such as bringing a fish to a saint. It is also possible the exclusion of some individual animal—particularly in the case of the hare—may be owing to the animal's sinister reputation in folklore[112] or to some association with magic. The latter might also apply to the rowan, a very beautiful plant, important to the druids, but rarely mentioned in the lives the saints. The one reference to rowans in Plummer's English translations of the lives of the saints in old Irish is to the druids of King Aeden using hurdles of rowan to prophesy."[113] Although the Celtic hagiographers certainly mention many plants and animals formerly sacred or employed in magic, perhaps some species had such negative or pagan associations that they could not properly be associated with the saints. Before accepting Plummer's suggestion that frequent mention may result from pre-Christian ritual significance, we must also be certain the opposite is not the case.

When Christianity arrived in Ireland, there had already been several periods of major forest clearing. Mitchell suggests all the woodland may have been gone from parts of the Boyne valley by the Megalithic period. Farming expanded during the Neolithic, and there were major intervals of woodland removal during the Bronze Age and the Iron Age. Grazing stock had begun to take a toll on species such as elm as early as 5500 B.C.[114] Degraded secondary woodland probably dominated the Irish landscape by the Iron Age.[115]

Two major agricultural innovations may be associated with two periods of forest removal early in the Common Era. About 300 C.E., the vertical knife or coulter was added to the Irish plow. About 600 C.E., the mould board plow was introduced. Ironically, the latter was probably brought to Britain by pagan Anglo-Saxons who were converted by Christian missionaries, many of whom were Irish. These missionaries in turn helped to introduce the new plow to Ireland.[116] The great flowering of monasticism was thus contemporary with a large expansion of cultivation. Columba's reported concern for axes in the woods of Doire may well have been a response to accelerated forest clearing.

Although the new plow probably would have eventually arrived without the monks, the monks did encourage the trend to expanded land clearing by bringing "productive activity into many hitherto unsettled areas, especially as many of the early communities deliberately selected sites of extreme remoteness such as offshore islands and inaccessible glens. Scores of monastic settlements were also built in lowland areas which previously had been forested and little occupied."[117] The impassioned saints, in their search for solitude, accidentally helped to open "the wilderness" to agriculture.

Celtic saints rescued animals from natural disasters, such as the attack of a predator or a fall into a crevice. The most frequent motifs, however, were relief from starvation and hunters. Since a number of the animals mentioned in the hagiographies were eventually hunted to extinction (i.e., the wolf) or severely depleted in numbers in the British isles, it is possible the low frequency of stories concerning saints killing vicious animals and the high frequency of protection stories may be a reflection of the saints' unconscious recognition that the animals were in peril. The removal of the forests would also have degraded wildlife habitat. Although wolves can starve because of the natural depletion of prey, human interference with forest cover and the supply of prey could also have left them "weak and fawning." The monk who fed a calf to a starving wolf, may well have recognized a problem in wildlife conservation, but probably did not comprehend its ultimate cause.

Although they wore animal skins, sometimes ate flesh when God provided, and cut firewood, the saints presented a strongly preservationist ethic for creatures residing on holy ground around their hermitages. The literature clearly discouraged the monks from coveting the products of nature. An otter, "great in its kindness," each day brought a salmon to some monks of St. Coemgen. One of the brothers decided the otter would look better as a glove. The otter recognized his danger and ceased to come, leaving the monks without the salmon. In the story, the brother realized his mistake and repented.[118] The monks also favored a vegetable diet, and although this was not absolute, it may have fostered their protectionist attitude towards wildlife.[119] In the Christian literature, the motive for protection, however, was usually compassion.

Pre-Christian Celtic religion may also have protected animals and trees, but the motives were likely to have been different, such as avoiding interference with the spiritual power of the animal. Is there any evidence for protection in the pre-Christian literature? A short investigation of Irish tales and histories most likely to contain Iron Age materials produced frequent references to druids and nature magic, but very little to parallel the protectionist activities of the saints. In such sources as the Ulster Cycle, a series of heroic tales including "The Tain," the tragedy of Deirdriu (Deirdre), and the Feast of Briciu, druids are presented as advisors to the warrior class.[120] They have large schools for teaching their arts and frequently serve as seers, prophets, or interpreters of dreams. The druids do change shape, turning themselves into wild animals, and in many tales a druid rod is used to turn a person into an animal of some type, usually a deer or wild boar, but also a swan, salmon, wolf, hunting dog, etc. The druids in these stories are not invincible like the saints and are frequently killed by the heroes. The saints acquired, in fact, more of the characteristics of the heroes than of the druids, and some heroic acts may actually have influenced the hagiographer's animal stories. The great Ulster hero, Cuchulain, catches a stag, for example, and has his charioteer step out on the stag's antlers.[121] The saints' cursing and turning people into animals are certainly rooted in druidic myth.

The old tales do record the cutting of forests. Queen Medb, in her pursuit of an Ulster bull, has a forest cut to facilitate a prophecy.[122] King Lugh bids the husband of his foster mother "to clear away the wood of Cuan, the way there could be a gathering of people around her grave. So he called the men of Ireland to cut down the wood with their wide-blade knives and bill-hooks and hatchets, and within a month the whole wood was cut down. And Lugh buried her in the plain of Midhe, and raised a mound over her."[12] The old tales are also full of hunting. Finn, perhaps the greatest of the hunter heroes, was also "a king and a seer and a poet; and a druid and a knowledgeable man."[124] Abstention from hunting seems most frequently to be a response to the fear that the animal was actually a person. Finn, for example, protected his son Oisin's mother who had been changed into a doe.[125] The opposition of saints to hunters is an interesting motif and does not seem to have a precursor among either the ancient heroes or the druids. In the (post-Christian conversion) tales of Oisin, who went to the Land of the Forever Young and returned several hundred years later to meet St. Patrick, Oisin argued against Patrick's attempts to convert him. Oisin lamented the loss of the Finna (Finn's men) and their hunting dogs and the passing of "the noise of the hunt on Slieve Crot, the sound of fawns round Slieve Cua; . . . the great deer of Slieve Luchra, . . . the hares of Slieve Cuilinn . . . ." Oisin, rejecting Patrick's call to faith, claimed: "The leap of the buck would be better to me, or the sight of badgers between two valleys, than all your mouth is promising me, and all the delights I could get in your heaven."[126] This tale portrays the hunter as the appreciator of nature, and the saint as a killjoy.

A countertheme to the lament of Oisin is the Christian saints' rescue of people turned into animals. In the famous tragedy, "The Children of Lir," Lir's four children are turned into swans by a jealous stepmother. After nine hundred years of life in the wild, the swans were living on Inis Gluaire—the Lake of the Birds—when St. Mochaomhog found them. The saint had silver collars made for them and "brought them to his own dwelling place, and they used to the hearing Mass with him." When someone tried to take them from the saint, they turned back into very aged human beings, and asked for baptism before they died.[127]

Although there is no way to be certain from the limited materials now available, the "antihunting" and "antiforest cutting" motifs in the Celtic literature do not appear to be druidic in origin, at least in their written forms. There is also little evidence they originated with the desert fathers. A best hypothesis is the saints protectionist role was rooted in Celtic respect for nature and the "animal friendship" ideal of the desert fathers, but formed as a response to the increasing need for preservation of wildlands through the Iron Age and into post Roman times. The "hunting and lumbering protection tales," despite roots in both pre- and post-Christian Celtic religious thought, probably reflect problems and attitudes common after the Christian conversion of the Celts.

The biblical writers could not imagine the desert wilderness disappearing under human manipulation. Their experience, in fact, was of a periodic resurgence of the wild during periods of warfare and social disruption. Isaiah predicted the jackals would reclaim Babylon. The desert was not something that could be tamed or conquered. In the last days the wilderness could become a wrathful agent of God.

The Celtic monks were experiencing something new—the complete displacement of larger wildlife and of mature forest stands by agriculture. Population growth accompanied by unregulated hunting were also taking their tolls. They added to their spiritual ideals a concern for harassed wildlife and for the vulnerable oaks, whose commercial value increased as the forests shrank. Unfortunately, it was an individual concern for individual animals and trees and, thus, was ineffective in curbing the cultural changes. The compassion of the saints, however, produced one of the first preservationist ethics on record.

From a theological perspective, the lack of biblical precedent does not make the response of the saints illegitimate. The Hebrew Law was concerned with protecting natural resources, but was not structured to deal with such massive changes in land use. The biblical declarations that Yahweh provides prey for the lions and habitat for the wild ass suggest God is also concerned about wild boar and oak groves. The prophecies of Isaiah give a special place to the wolf. The saints, in recognizing the value of these wild creatures and seeking to protect them, were taking a necessary first step in dealing with some major environmental issues that have become increasingly important as humankind has occupied more and more of the earth.

A Fawn across His Shoulders, a Blackbird in His Hand, a Crane at His Breast

In summary, the *Celtic model* of wilderness spirituality emphasized:

1. Withdrawal into wildlands for spiritual exercise. This included prayer in lakes (actually in the water), on lake or ocean shores, in the woodlands, and in caves. Hermitages and monastic communities were established on islands, in forests, and by bodies of water.
2. Withdrawal into the monastic community for a lifetime. There is evidence withdrawal away from the monastic settlement deeper into the wilds for prayer could occupy shorter periods. (Saints would go off by themselves into the forest to pray.) Coemgen and others practiced complete isolation for extended intervals (in the case of Coemgen, for seven years).
3. Extremely positive attitudes towards wild nature. Images from the wild are an important component of Celtic Christian poetry.
4. A deep understanding of God as Creator and Provider. The Celts expressed this through natural imagery in their Christian literature.
5. The demonic in wild nature was of relatively little importance in wilderness spiritual practice. Encounter with wild creatures or with angels was more frequent. Of the wild creatures only the monsters (usually lake dwellers) were considered evil beings.
6. An extremely diverse concern for wild nature. Not just large mammals, but birds, fishes, and plants attracted their attention.
7. Not only living in peace with wild creatures, but also protecting animals from hunters and natural catastrophes. A great sign of saintliness was never to have harmed a bird. The saints also protected forest stands and trees near their churches. Being kind to wild animals was considered a result of compassion and other Christian virtues. On several occasions a saint rescued an animal, while the saint had withdrawn for prayer.

Since we have already investigated the Bible and the desert fathers, the Celts have little new to teach us about ascetic practice, but they do extend the types of sites used out of the environments favored in the Levant. The northern monks, in fact, successfully translated the *ascetic* and *contemplative models* into much wetter habitats. The ease with which they moved contemplative prayer into the forest provides further evidence that isolation and wild surroundings, rather than deserts in particular, forward spiritual exercise.

From a contemporary Christian perspective the Celtic literature challenges us to think about our own relationship to wild nature in terms of Christian virtues. Should mercy and compassion towards the wild be spiritual goals? Should we make a conscious effort to interact with wild nature in such a way that we cause minimum disturbance? Is it a sign of holiness to protect creatures fleeing from human caused destruction? Do we honor God by preserving the beauties of the wild around our churches and spiritual sanctuaries? Do we see God's providence in a handful of berries or a bird's nest on the roof?

The Celtic monks offer an interesting example of spiritual balance. They could, on the one hand, appreciate the waves breaking on the rocks and, on the other, turn themselves towards God. They could wonder at whales, harvest some seaweed, praise the Lord of Stars, and serve the poor. They had a comfort and an appreciation of their environment that we tend to lack, and their feeling of belonging seems to have been matched by an ability both to give freely and receive joyfully. Their love for wild nature aided their Christian simplicity, and their nature poetry enhanced their understanding of divine love. We would do well to praise God as Creator as beautifully as they did.

12
A Cave outside the City
(St. Francis)

Home or the Sierras?

I thought about my alternatives for a while, and then picked up the phone to call the wilderness trip coordinator at Fuller Seminary.

" I am sorry; I have to cancel the trip to the Sierras," I told her, knowing full well it would cause organizational problems on her end.

"But you've already paid your deposit," she objected.

" I know, and I'll have to lose it. I have a Sri Lankan high school student coming to stay with me for the year. Her family has been caught in the ethnic violence, and I offered to help get her a student visa. We just made the arrangements, and she is arriving a week before the backpacking trip is supposed to start. I just can't leave until she is settled into school."

"Can't you get someone else to take care of her; it's only ten days," suggested the trip coordinator.

"It would be difficult, and I just can't bring myself to take off for California immediately after she arrives. It's been a tough decision because I have already finished all the pretrip readings, and it's a rare opportunity for me to be able to go backpacking with a Christian group interested in spiritual growth. But, I think the student's needs come first . . . ."

"Well if I can't convince you to come, I guess I'll just have to take your name off the list."

"I'm sorry," I apologized, " I didn't have much warning about this."

* * *

For the Christian, the question of how much time should be spent in personal prayer and other spiritual disciplines versus how much should be spent in service is often problematic. Through Christian history, the issue has been the subject of much debate. One of the greatest objections to the wilderness vocation of the desert fathers is it keeps the practitioners from performing numerous services to others, including bringing the word to the unbelieving, healing the sick, and caring for the hungry. Monasticism as a whole can hardly be accused of ignoring ministry, however. One of the greatest lovers of nature in the entire history of Christianity and a regular practitioner of prayer and fasting in the wilderness was also known for his attention to lepers, his love for the poor, and his public preaching. St. Francis of Assisi, who gave the birds a sermon, balanced a deep desire for personal knowledge of Christ with a passionate concern for the least of God's creatures—and the least of God's people.

The Cultivated Countryside

In *Wilderness and the American Mind,* Nash remarks: "Among medieval Christians St. Francis of Assisi is the exception that proves the rule. He stood alone in a posture of humility and respect towards the natural world."[1] This conclusion overlooks the continuous interest of Christian monks in wild nature from the time of Antony through the Middle Ages, a span of nearly one thousand years. Not only did the desert fathers influence the Celts, the Celts in turn influenced the Italians, including the Franciscans. St. Columban (also called Columbanus, an Irish saint who lived later than Columba) traveled in France and crossed the Alps. In 612 C.E., he established a monastery at Bobbio in Italy, "which in subsequent generations came to house a library of manuscript collections scarcely rivaled during the Middle Ages and Renaissance."[2] Irish writings were widely circulated during the Middle Ages, and some, such as *The Voyage of Brendan,* were translated into other European languages.[3] Edward Armstrong, in *Saint Francis: Nature Mystic,* has shown that almost all the Franciscan legends have precursors in Irish saints' tales and other monastic sources. Today, St. Francis of Assisi is the best known of the nature-loving monks. Yet, Francis was hardly an isolated example; he was, instead, the ultimate expression of traditions which had been growing and interweaving for centuries.

Before we consider Francis to be a mere imitator of older monastic ideals, however, we need to evaluate his environmental views in light of his total message. The founding of the Franciscan order, together with the founding of the Dominicans, marked a major transition in the structure and ministry of monastic communities and represented an important adaptation to changing social conditions. If we assume Franciscan nature lore was just a romantic throwback to another simpler era, we will have trouble explaining its significant impact on the natural philosophy, art, and science of the Renaissance.

Francis was born in 1181 or 1182 C.E.[4] The son of a prosperous merchant, he was raised in Assisi, a fair-sized town (for the Middle Ages) of more than twenty-two thousand souls.[5] Francis grew to maturity during a period of economic and demographic growth for most of Western Europe. Commercial ventures, including his father's textile business, flourished as industry expanded. Rising out of the feudal and rural social environment of the preceding centuries, the cities of the late Middle Ages gained population, territory, and political influence. The new wave of commercial activity favored literacy among the laity of the church, as some knowledge of reading and arithmetic was necessary to mercantile activity. Without the constraints of the rural manor, the commoner had greater access to new ideas and to political expression and had also become well educated enough to be suspicious of the clergy and the monastic establishment.[6] Looking back from our megacities and complex commercial culture, we tend to see Francis as a quaint wanderer of country fields and cart tracks. In actuality, Francis was a product of the growing urbanization and intellectual sophistication preceding the Renaissance.

Francis founded a mendicant order. His monks did not remain in cloistered communities, but traveled from town to town, preaching the gospel and ministering to the poor. Initially, the Roman church establishment was wary of the mendicants who operated outside established ecclesiastical structures and preached in the market places. The success of the mendicants, in face of resistance from the church hierarchy, was rooted in the growing need for their services, particularly in the towns and cities. Religious skepticism was already common in Francis's day; thus attendance at Latin masses no longer ensured belief. The urban areas were beginning to attract not only beggars who were unfit for work, but also the poor who had been displaced from their land or could not otherwise make a living in the country. Francis's followers were free of the rural parochial structure and could speak to the people in the vernacular. They took vows of poverty, thus avoiding the temptation to use a religious position to gain wealth or political status. Like many of the people to whom they ministered, they begged for a living. Disconnected from land holdings and specific territories, such as parishes, they could reach out to the cynics, the literate laymen, and the poor where they were gathering in the newly flourishing cities.

The Fearful Wolf

The landscape Francis experienced is evident in the early biographies. Francis moved from the urban centers, through the cultivated countryside, and withdrew to the woods, only to return to town again. Francis met his creatures in various places, including marshes and forests, but many of the animals came around his dwelling or were captured by someone and offered to him. A cricket used to perch on a figtree beside his cell,[7] a falcon made a nest where he was staying at the hermitage of La Verna,[8] and swallows at the city of Alviano competed with and defeated him while he was trying to speak.[9] On more than one occasion Francis released a hare or a rabbit caught in trap, and he also freed fishes and waterfowl from the hands of fishermen.[10] Francis talked a boy from Sienna into releasing a flock of turtle doves he had trapped and was planning to sell.[11] In the Franciscan stories, the animals are primarily creatures of the domesticated countryside, nonthreatening and easily controlled. They reflect an environment dominated by cultivation and subject to human exploitation.

In the Franciscan nature legends, tales of song birds and rabbits greatly outnumber narratives about large wild mammals, such as deer and boar. The literature does include several accounts of wolves, but these animals, such as the wolf of Gubbio or the wolves of Greccio, are bothering towns or farmers, and the medieval hagiographers have cast them as unrelenting villains. According to *The Little Flowers of St. Francis,* the wolf of Gubbio

> was a fearful wolf, enormous in size and most ferocious in the savagery of his hunger. It had devoured not only animals but men and women too, so much that it held all the people in such terror that they all went armed whenever they went into the countryside as if they were off to grim war. Even armed, they were not able to escape the tearing teeth and ravening rage of the wolf, if by mischance they met him. Such terror gripped them all that scarcely anyone dared to go outside the city gate.[12]

This description is hardly accurate natural history. Unless rabid or cornered, wolves rarely attack people and are no match for an armed adult who knows how to use a weapon. In contrast to all the meek animals in the Franciscan stories, the wolf of Gubbio appears as a wanton killer. The wolves of Greccio, equally evil, were also "devouring not only animals but even men."[13] Francis spoke to the wolf of Gubbio as if the wolf were a violent criminal, but offered pardon:

> Brother wolf, you wreak much harm in these parts and have done some dreadful deeds, destroying creatures of God without mercy. Not only brute beasts do you kill but, a deed of more hateful boldness, you kill and devour men and women made in the image of God. So you deserve an awful death, to be hacked like any footpad or loathly murderer. That is why all justly cry out and murmur against you and the whole city is your foe. But, brother wolf, I want you and them to make peace so that they may be no more harmed by you, nor the hounds further pursue you.[14]

Francis makes a covenant with the wolf of Gubbio, promising the wolf that the townspeople will feed him in return for more peaceful behavior. The townspeople agree to care for the wolf forever. The wolf "went from door to door begging. Harming no one, and harmed by no one, he lived like a state ward. It was wonderful that no dog barked at him. At length he grew old and died, and the citizens mourned him, for by his peaceful and kind forbearance, he recalled to mind the worth and holiness of Saint Francis whenever he went to town."[15] Just as Gerasimus's lion and Cirian's wild boar became monks, so did the wolf of Gubbio. Francis converted him in the best monastic tradition, and the wolf, following the example of the Franciscans, became a humble, mild-mannered beggar. The wolf also became a worthy citizen of the town and, unable to care for himself without doing violence, learned to content himself with handouts.

The effect of cultivation and urbanization may be detectable in this story. The wolves and lions of the desert fathers stayed with them in their cells. The Celtic monks had no towns in which to leave a wolf, and, in most of the Celtic stories, the wolves apparently departed after they were fed. The wolf of Gubbio did not return as a penitent to the wild, nor did he remain with St. Francis; he adjusted to urban life. On one hand, the medieval hagiographer may, with some understanding of natural history, have thought it "unsafe" for the wolf to leave the town. In a heavily managed countryside, the unfed wolf will always find livestock a temptation. The importance of the farmers and townspeople in the Franciscan wolf legends points to a landscape where large animals could easily come into conflict with human interests. On the other hand, the wolf was in the same straits as a hungry person in the late Middle Ages. The discussion may represent an anthropocentric analysis of the state of the poor and their potential for violence. The poor, like the wolf, were drawn to the city.

Loving the Lower Creatures

St. Francis was not interested in only *wild* nature, he was interested in *all* nature. He attended to the cultivated fields, just as he had attended to the wolf. "When he found an abundance of flowers, he preached to them and invited them to praise the Lord as though they were endowed with reason. In the same way he exhorted with the sincerest purity cornfields and vineyards, stones and forests and all the beautiful things of the fields, fountains of water and the green things of the gardens, earth and fire, air and wind, to love God and serve him willingly."[16]

Francis treated the nonhuman portions of the cosmos as equals. The most diminutive organism—be it a cricket or a lamb—was worthy in his eyes. According to his first biographer, Thomas of Celano:

For who could ever give expression to the very great affection he bore for all things that are God's? Who would be able to narrate the sweetness he enjoyed while contemplating in creatures the wisdom of their Creator, his power and his goodness? Indeed, he was often filled with a wonderful and ineffable joy from this consideration while he looked upon the sun, while he beheld the moon and while he gazed upon the stars and the firmament. O simple piety and pious simplicity! Toward little worms even he glowed with a very great love, for he had read this saying about the Savior: I am a worm, not a man. Therefore he picked them up from the road and placed them in a safe place, lest they be crushed by the feet of the passerby. What shall I say of the lower creatures, when he would see to it that bees would be provided with honey in the winter, or the best wine, lest they should die from the cold?[17]

His love took Francis beyond the care of living organisms only and into care for the inanimate: "He spared lights, lamps and candles, not wishing to extinguish their brightness with his hand, for he regarded them as a symbol of Eternal Light. He walked reverently upon stones, because of him who was called the Rock."[18] Although allowing candles to burn is not the best conservation practice, Francis understood the need to protect living creatures and to leave, in the midst of cultivation, some habitat for the wild:

He forbade the brothers cut down the whole tree when they cut wood, so that it might have hope of sprouting again. He commanded the gardener to leave the border around the garden undug, so that in their proper times the greenness of the grass and the beauty of the flowers might announce the beauty of the Father of all things. He commanded that a little place be set aside in the garden for sweet-smelling and flowering plants, so that they would bring those who look upon them to the memory of the Eternal Sweetness.[19]

Francis, like the psalmist, recognized that all creatures praise God. In his own composition "The Exhortation of Praise to God," he called on his friends:

Heaven and earth praise Him
All you rivers praise Him.

. . . .

All you creatures bless the Lord
All you birds of the heavens, praise the Lord
All you children praise the Lord.[20]

Francis considered all the elements of creation to be his brothers and sisters in the Lord. In the best known of his works, "The Canticle of Brother Sun," he valued all creation equally:

Most High, all-powerful, good Lord,
Yours are the praises, the glory, the honor, and all blessings.
To you alone, Most High, do they belong,
and no man is worthy to mention your name.

> Praised by You, my Lord, with all your creatures,
> especially Sir Brother Sun,
> Who is the day and through whom You give us light.
> And he is beautiful and radiant with great splendor;
> and bears a likeness of You, Most High One.
> Praised be you, my Lord, through Sister Moon and the stars,
> in heaven You formed them clear and precious and beautiful.
> Praised be You, my Lord, through Brother Wind,
> and through the air, cloudy and serene, and every kind of weather
> through which you give sustenance to Your creatures.
> Praised be You, my Lord, through Sister Water,
> which is useful, and humble and precious and chaste.
> Praised be you, my Lord, through our Sister Mother Earth,
> who sustains and governs us,
> and who produces varied fruits with colored flowers and herbs.[21]

The Canticle expresses human dependence on the rest of the creation and demonstrates Francis's insight into the interdependence of all things. The end of the poem praises bodily death for the righteous as the final divinely ordained member of the universal order.

Francis not only protected his fellow creatures, he extended his ministry to them and showed concern for their spiritual well being. Celano reports that the hay in the manger of a life-sized creche built by Francis cured the animals that ate it of various illnesses.[22] Preaching was a critical part of Francis's human ministry, and he exhorted the birds as well:

He saw some trees beside the road in which was a multitude of different birds as had never before been seen in those parts. Another huge flock was in the open fields beside the trees. Looking at the mass of them wonderingly, with the Spirit of God upon him, Saint Francis said to his companions: "Wait for me here on the road. I shall go and preach to our sisters the birds," and he went into the field where the birds were on the ground. As soon as he began to preach all the birds who were in the trees came down to him, and, like those on the level ground, remained quiet, though he moved among them brushing some with his gown. . . . To these birds Saint Francis said: "My sisters the birds, you have much from God and should always praise him for the free flight you have, for your double and triple plumage, colored and decorated vesture, for your nourishment set out for you without care, for your song accorded you by your Creator, for your number increased for you by the blessing of God, for your race was preserved in the ark, for the element of the air set apart for you. You neither sow nor reap and God feeds you, gave you rivers and streams at which to drink, mountains, hills, rocks and crags to hide in, tall trees in which to build your nests; and since you cannot spin or weave he gave the garments you need to you and your chicks. It follows that the creator who so blessed you loves you much. So take care, sisters mine, the birds, not to be ungrateful but be zealous always to praise God."

> At these words of the most holy father, all the birds began to open their beaks, stretch their wings and necks and reverently bend their heads to the ground, with songs and actions to make it clear, in manifold ways, that the words pleased them.[23]

Francis gave a repeated message of the value before God of those in the lower strata of the hierarchal society of his day. He was "filled with compassion not only toward men in need, but even toward dumb animals, reptiles, birds and other creatures, sensible and insensible."[24] We should recognize, that by his actions, Francis not only demonstrated his interest in nature, he was calling for mercy for the poor also. If birds were worthy of Christian attention, so were beggars and lepers.

A Cave outside the City

In terms of spiritual practice, the early hagiographies describe Francis and his followers as repeatedly, if not preferentially, pursuing prayer and contemplation outdoors. The biographers most frequently mention woods, although Francis prayed by lakes, on mountains, and in caves. On one occasion an angel came to Brother Masseo and asked him to disturb Francis while the saint was "meditating in the wood."[25] In another encounter with the divine, Brother Bernard was "in a wood totally lost in divine contemplation and united with God" when Francis called to him; Brother Bernard could not answer him because he was "in such ecstasy that he simply did not hear what [Francis] said."[26] Francis, seeking a quiet place for a Lenten fast, crossed over to an island in a lake and "crawled into a dense thicket where the vines made a kind of a shelter, and stayed there through the whole forty days, without eating or drinking."[27] Saint Bonaventure reported that Francis began to "seek out solitary places" immediately after his conversion, went out "to mediate in the fields," and made "his way through a certain forest, merrily singing praises to the Lord in the French language."[28]

A nobleman gave Mount Alverna, "a very sacred and very solitary mountain"[29] in Tuscany, to Francis. Francis and two companions climbed to the peak.

> During the ascent with his blessed companions, as they rested a little while under an oak tree, a flock of birds of various kinds descended upon Saint Francis with joyous song and fluttering wings. Some settled on his head, some on his shoulders, some on his knees, some on the hands of the holy father. At such a strange and wonderful sight Saint Francis said to his companions: "I believe, my very beloved brethren, that it is the will of our Lord Jesus Christ, that we should occupy a place on this lonely mountain where our sisters the birds show such joy at our coming." And rising up rejoicing greatly in spirit, he continued on. . . . Having sought out a lonely spot where he could pray away from the others, he built a very modest cell on the mountain side and gave orders that none of his companions should come near him . . . except Brother Leo . . . [whom] he instructed to come to him only once a day bringing bread and water.[30]

Francis received the stigmata on Alverna, where "on a certain morning about the feast of the Exaltation of the Cross, while Francis was praying on the mountainside, he saw a Seraph with six fiery and shining wings descend from the height of heaven."[31] The Seraph bore a figure of Christ crucified, and "he appeared at night with such brilliance that he lit up the various mountains and valleys round about more brightly than if it had been the light of the sun."[32]

Unlike some of the desert fathers, Francis had no inclination to stay in one place, and he never completely withdrew to the contemplative life. Francis found his first wilderness in a cavern among "the black firs"[33] on Mount Subasio outside Assisi. For Francis, these initial periods of withdrawal were not for a lifetime but, rather, for a few hours of prayer, during which he sought, he struggled, and he wept. Even with his growing religious fervor, these forays into solitude were always balanced by returns to the harsh social realities of the medieval world. Francis would pass a leper hospital on his walks or encounter the poor at the city gates. He traveled to Rome and experienced the city as a beggar. Francis hid in wild nature, but only for a time. Unlike Antony, Francis used "the Inner Mountain," not as a permanent residence, but as a relief and source of inspiration. Except during a period of illness immediately before his death, Francis always returned from the natural and solitary places to preach in the markets or to minister to the poor. After receiving the Stigmata on Mount Alverna, Francis went on a missionary trip, speaking in several towns, and demonstrated his unflagging interest in the sick and destitute by caring for a leper.[34]

Saint Bonaventure described this alternation between ministry to the people and withdrawal to the woods and the mountains:

> The angelic man Francis had made it his habit never to relax in his pursuit of the good. Rather, like the heavenly spirits on Jacob's ladder he either ascended to God or descended to his neighbor. For he had wisely learned so to divide the time given to him for merit that he expended part of it in working for his neighbor's benefit and devoted the other part to the peaceful ecstasy of contemplation. Therefore when in his compassion he had worked for the salvation of others, he would then leave behind the restlessness of the crowds and seek out hidden places of quiet and solitude, where he could spend his time more freely with the Lord and cleanse himself of any dust that might have adhered to him from his involvement with men.[35]

Francis and his mendicant monks tried to balance the holy life, moving between the needs of the people and adoration of God. Many of their visionary experiences or contacts with the divine were in isolation. In the end, however, they returned to the streets and cities and poured compassion on their fellow human beings.

Of All Animals He Preferred the Gentle

St. Francis was the product of a European city of the Middle Ages. His surroundings were cultivated countryside, tamed by plows and feudal agriculture. Medieval farmers were pushing into every corner they had the means to cultivate. Italian ship builders had cut the Apennines for wood, as had the Romans. Sheep and goats, foraging on the Umbrian mountains since ancient times, had turned forests into meadows.[36] Wild boar and wolves had been driven back into the higher, more rugged peaks.[37]

An heir of the Celts in his treatment of nature, Francis was responding to a far different social environment. Although large vertebrates were still present in Italy, relatively few of the Franciscan legends concern them. When mentioned by Francis's biographers, the larger predators are always in conflict with the human. Unlike the desert fathers, Francis did not spend his time with noble or strong beasts, and Celano noted that "among all the animals he preferred the gentle."[38]

St. Francis did not exercise power over nature to get animals to do favors for him, such as draw a wagon or hold his book, as did the Celts, although he did quiet birds and tame wolves. Animals appreciated Francis, and they responded by flying or leaping around him, or by jumping into his arms (themes found in the Celtic literature also), rather than by laboring for him. Francis was nonexploitative and respectful in his treatment of the wild. He extended the mercy and compassion of the Irish saints to all of creation. Francis was also a preservationist and, not having many large animals to save, rescued small ones.

For Francis, nature became a source of inspiration and beauty, circling in and out of his life as he wandered from town to town. Francis saw all nature as a blessed creation of God. He agreed very much with the psalmists that even inanimate creation praises the Creator. Nothing was too small or insignificant to be of interest. The followers of the Italian saint could appreciate a swallow in a cottage roof, or a mouse in a cupboard. The Franciscans began to look at nature in detail and, in so doing, encouraged the careful observation which helped to develop the visual arts of the Renaissance and set the stage for modern biological science. In medieval painting, humans and their livestock are larger than trees and mountains. Landscapes are idealized and not very realistic. Franciscan attention to nature brought a better understanding of how nature actually looks and how it operates. Franciscan thought did not automatically make the human the overwhelming centerpiece of the universe. Crickets, worms, larks, and rocks all had something to say about Christ. Francis and his mendicant monks became, then, a transition to a contemporary view of nature where even the smallest organisms are worthy of study and may be linked to human welfare.

Ironically, as was the case with the Celtic monks, Francis's love for nature may have helped to forward human domination of the natural. Western scholarly traditions had encouraged Platonic body/soul dualism and the classical Greek tradition of aloofness from nature. Francis disregarded the strict hierarchies of the medieval universe and reordered the cosmos, raising nature in spiritual value. Saint Bonaventure "encouraged Christians to study nature in order to learn of God."[39] The Franciscan declaration that nature was worthy of human attention, and therefore of scholarly study, forwarded systematic observation of nature and the development of the empiricism which produced modern science. Not coincidentally, a number of the greatest medieval scientists were Franciscans. The tremendous growth in the natural and physical sciences from the end of the Middle Ages onward has helped to give western culture increasing control over nature, and, although our knowledge has produced many benefits, it has also caused great disruption of natural environments.

Without the love and compassion expressed by Francis, an intense human interest in nature quickly becomes dangerous. Francis viewed other creatures in terms of their relationship to God, rather than in terms of their usefulness to people. Our modern ignorance of the will of God for creation has left us disposed to interacting with the cosmos entirely in terms of our own immediate material desires. Francis would never have approved.

In regard to spirituality, Francis maintained the emphasis on the contemplative life, while redefining monastic public ministry. Of the monks we have investigated, he was the closest to the New Testament *contemplative model* in his approach. Withdrawal to the wilderness was critical to Francis's early Christian development. The time spent in the cave helped him to define his values and his ministry and brought him closer to Christ. After this time of preparation, Francis alternated between ministry and prayer, much as Christ had, but, if we believe Francis's early biographers, the temporal balance was different. Francis spent relatively longer intervals in contemplation and fasting than Christ and his disciples had. Francis took his followers with him to the solitary places, much as Christ had done, but Francis had little inclination to rest. The traditions of strict asceticism among the holy still reigned, driving Francis into long fasts, which eventually, in combination with his repeated missionary journeys, may have impaired his health.

Francis and his companions saw both angels and demons, as well as visions of Christ, in their wild retreats. The otherworldly experiences sometimes affirmed the holiness of the friars, sometimes instructed the faithful, and frequently demonstrated Franciscans values. On a few occasions, the visions presented the person of Christ or provided directions for ministry. The influence of the desert fathers was still very evident, although the Franciscans had more of a taste for the angelic.

In summary, we can identify several characteristics of the *Franciscan model* of wilderness spiritual experience:

1. Withdrawal into wild nature was encouraged for spiritual exercise. This occurred in woods, in caves, on islands, and on mountains, although Francis was known to walk through the fields just praising God. His monks had a mountain where they went to fast and pray.

2. Withdrawal into wild nature and the solitary places was temporary. The monks would return to preach and serve the poor, although withdrawal did include forty day fasts and longer periods of contemplation.

3. Withdrawal into wild sites was used both for spiritual preparation for ministry and for renewing communication with God in the midst of ministry.

4. Francis treated all nature as "brother and sister." Everything had worth before God. Franciscan interests in nature were even more universal than the broad concerns of the Celts. Francis preached even to the birds, an affirmation of their spiritual worth before the Creator.

5. Francis rescued animals of all sorts from human caused destruction and also mediated conflicts between wild animals and humans. Francis emphasized mercy and compassion towards nature; animals in return expressed their love for Francis by approaching him joyfully.

Of all the models of wilderness spirituality we have investigated, the *Franciscan model* is one of the most applicable to the contemporary Christian situation, perhaps because it was developed as western Europe was urbanizing. The Franciscan model recognizes the critical need for Christian ministry and the importance of carrying the words and love of Christ to others, yet allows time for reflection, gaining self understanding, seeking God, and appreciating the wonders of creation. In his abandonment of "the world," Francis embraced the cosmos and found signs of God's handiwork everywhere he turned.

Francis's use of the cave when he was first considering his spiritual state helped to bring him to an early Christian maturity and led him to a depth of understanding of the needs of his own culture and his own time. When ministry is so oriented towards action that such periods of reflection are lost, the result will often be an impassioned attack on the wrong problem or an intensive effort with the wrong goals. Francis had a capable spiritual guide or an advisor during this early period, and he had solitude. The result was tremendous missionary vision and an unwavering dedication to Christ.

Franciscan attention to the whole of nature was a needed return to biblical values. Francis expected the birds and trees to praise God. Like Christ, he knew God cared for the sparrows. Francis was born at a time when humankind had just managed to displace wild nature completely from much of the European landscape. He called attention to worthy citizens of creation whom human technological and economic advancement were trampling under foot. Today we need to ask ourselves if we have acquired Francis's interest in all creation and his recognition of the importance of even the smaller creatures (and the accompanying passion for the microscopic and the molecular), while abandoning, with many of our other religious values, his understanding of that same creation as beloved of God.

13

The Limits of Western Wilderness
(The Reformation)

Whither the Wilderness

Wilderness spirituality has its limitations. Wild nature isn't always available, and we can't always "withdraw" when we wish. Sometimes other Christian concerns, such as home and family or interpersonal ministry, interfere. For the Christian, the questions of how much time should be spent in personal prayer and other spiritual disciplines and of how much should be spent in service can be problematic. In the past, issues such as this have not been confined to individuals trying to lead righteous lives, but have been matters of debate for all Christendom.

Having reviewed biblical wilderness texts and three movements in the early and medieval church—all of which had a strong appreciation for wild nature—we now should ask: Why aren't these attitudes and practices common among Christians today? Why do modern Christians tend to ignore these traditions, or simply deny that they ever existed? Why would a liberally educated historian such as Roderick Nash assume Francis of Assisi was an anomaly? The answer lies partially in social, landscape, and theological changes that affected early modern Europe. Western wilderness traditions were very strongly tied to monasticism and the monastic vocation, which were largely rejected by the Reformation and underwent considerable change within Roman Catholicism. (The Eastern church has maintained somewhat greater continuity in wilderness practice and greater appreciation for the way of the desert.) Further, many of the attitudes towards nature found in both the Bible and in pre-Enlightenment Christian literatures have been displaced by a scientific or mechanistic view of nature. Nature cannot respond to holy St. Coemgen (Kevin) if it is entirely controlled by physical forces. Further, from the Renaissance onward, western science has emphasized the ability to control nature, a philosophy that is, in itself, in intrinsic conflict with wildness. Coemgen's holiness, which brought the birds to him, has been replaced by intellect, and, on far too many occasions, brute force.

Politics, Plague, and Famine

Through an analysis of the social circumstances surrounding St. Francis, we can begin to understand why "wilderness spirituality," although continuing as an ideal in the western church, began to decline well before the Protestant Reformation: there was simply no place to practice it, except for a few mountain peaks and some of the outer islands. Europe had slowly opened most of her forests and occupied almost all lands that could be farmed or grazed. Overcoming the disruption that followed "the fall of the Roman Empire," the population of the continent continued to grow, as did the need for food and fuel. The oak forests, so dear to Columba, were cut. Europe was so heavily populated by the fourteenth century that famines fell on several countries between 1309 and 1317. With the advent of the Black Death in 1348, there was a major population collapse.[1] The church of the late Middle Ages, in the midst of a terrifying encounter with human mortality, lost most of its concern for the bucolic monastic life.

The routes to the wilderness were restricted not only in Europe; the rise of Islam reduced access to the deserts of the Holy Land also. Rather than sending monks to the wilderness of Judea, Europe was contemplating how to keep the Moors and Turks from capturing Madrid, Vienna, and even Venice. Friendship with lions and wolves did not provide adequate protection from sultans who proselytized with the scimitar. The spirituality of the desert fathers had rejected the decadence of the Roman culture, but it thrived on the peace the Empire provided. In 614 C.E., the Persians, the ancient enemies of the Greeks and Romans, captured Jerusalem, ending the *Pax Romana* for the desert monks.[2] After the death of Mohammed in 632 C.E., Islamic fervor spread through the Levant, thwarting most opportunities for further Christian movement into the desert.

Wilderness spirituality continued in various forms, among the Franciscans and groups such as the Cistercians. Yet the great "wilderness" initiatives in monastic circles ended with the Renaissance. Europe was turning outward to look at the rest of the world, and the later monastic orders would be "soldiers," such as the Jesuits, called to meet the challenge of the Reformation and expanding mission fields. Monastic mysticism, meanwhile, became an even more internal matter, influenced by the Neoplatonism of Meister Eckhart and the interior castle of Teresa of Avila.

Through the end of the Middle Ages and into the Renaissance, the church and the monastic orders themselves became havens of decadence. Many of the monasteries were wealthy, housing chubby abbots and, in some unfortunate cases, feeding their concubines. Fraught with war and politics, the church sometimes seemed to be barely interested in religious pursuits. The papacy had become a position of earthly power, rather than a post for a spiritual shepherd. The need for change was widely recognized, yet concerned Christian leaders remained unable to execute the needed major house cleaning. On the eve of the Reformation, the western church had merged so thoroughly with the secular culture and was so busy laying up treasures on earth that the collection of tithes and indulgences had become a primary concern. The kingdom of heaven owned Sunday morning, and the kingdoms of the earth owned the remainder of the week.[3]

Devil's Lies and Deadly Venom

Wilderness spirituality hit a low point in the early Reformation period. This may at first seem surprising since renewal and revival movements in the early church so often had "wilderness" elements. The social and theological circumstances of the Reformation are very complex, but a few major factors stand out as possible roots for early Protestant disinterest in wilderness.

First, the Reformation was, in many ways, an urban movement centered in the very independent cities of middle Europe. Many of the leaders were educated burgers. As the Reformation spread, entire towns and principalities withdrew from the authority of Rome, giving the municipalities an important role in the theological battles.

Second, the reformers viewed monasticism and its practices as an important source of the *problems* within the church. The reformers' reaction against the evils of the monastic institutions included an attack on their spiritual foundations. Leading reformers deemed not only celibacy but also withdrawal unnecessary—if not of Satanic origin.

Third, Luther abandoned the concept of a specialized spiritual vocation and insisted that each Christian had a calling from God. Being a good blacksmith was as worthy as being a good priest. The person who lived in the desert and sought holiness was, therefore, wasting his or her time—if not acting in a self-serving fashion and taking a fast route to hell. Service in a secular profession and participation in the community replaced Antony's wilderness vocation.

Fourth, Luther's doctrine of *sola Scriptura* (all matters of Christian faith should be based on the canon of Scripture) tended to place emphasis on reading the Bible, while removing emphasis from many other sorts of spiritual practices. Although prayer and, to a lesser extent, fasting were still very important in rebel circles, many ascetic practices were abandoned. Preaching and community worship became the centers of Protestant spirituality.

Fifth, the struggle over the marriage of priests and former monks mixed the Protestant interest in the community (and perhaps Reformed bourgeois values) with a new attention to marriage and the family. Luther, who had been a monk, married a former nun and became a dedicated father. Long sojourns in isolated spots did not fit the early Protestant ideal of finding a godly spouse, pursuing a respectable vocation, and, if blessed with offspring, teaching them to read the Holy Word around the dinner table.

Perhaps one of the best ways to understand the changes taking place during the Reformation is to look at the writings of the reformers, particularly Luther and Calvin, and evaluate their attitudes towards "wilderness spirituality." The tirade against monasticism begins with Luther, who was concerned not only about decadence within the monastic orders, but also about the concepts of salvation by works and righteousness based on human merits.[4] In *The Judgment of Martin Luther on Monastic Vows,* Luther wrote: "But these godless people (monks) not only want the monastic life to be regarded as a way of life in which a man may lead the good life, but rather as the way of life which is good in itself."[5] He thus rejected the idea of the monastic vocation. Luther also condemned many types of spiritual exercises and, coincidentally, the value of the wilderness sojourn:

> And if any monk, devil or schismatic spirit . . . comes to frighten you with the words: "God is a stern angry judge!" and then directs you elsewhere . . . or if they order you to go on so and so many pilgrimages, to enter a cloister or to flee into the wilderness, etc., until you have rendered satisfaction for your sin and merited mercy—then on the basis of this you can judge and say that such doctrine and such notions are the devil's lies and deadly venom.[6]

Luther's final blow to monasticism was to deem it sinful and dangerous, when he wrote: ". . . the monastic institution has no divine authority but . . . it is actually contrary to the Christian faith and evangelical freedom."[7]

Luther discouraged outdoor spiritual practice away from the home church when he stated: "The chapels in forests and the churches in fields, . . . which have become the goal of pilgrimages must be leveled. Oh, what a terrible and heavy reckoning those bishops will have to give who permit this devilish deceit and profit by it. . . . They do not see the devil is behind it all, to strengthen greed, to create a false and fictitious faith, to weaken the parish churches, to multiply taverns and harlotry, to lose money and working time to no purpose, and to lead the ordinary people by the nose."[8] Luther, of course, was objecting to all the miracle-mongering associated with rural shrines and the revelry associated with pilgrimages, rather than to the type of wilderness experience associated with Elijah. He was also, coincidentally, removing one of the few contacts the common people had with spiritual exercise outside the community churches.

No Slight Evil Example

John Calvin took up Luther's banner, and, although Calvin found ancient monasticism to be acceptable and righteous, he found the monks of his time to be contemptible. He, like Luther, thought Christ's rule of life is for everyone. In *The Institutes of the Christian Religion,* Calvin summarized his position by rejecting monastic vows and suggesting that his contemporaries would do better by remaining in the Christian community and everyday vocations:

> I frankly admit that even in the ancient form (of monasticism) which Augustine commends there is something that I do not like very much. I grant that they were not superstitious in the outward exercise of a quite rigid discipline, yet I say that they were not without immoderate affection and perverse zeal. It was a beautiful thing to forsake all their possessions and be without earthly care. But God prefers devoted care in a ruling household, where the devout householder, clear and free of all greed, ambition, and other lusts of the flesh, keeps before him the purpose of serving God in definite calling. It is a beautiful thing to philosophize in retirement, far from intercourse with men. But it is not the part of Christian meekness, as if hatred of the human race, to flee to the desert and the wilderness and at the same time to forsake those duties which the Lord especially commanded. Though we grant there was nothing else evil in that profession, it was surely no slight evil that it brought a useless and dangerous example into the church.[9]

Calvin suggested that extreme asceticism and other forms of "exclusive" spiritual practice are wrong. He did not consider forsaking the world as sin, but he deemed it less desirable than serving God in a profession in Christian community. He proposed that fleeing to the desert was actually abandoning Christian service. Calvin's objection was not to temporary withdrawal for renewal or to seeking solitude for prayer but to the wilderness *as a spiritual vocation.* The effect of this sort of writing, however, has often been to make individual Christians afraid that, if they withdraw for a solitary time of prayer or reflection, they will be shirking community responsibilities and God's true call.

Calvin, like Luther, attacked individual religious exercises and, like Luther, coincidentally discouraged use of wilderness or natural sites. In eliminating the veneration of saints, for example, Calvin also spoke against pilgrimages and pointed out that "Christ abolished all distinction of places, when he said, 'The hour cometh, when not at this mountain, or Jerusalem, but everywhere shall the true worshipers worship God in spirit and in truth.' (John iv. 21, 23.)"[10] For Calvin, the pilgrimage to the holy site was a revival of Judaism, "because Jerusalem was the appointed place of worship," and those who went on pilgrimages were "men, after the example of idolaters, (who) erect for themselves groves . . . at pleasure."[11] In addition to the trip to the painted altar, Calvin's objections to the veneration of holy places tended to discourage spiritual traveling for any purpose other than evangelizing the unbeliever or taking the word to the Christian community.

Another example of an attack on a religious exercise that might even more strongly affect a Christian's use of wild sites was Calvin's commentary in his *Harmony of the Evangelists* on fasting for forty days in imitation of Christ in the wilderness. Calvin suggested Christ's forty-day fast was a special event, "not to give an example of temperance, but to acquire greater authority, by being separated from the ordinary condition of men, and coming forth, as an angel from heaven, not as a man from the earth."[12] The fast, therefore, should not be imitated. The fasts of Moses and Elijah were also related to special employment by God and would not justify such a practice for most Christians.

The practice in Calvin's time apparently was to fast once a day for forty days, which, as Calvin made clear, has little to do with the difficulties of a complete fast for forty days. Calvin noted further that Christ and Moses only observed solemn fasts once during their lives. Calvin then blasted those who wished to copy them:

> I wish we could only say that they had amused themselves, like apes, by such fooleries. It was a wicked and abominable mockery of Christ, to attempt, by this contrivance of fasting, to conform themselves to him as their model. To believe that such fasting is meritorious work, and that it is a part of godliness and of the worship of God, is a very base superstition.
>
> But above all, it is an intolerable outrage on God, whose extraordinary miracle they throw into the shade; secondly, on Christ, whose distinctive badge they steal from him, that they may clothe themselves with his spoils; thirdly, on the Gospel, which loses not a little of its authority, if this fasting of Christ is not acknowledged to be his seal. God exhibited a singular miracle, when he relieved his Son from the necessity of eating: and when they attempted the same thing by their own power, what is it but a mad and daring ambition to be equal with God? . . . Away then, with that ridiculous imitation, which overturns the purpose of God, and the whole order of his works. Let it be observed, that I do not speak of fastings in general, the practice of which I could wish were more general among us, provided it were pure.[13]

Again the target was not the wilderness, but a type of wilderness spirituality, that was at issue. Calvin did not address directly monastic wilderness practice in this commentary, but it is doubtful one of his well-reformed followers would have chosen St. Antony as a spiritual mentor.

It should also be noted that the reformers tended to interpret biblical passages concerning wilderness or wild nature allegorically or metaphorically. Although not the first interpreters to do this, the reformers influenced several generations of Protestants, including the Puritans in the New World. Luther, in his *Lectures on Isaiah 1–39*, for example, interpreted the animals of Isaiah 11:6–9 ("The wolf shall dwell with the lamb, and the leopard shall lie down with the kid, and the calf and the lion and the fatling together . . .") as personages in the church. The wolf is a false teacher, the lambs are Christians, the leopards are persecuting tyrants, the goats are a clean offering or martyrs, and the lions are the rich.[14] This interpretation is quite different from that of the desert fathers, who, familiar with the creatures involved and convinced the kingdom was at hand, would have taken the passage more literally.

An example from the work of Calvin, is the interpretation of John the Baptist as "a voice crying in the wilderness." Calvin wrote:

> The word wilderness is here used metaphorically for desolation, or the frightful ruin of the nation, such as existed in the time of captivity. It was so dismally shattered, that it might well be compared to a wilderness. . . . The prophet magnifies the grace of God. "Though the people," says he, "have been driven far from their country, and even excluded from the society of men, yet the voice of God will yet be heard in the wilderness, to revive the dead with joyful consolation." When John began to preach, Jerusalem was in a sense a wilderness: for all had been reduced to wild and frightful condition. But the very sight of a visible wilderness must have had a powerful effect on stupid and hardened men, leading them to perceive that they were in a state of death, and to accept the promise of salvation, which had been held out to them.[15]

Although John the Baptist's stay in the wilderness was both a metaphor and a fulfillment of prophecy, it was also a reality of ministry. Calvin's commentary takes a negative tone and associates the wilderness with sin-ridden Jerusalem. As suggested previously, wilderness at the time of the Baptist, may have been an actual place where one could escape from sinful cultural influences.[16] Calvin proposes here that the wilderness may have improved the vision of those who went down to John outside of Jerusalem, but, since Calvin had never actually seen the Jordan valley, his interpretation may reflect primarily the cultural views of his time when he suggests that the desolate state of the wilderness shocked sinners into conversion. (It seems unlikely that natives of Palestine, who were used to arid landscapes and had to walk through them to get from Jerusalem to Jericho, would have been shocked or startled by the desert wilderness.)

In defense of the reformers, it must be stated that they did not lack either a creation theology or an interest in nature. Calvin wrote an extensive commentary on Psalm 104,[17] for example, and, in his *Institute of the Christian Religion,* included a chapter entitled "The knowledge of God shines forth in the fashioning of the universe and the continuing government of it."[18] The reformers were open to the study of the natural sciences and perhaps less worried about the potential worldliness of an interest in creation than many of their Roman colleagues. Yet wilderness was, at first, only a place to flee under persecution, and much of the appreciation of the early church fathers for solitude was lost in the battle against the monastic establishment.

The linkage of wilderness spiritual experience to a lifelong, celibate vocation and to asceticism was, for the first Protestants, its downfall. In shifting the emphasis back to family and community, they threw out much that was associated with isolation, personal spiritual practice, or the ascetic pursuit of holiness. The primary disadvantage of this strategy was the coincidental removal of practices that aid in fighting sin and temptation at the cultural level. The strategy also discouraged spiritual practices that had for centuries been associated with deeper perception of the person of God. In many congregations, the concept of spiritual rest became tightly tied to the Sabbath and to church related social and educational activities, while short withdrawal for prayer and contemplation or dealing with internal turmoil fell from the community repertoire.

The five key issues for the reformers remain important topics for discussion among the various Christian denominations. These are, in summary:

1. The association of wilderness withdrawal with salvation by works. Someone enters the wilderness to become "super holy." Luther could hardly have accused Christ or John the Baptist of entering the wilderness to prove or earn their salvation, yet there can be no doubt the desert fathers thought desert spiritual discipline produced divine rewards. The real question here is not wilderness, *per se* but the role of ascetic practice, which is biblical but was extended by the early church. The reason one enters the wilderness becomes critical. Is it flight, an answer to a divine call or an attempt to prove oneself?

2. The association of wilderness with religious vocations, including wilderness monasticism. The issue, in terms of wilderness, is less, "Should one go?," but more, "How long should one stay?" Since the concept of a complete wilderness vocation is postbiblical, one's acceptance or rejection of desert monasticism depends on whether one accepts early church traditions as establishing precedents for continuing Christian practice. One can seek the wilderness for rest or contemplation without remaining a lifetime.

3. The association of withdrawal with disruption of the Christian community and interference with ministry. Although this certainly is a risk of entering the wilderness, both the biblical record and early church literatures indicate that wilderness sojourns were

frequent precursors to difficult ministries and to initiating cleansing
or transformation within the community. The question becomes,
"When is aloneness appropriate?" and again, "How long should
one stay in the wilderness?"

4. The association of wilderness spiritual practice with improper spiritual discipline and improper imitation of Christ. Calvin was quite right
when he pointed out evening fasting on top of a full lunch is a poor
replicate of Christ's forty days in the wilderness. Fasting, however, is
certainly condoned by the Bible, and so are other potential wilderness
disciplines, such as prayer. When and how one should imitate Christ
can be a very difficult issue. Today, Christians argue over the
meaning of miraculous healing in the Gospels. Christ healed the
sick—should we? The wilderness case is fraught with contrasts.
Christ did not, in personal spiritual practice, participate in any
activities foreign to the Judaism of his time. He prayed, he fasted,
he read the Scriptures. The unusual thing about Christ's forty days
in the wilderness was the length of the fast and the confrontation
with Satan, not the fasting itself. Christ's stamina has scriptural
precedents. Elijah, after taking food from an angel, "went on the
strength of that food forty days and forty nights to Horeb the mount
of God." These events are both clearly in the realm of the
miraculous and would be difficult for us to imitate. Fasting, however,
is condoned by the Scriptures. The question becomes, "When is
imitation of Christ and other biblical figures, such as David,
appropriate?"

5. The association of wilderness or outdoor retreats with holy sites.
Historically, in Christianity, wilderness spirituality does not
require holy sites, but, in an environment where veneration of saints
is practiced, wilderness spirituality tends to generate new holy
locations. Since there is clear evidence of holy sites, such as Horeb,
in the Old Testament, and some evidence of special locations or
landscapes, such as the mount of transfiguration, in the New
Testament, we must be careful to distinguish sites chosen by God
for interaction with human kind from sites of historic events and
from preferred landscapes for spiritual practices, such as prayer.
The question becomes, "Am I trying to go to the mountain where
Moses saw God?" or "Am I trying to choose the best place for solitary
spiritual exercise?"

Ironically, although the theology of the reformers has had a tremendous impact on Christian thinking about spirituality, some of their objections to monastic
practice are now almost archaic. In past ages, for example, extreme asceticism

has been a genuine theological and spiritual concern. Today's Christians have little feeling for spiritual activity, other than prayer or Bible study, and would not think of a wilderness sojourn as a means of gathering "holiness points." Even in Roman Catholic and Eastern Orthodox circles, asceticism and solitude are out of fashion. Most modern Christians have no association of wilderness with holiness, so the temptation to salvation by works in a wilderness setting is relatively minor. The average Christian who seeks solitude in the wilderness is much more likely to act out of need than out of pride or social competitiveness.

Western Christians are so rushed in their daily lives that the lack of time for spiritual exercise is more of a current dilemma than the potential for too much withdrawal. Modern, highly mobile society has also destroyed our sense of place. Pilgrimages to country shrines have never been a major social activity in Anglo-America, and, throughout middle-class culture, they have, to a large extent, been displaced by the family vacation. Association of a wilderness location with the divine may cause an emotional attachment to place, but real veneration is, for the contemporary Christian, unlikely.

We can conclude that Luther and Calvin, in a sincere effort to bring biblical values back into Christendom, probably discouraged some biblically based wilderness spiritual practices in the process of attacking the monastic vocation. There can be no doubt that withdrawal into the wilds can be self-serving and can encourage extreme asceticism. There can also be no doubt that contemporary Christianity is weak in spiritual discipline and sets little time aside for individual pursuit of God. Somewhere between monastic excess and reformation rejection, we must return to God's intent for creation and tie together the threads of a fractured wilderness tradition.

14

Spiritual Journeys –
On-and Off-Trail
(Contemporary Questions)

The Appalachian Trail

"I enjoy being alone," the solo hiker said. "My parents are worried about me, and my brother thinks I ought to be hiking with someone else."

The young woman grinned. "But I know when I met you sitting in front of that trail shelter, I talked your ear off. I think I need to talk when I haven't seen anyone else for two days."

"You did talk for about an hour straight," I replied. "I didn't mind. I had already eaten dinner and was just lying on the grass relaxing when you came along." We had met the day before, at an Appalachian Trail shelter on a mountain peak. Although we did not hike together coming down off the ridge, we kept passing each other over a twelve-mile stretch and finally ended up staying at the same bed and breakfast place in a little mountain town.

"Going out alone does present safety problems. Somebody needs to know where you are," I suggested. "Why did you start doing the trail anyway?" I leaned back on the couch. The sitting room was furnished with Victorian chairs, surrounding a cast-iron stove. The wall decorations were a mixture of 1960s counterculture and pre-1930s mountaineer memorabilia. It was a great place to discuss walking the Blue Ridge.

"I started the trail to stop drinking," she replied.

"Were you drinking that much?" I asked mildly surprised.

"Yeah, all day. When I was painting and refinishing, we'd stop working for awhile in midafternoon and do some shots. Then we'd go back to work. We'd drink all evening too."

"That's bad. You seem too young to get into that."

"Lately, I've been tending bar. All these people in their forties and fifties come in. I enjoy talking to them. A lot of them divorced . . . and lonely. And I feel bad giving them alcohol. Watching things getting worse for them instead of better."

"Do you do any drugs?" I inquired.

"Yeah, mushrooms, hash, some coke. I've been smoking marihuana since I was seven years old, and tripping since I was eleven."

I shook my head.

She looked down, contemplative. "But there's none of that out on the trail. I can't understand people who carry all kinds of liquor to the shelters just to party," she said. "The people who come out to party never see anything. They just get drunk and talk to each other. They don't see any flowers, they don't see any wildlife. They don't see the ground. I like to look at all the different kinds of ground when I'm walking. I like the thick pine needles . . ."

"The last year's been the worst year of my life," she reflected. "This is so much better. I love it out here. . . . When I'm hiking by myself, I have the whole day to think. Things become so clear. I look back and I say, did I really do all that stuff?"

* * *

Contemporary problems and contemporary spiritual states require con-temporary transitions. Our generation has its own peculiar conflicts to face, its own ministries to initiate, and its own basic spiritual needs to fill. A literal interpretation or replication of Christian wilderness traditions can never serve us completely. Yet the traditions do provide precedents and suggest certain paths may be more fruitful than others. They also provide a series of alternate approaches to resolving specific spiritual problems.

The origins of the Hebrew wilderness encounter with God were in a nomadic pastoral culture very different from our own. Not only have the level of technological advance and the transportation and communication systems changed since the time of Abraham, but the need for revelation has changed also. God's character and will as displayed in the Law, the writings of the prophets, and the history of Israel are available for study. Christ has come. We can investigate the traditions—if we frame general inquiries and avoid trying to repeat history that has already fulfilled its divinely ordained purpose. The traditions cannot tell us when to seek the wilderness, nor can they tell us exactly what to do if we enter it. They can, however, provide time-tested patterns of wilderness practice and basic wisdom concerning the value of the desert.

A Venerable Tradition

The Christian tradition of wilderness spirituality is extremely diverse. Spanning five thousand years and numerous geographic locations and cultures, it began with desert herders—before Israel was a nation—and continued through empires, exiles, and dispersion. Influenced by prophecy in the divided kingdoms, the tradition anticipated the coming of the Messiah and the establishment of a new order and, eventually, set the stage for the New Testament. Embraced in the desert, the wilderness tradition followed the spread of Christianity into the forests and caves of northern Europe where it flourished and grew in the great blossoming of medieval monasticism. Wilderness sojourns have been despised and honored, fought and zealously pursued. Out of wilderness journeys have come promises and covenants, the Law and the tabernacle, poetry and theology, visions and revelation, and new ministries and the New Kingdom. Faced with the centrality of wilderness in Christian tradition, we must ask: "Has it any relevance for us today?" And, if so, what inhibits us from entering the wilderness?

There can be little doubt we seem to have trouble developing an approach to wilderness spirituality that is both truly Christian and suitable to our time and culture. The problem does not lie in contemporary disinterest in nature—we are very enthusiastic about the environment, whether or not we care for it properly. As suggested in the previous chapter, a more likely source of difficulties is our unwillingness to cope with spirituality in a disciplined way. The wilderness sojourns of the Bible are exactly what modern culture does not do well. The traditional Christian wilderness experience is an intense confrontation with God. Visionary and other-worldly in substance, the material and the cultural dissolve in the fire of divine splendor. The voice of God, roaring from the mountain or whispering through the depths of the soul, overwhelms all other communications. Since the Enlightenment, western theology has increasingly forced this type of encounter into the category of the mythological or of the imagined. The desert travels of Hagar or Elijah become difficult to understand because industrialized humankind does not expect to participate personally in these sorts of adventures—except in a symbolic sense. The Christian wilderness heritage is, thus, the portion of the Christian spiritual heritage we have the greatest tendency to reject as primitive, contrived, or impractical.

Western emphasis on scientific inquiry and on material knowledge of structure and function teaches us not to ask certain kinds of questions and not to see or hear certain kinds of things. The seeker does not jump the fences of the material. Consensus and mutual verification define truth. From the time of the patriarchs onward, Hebrew wilderness experience had no such limitations—the questions asked were as broad as the range of spiritual concerns, and God sometimes presented answers when no inquiry had been made. The continued presence and blessing of God and the response of the wilderness itself marked the holy nature of an encounter or communication. We technocrats walk out into the wilderness and fall immediately back to the comprehensible. We would never, like Jacob, wrestle with a man whose name we do not know.

Social convention limits contemporary wilderness experience to communing with nature. There is of course nothing wrong with contemplating the beauty of the mountains or appreciating the vastness of the oceans. Unfortunately, we gaze up at the clouds from our mechanical world and suspect the heavens may hide something holy, but we find ourselves unable to look beyond and see God in a flaming chariot descending on the towering, desert peak. Our rational infatuation with space and time has become a shoddy substitute for deeper perception of the essence of creation.

Systematic theology is still the favored technique of western biblical interpreters. Tightly logical, in a Greek sense, and apt to abstract overall principles from scriptural texts, this analytical methodology tends to ignore or allegorize both physical and social environments. Through the centuries, academic distillation has divorced the wilderness spiritual experience from its setting. This is particularly true of a theology that holds the wilderness experience to be an interior conversion, caused by the stress of separation from God. A literal sojourn in an isolated place is thus no longer necessary. Any social, spiritual, or environmental crisis can initiate a "wilderness experience." Christians say they are "in the wilderness" when they are having professional conflicts on the job, or when their children are misbehaving.

The growing strength of biblical theology and of historical scholarship that purposefully investigates the environmental and social milieu of ancient writings should greatly aid contemporary understanding of Hebrew and Christian wilderness traditions. Before allegorizing or systematizing, the spiritual traveler should personally "walk through" the passages and join the children of Israel at fiery mountain, rejoice with Hagar at the spring, struggle to climb the slippery rocks with Jonathan, and pray with Christ in the lonely places at the break of day. One must also reach into the text to see the mighty God they saw and speak to the sovereign Lord to whom they spoke.

The biblical wilderness is indeed symbolic, but it had already acquired its special associations with the history of Israel when Isaiah wrote of the trees rejoicing or of John baptized in the Jordan. From the beginning, the mountain and the desert were places where the extraordinary transpired and this world met another dimension of being. Modern christendom has confined the divine so completely to church buildings and to organized meetings and groups that Christians can no longer relate to God the Creator who uses the entire Sinai for a stage and chooses a mountain as a platform for communication. "God in a box"—God restricted to the "religious" part of our lives—is not the God of the wilderness.

As the church has become increasingly institutionalized, many of the events that once occurred in the wilderness have been brought indoors. Spiritual transition has become ritualized in series of formal ceremonies, such as confirmation. Baptism is moved from the river to the marble font, the calling of disciples is shifted from the seaside to formal ordination services, and prayer is moved from the isolated place to the packed church pew. Many denominations emphasize encounters with Christ in organized community worship. Although the life of the community is critical to the health of the church and to the accomplishment of God's purposes, an *exclusive* emphasis on corporate or indoor activities (and often on more conversions and financial contributions) may discourage the development of wilderness strengths, such as deep insight into the character of the holy, dependence on God when under stress, and a discerning eye for idolatry in societal values.

In the west, a material culture has obtained so strong a hold on religious activity that entire denominations count success in numbers attracted, building programs consummated, and funds solicited. Caught in the pressures of becoming Christian achievers, individual believers cease to reflect on the real substance of their lives—or on the essence of their relationship with God. Biblical wilderness experience is associated with spiritual transition, acquisition of a deeper knowledge of God, strengthening for ministry, and liberation from cultural or personal bondage—all functions that are potentially threatening to material religiosity. Some contemporary reluctance to define Christian relationship to wilderness may, thus, arise from the deep fear that God might actually send the church back into the desert.

One of the more subtle reasons for contemporary difficulties in dealing with wilderness traditions is the assumption that the wilderness sojourn was restricted to the professional prophets, such as Moses and Elijah. This concept causes two problems. First, it makes wilderness experience appear exclusive and limited to those with a very specialized calling or a very unusual relationship with God. Second, since the Old Testament prophets seem to have disappeared, it makes wilderness appear to be an archaic site for spiritual activity. Since the time of Paul, prophecy has been considered a spiritual gift and the responsibility of the church as a whole. This does not mean, however, that the church does not need leaders who are moved by God to speak out. The role of the biblical prophet was far more than that of a colorful soothsayer. The prophet led the people out of Egypt, chastised the king, and inspired the people to repent. The prophet was the "soul" of the nation, crying out for compassion and righteousness before God. The church and individual Christians still have prophetic responsibility and may, in conjunction, still require a wilderness sojourn to prepare for the mission.

All of the biblical literature reviewed in this book suggests that the wilderness sojourner is able to "see" and "hear" in an exceptional sense. In the wilderness, both the individual and the community can perceive the divine—in many cases literally. The "seeing" and "hearing" are often associated with the comprehension of the being of God and an improved understanding of God's person. In the earlier literature, the usual vehicle for this understanding is the theophany and an announcement of the name or characteristics of God. In the later literature, the "still small voice" and contemplative exercise become increasingly important. The association of wilderness with the prophet or forthteller is thus natural since the prophet must hear or see before he or she speaks. The biblical visions in the wilderness go deeper, however—to the very person of God—and have never been limited to the religious professional.

As a place of exceptional perception, wilderness became a site of major revelation. This began with the promises to Hagar and Abraham and continued through the first prophetic call to Moses and the giving of the Law at Sinai to the coming of John the Baptist and the visions of the New Jerusalem. Many of the early revelations concerned promises and covenants, whereas the later ones were more likely to concern calling and ministry. Although the closing of the canon of scripture limited the importance of direct revelation from God after the end of the Apostolic Age, we can not assume the divine no longer desires to be seen or heard. The modern predisposition is to look for prophecy and revelation in written texts, such as the Bible, and not to seek personal encounter with God. This approach neglects the need for revelation to deal with individual and community decisions, matters of personal calling, difficult ethical questions, and even strategies for evangelization.

The perceptions developed in the wilderness have proven to be of unassessable value to the formation of the Christian faith. If we can accept the concept of revelation directly from God, or of revelation of the person of God, we are, however, still left with the questions, "Why would this happen so frequently in the wilderness?" "What is it that gives wilderness its spiritual importance in Christian tradition?" "What elements in the wilderness experience are critical?"

If we are to obtain a contemporary understanding of traditional patterns, we will have to determine what our predecessors valued and what dictated the routes for their journeys.

Solitude and Seeing

In analyzing the various components of the wilderness sojourns investigated in this book, the most important common denominator is solitude. The protagonists were separated from their family or their culture, or, at least, they were away from antagonists or ungodly influences. In the last three centuries, romanticism and transcendentalism have placed a great emphasis on the beauty in nature as a precursor to spiritual experience. The stories in Genesis contain no evidence, however, that the sites of divine encounters were in any way natural wonders or exceptionally beautiful. The spring Hagar found on the way to Shur was named for a property of holiness, something like "God sees," not for its scenic qualities, or even for the potability of its water. The wadi Jabbock does have a deep, spectacular canyon, but the Bible does not even mention this fact. The same pattern is found in the Exodus. It is the Divine Presence, the appearance of Yahweh, that makes the burning bush or Horeb remarkable, not some intrinsic natural quality.

Some of the patterns found in the Bible may be due to literary convention. Manifestations of the divine or visionary encounters with God are primarily found in the historic narratives, and descriptive adjectives for locations where these occur are rare in the historic texts, even in the New Testament. Christ went up on a "high mountain apart" for the Transfiguration. Matthew and Mark note the location's exceptional elevation and its isolation, and that is all they have to say about the site. Most of the discussions of beauty are found in Psalms and in the more poetic works of the prophets (i.e., Isaiah), which do not discuss literal wilderness sojourns. The beauty of creation tells us much about the character of God. There is, however, no evidence that the exceptional beauty of a wilderness location in any way initiates direct interaction with God or is related to the occurrence of theophanies.

One could argue that solitude was not available during the wilderness wanderings, the most extensive wilderness sojourn in the Bible. Since the spiritual experience concerned all the people of Israel, however, one could consider the entire nation to be alone in the wilderness, away from Egypt and other bad influences. Wilderness solitude most often concerns individuals, but it may also concern pairs (Abraham and Isaac), a small group (Christ and his disciples), or an entire chosen nation. In the Hebrew scriptures, spiritual principles that apply to individuals often apply to the nation, and vice versa.

What gives wilderness solitude its spiritual importance? And why solitude in wilderness, rather than somewhere else? These questions are more difficult to answer because the reason for solitude seems to have varied from case to case. Some overall patterns are discernible, however. First, in the case of visions and direct encounters with Yahweh or the otherworldly, only those who will understand the vision or who are directly concerned with the message are present. The theophany tends to be exclusive. Only Abraham and Isaac were on Moriah, only the children of Israel were present at Sinai, and only Christ and three disciples were present for the Transfiguration. In the case of theophanies in other types of locations, such as Isaiah or Zachariah in the temple seeing Yahweh and an angel respectively, the general pattern is still primarily the isolated visionary. The exceptions to the pattern are primarily from the New Testament and concern situations in which the presence of witnesses to verify the event may have been important. When the light forced Paul to the ground on Damascus Road, the presence of other travelers verified "the call" of the zealous persecutor. The New Testament also stresses evangelization, which may have encouraged the proliferation of group vision.

The Hebrew tradition is one of chosen individuals and nations. Literal seeing and hearing of God is reserved for a few select persons or groups at special places and times. The wilderness is not the only type of location where this occurs, but is the primary visionary landscape. The wild or untraveled territories exclude those not privy to the otherworldly communication and provide suitable space for interaction with the divine. In the early tradition, going back to the patriarchs and the Exodus, the theophanies (with the possible exception of Jacob's ladder) are always presented as real persons or presences, rather than just something seen in the mind's eye. The landscape thus had to be expansive enough to allow angels to walk by, springs to appear, wrestling matches to take place, and, most importantly, Yahweh to appear in glory.

Although the pattern does not apply to all wilderness sojourns, solitude provided freedom from cultural and spiritual antagonists. This begins with Hagar fleeing Sarah and Jacob trying to avoid Esau. Hagar's encounters with the divine would have been in a different context if they had been around Abraham's tents. Both Abraham and Sarah might have interfered or reacted unfavorably. After the patriarchs, the theme of freedom from antagonists takes several different directions. The children of Israel are physically protected from Pharaoh by the Reed Sea and the desert, and they are also freed from the fleshpots of Egypt (whether they wanted to be or not). The wilderness forces the people to cleanse themselves of inappropriate behaviors and to push the idolatry out of their culture. Part of gaining their identity as a nation is the recognition that they are no longer part of Pharaoh's empire. The concept that wilderness separates the sojourner from evil cultural influences and aids in eliminating sin continues through the later prophets, the coming of Christ, and the development of Christian monasticism. John the Baptist used the wilderness to forward ascetic practice and called the people to come to the Jordan and repent of their sins. The wilderness offered lack of interference by the Romans or the Jewish religious leaders. Hosea portrayed the time in the wilderness as a time for Israel to leave false gods and return to her true lover. Isaiah saw the wilderness as the road to freedom from captivity. Wilderness both removes material temptations and false gods and provides time for reflection. One can look back on "the nation" from a distance and contemplate what belongs to God and what does not.

In the stories of David and Jonathan and in the case of Elijah fleeing Jezebel, the wilderness provided literal protection from enemies in sincere pursuit. The wilderness did not free David from sin as much as it freed him from Saul. Lying subtly beneath this theme are the notions that the God-dependent and the faithful can survive better than the unfaithful in the difficult environment, and that creation itself favors the chosen of God. The physically protective aspects of wilderness do not directly precipitate theophanies or the more spectacular encounters with the divine. They are related, however, to the development of a heart knowledge of God, when the Lord is recognized as the ultimate source of the relief from danger.

Also associated with freedom from daily pressures and antagonists is the opportunity for rest. This theme appears to be a relatively late concept and is better developed in the New Testament than in the Hebrew scriptures. In the Gospels, Christ voluntarily entered the desert or solitary places for short periods of preparation, rest, and prayer. This new direction in wilderness spirituality was more a function of Christ's style of ministry than a conscious effort at creating a new form of wilderness renewal. The Gospel writers, particularly Mark, emphasize Christ's desire to withdraw and, in doing so, to portray the human side of his nature. Christ, more than any of the Old Testament prophets, was beset by his followers. No prophet had ever preached and healed the way Christ did. No prophet had ever had such a need to retreat from the crowds. Christ often stayed at other people's homes while traveling and was usually accompanied by his disciples. Lacking personal space while on his journeys and constantly sought by the sick, he had to remove himself from the towns just to obtain quiet. He couldn't just take a stroll to the edge of a village and expect to be left alone.

During the Roman period, Palestine had a greater population than at any other time in its history—except for the modern period. The land was under intensive agricultural management. The roads were heavily used by merchants and military personnel. On one hand, getting away from other people for contemplative prayer was much more difficult than it had been during earlier eras. On the other hand, owing to the well-developed road network, getting to the wilderness was not as difficult as in the days of Elijah, and news of what was happening in the wilderness traveled faster than it once had. The social situation was, in many ways, very modern, and it is not surprising that Christ chose to take short "vacations" for prayer and rest.

In the case of spiritual rest in wild nature, it is again solitude and not beauty that appears to be the primary attraction—at least for those in ministry. The historic biblical texts mention isolation or the distance traveled, but do not mention the aesthetic qualities of the sites chosen for retiring. One can not neglect, however, the potential value of attractive or quiet settings for spiritual restoration. Healing or regeneration with wild nature as a site or mediator seems to be primarily expressed in the poetic literature and, thus, is more strongly associated with characteristics of the Creator expressed through the creation than with direct encounter with the divine.

The Bible dedicates much less text to the potential "healing" or "restorative" properties of wilderness than the modern reader might expect. There are three possible reasons for this. First, the psalms and other poetic writings usually employ a direct call for help directed to a personal God. A complex system of intermediary agents is unnecessary. The Bible uses natural settings as spiritual sites, but the setting itself is never the core element in the spiritual interaction. Second, the Hebrew scriptures are products of a rural and very pastoral environment. The spatial confinement of postindustrial, urbanized society was not present. The European Romantic poets, who did a great deal to forward our modern wilderness ideal, found themselves seeking the simple country life and wandering the fields and fells to get away from crowded, mercantile centers. David, in contrast, hardly found looking after sheep a withdrawal from uncomfortable environments; shepherding was, instead, the center of his cultural experience. The ancient Hebrew spent a great deal of time outdoors (the women as well as the men worked in the fields), and did not need to seek the healing properties of fresh air or exercise. By New Testament times, personal rest and restoration in wild settings was beginning to gain importance, but the Hellenistic environment was still very rural compared to our contemporary milieu. Today, we may have more need to consciously seek association with other living creatures and with environments with little human influence.

The relationship between solitude and "seeing" and "hearing" may not be immediately evident, but the removal of the human concerns is apparently important to deeper spiritual perception. Perhaps God cannot be heard over day-to-day conflicts and cannot be seen through a curtain of household responsibilities. Even in the wilderness, Egyptian culture proved a substantial block to Israel's perceptions of divine will. The desert is free of human structures and human territories, and the wilderness peak does not support massive networks of human communication lines. If nothing else, Israel could perceive, in the absence of palaces and princes, how deeply rooted Egyptian ways had become.

Lest we treat wilderness as a neutral platform for otherworldly action or as a barren stage with no props or backdrop, we need to remember nature itself was a player in many of the wilderness sojourns. God moved repeatedly through creation to demonstrate the divine will or to respond to the needs of the protagonists in the biblical narratives. The response of creation was integral to construction of a holy history, from the appearance of a spring, to the bogging of the chariot wheels, to the darkness on the day of the crucifixion. The biblical writers portray wild nature as very much in God's hands. Isaiah's use of wilderness imagery to describe the return from exile is far more than artistic license. The return of the chosen people to the Promised Land is of universal importance, and creation responds joyously to this mighty act of God because creation will share in the blessing. If the removal of human clutter pulls the veil away from the relationship between humankind and God, it also allows us to perceive clearly how the divine undergrids our relationship with creation. Creation itself may prompt some of the "seeing," not so much through its natural virtues as through its continuing interaction with Yahweh.

Stress and Sweet Water

In the company of solitude, the Hebrew wilderness experience is often marked by stress. The troubles may be environmental, such as lack of water, but they may also come directly from interaction with the otherworldly, such as Abraham's conviction he would have to sacrifice his son at Moriah, Jacob's struggle with the angelic figure at the ford, or Moses refusing to commit himself at the burning bush. Wilderness sojourners may bring their problems with them, like Elijah sitting under the broom tree, or may find human troubles chasing them, like David running from Saul. Stress encourages "seeing" and "hearing" because it forces the sojourner to turn to God. The God-dependence generated by wilderness trials can be physical, such as the children of Israel gathering the manna, or it can evolve into social responses, such as the courage displayed by David and Jonathan. The stress produces not only a knowledge of the person of God, but also a knowledge of the self in relationship to God.

The theme of stress is counterbalanced by the constant operation of God's providence during the wilderness journeys. Since this providence often takes the form of physical provision, one must avoid viewing God's action as primarily material and recognize that the sudden flow of water or the unexpected arrival of an angel always point to the more basic purposes and interests of Yahweh. Hagar wandered in the desert, nearly died of thirst, and received, with a life-saving spring, a new understanding of the person of God. God let the children of Israel wander in the arid lands of Sinai, and, with the manna, they received a national calling and the faith to walk into the Promised Land. David proved himself against the lion and the bear and inherited not only the Lord's protection and an earthly kingdom, but also a heart crying for God. John prepared himself for a short but very critical ministry by subsisting on wild foods, and he fulfilled his destiny not only by baptizing the people in the River Jordan, but also by seeing the long-awaited Messiah. In dozens of instances where people were suffering or struggling in the wilderness, God made miraculous provision or intervened to remove physical threats. Yahweh sweetened the waters of Marah, helped Jonathan over the crags, and sent hot cakes for Elijah. Christ provided bread and fish for the crowd and stilled the storm. In each case there was also a new revelation—a critical personal vision or a message to the nation that would determine the course of holy history.

The human relationship to God is reflected in the human relationship to the wilderness environment. Directed by God, wild nature supports or assists the faithful and destroys the unbelieving or apostate. This may occur on an individual basis, such as Daniel in the lions' den, on a community basis, such as the environmental blessings and hardships of the wilderness wanderings, or on an international or universal basis, such as Isaiah's prophecies of the response of wild nature to the righteousness of the shoot of Jesse. Most of the biblical examples of this relationship in the historic literature describe individual incidents—people do something offensive to or blessed by God, and creation responds in a single event. The later prophets, however, project peace with wild nature as a continuing state. The relief of stress in the wilderness is thus converted to a universal condition, and wild nature loses its role as a site for a temporary sojourn and becomes a part of the New Kingdom.

One of the interesting patterns found in the wilderness motifs throughout the biblical literature is the variation in the way the environment is portrayed (sometimes as very harsh, sometimes as very helpful). Despite the clear dangers of the drier deserts, however, no one who is faithful to God ever is reported as dying from exposure or thirst. The negative aspects of the wilderness—its lack of water, its impassable terrain—are used to demonstrate either the power of God or the holiness of God's people. The faithful or anointed may be led out for testing and, like Hagar and Ishmael, come close to death, but in the end, God never fails them. Another spring always appears. The biblical writers probably knew of people who had died in the desert, but they only report mortality of those who disobeyed, such as the children of Israel who ate the quails during the exodus, the disobedient prophet killed by a lion, and the boys who taunted Elisha. The wilderness never operates as a random source of death, independent of the victim's status before God. The wilderness can be a threatening environment, but it is always an environment under God's control.

Walking to the River

A third key element in wilderness spiritual experience is the possibility for spiritual transition or transformation; Hagar went into the wilderness as an outcast and came back with promises. Jacob in crossing the wadi gained a new name and, therefore, a new spiritual identity. The children of Israel entered the wilderness as a group of disorganized slaves and came out with the Law, a system of worship, and a national identity. David served in the wilderness as a shepherd boy and emerged as a king and a man of God. Elijah fled into the Negeb ready to die and returned to pursue a new mission. John called the people to the wilderness to repent and sent them up from the Jordan ready to receive the Christ. The eunuch drove his chariot through the desert and discovered Jesus. The writings of Hosea and Isaiah, as well as John's baptism, presents the wilderness as a place of renewal where one can return to the true love and service of God.

Transformation grows out of the other elements of the wilderness experience. Isolation encourages introspection and self-evaluation, as well as an accurate perception of the holy. If someone begins to "see" and "hear" God, he or she will begin to understand who God is, how God works, and what God desires. As a result that person will move towards God and, in the process, begin to change. If the person or community is under stress, they are more likely to turn to God for help or to appreciate God's provision. If they are freed from negative cultural influences, they are more likely to respond to a divine call or to execute the Lord's will. Removal of material temptation accelerates the pursuit of the holy. Encounters with God, the experience of personal revelation, coping with difficulties, and an understanding of creation—all key elements in traditional wilderness experience—all foster spiritual growth and change.

The very process of going into the wilderness and returning may aid transformation and renewal. The wilderness sojourner temporarily breaks cultural and social ties. This allows a period of adjustment and a new start on the return from the "nonhuman" or "spiritual" environment. The journey to the Jordan for baptism, for example, caused a symbolic break with the ruling powers and the Jewish social order and forced John's followers to leave their homes for a short period. If John had baptized in front of the temple, the people would not have had to exert as much effort to get to the site and would have been conforming symbolically to the existing cult. The time taken to reach the wilderness in ancient times may also have helped nourish transition. The longer sojourns, such as Elijah's trip to Horeb, allowed time to reflect on the internal and external roots of the problems or to examine one's own spiritual state. Such trips also provided an extensive or open ended period for interaction with God and a lengthy return journey for thinking through the implementation of new missions.

Transformation is so much a characteristic of wilderness experience that Isaiah suggests that the wilderness itself may ultimately be renewed. As the people are freed and a new nation arises, Yahweh will reshape the rejoicing desert landscape. Transformation of nature through God's power is a very foreign idea to the contemporary mind, yet, when we think of a heavily mined, carelessly logged over, or otherwise degraded landscape, the possibility of spiritual regeneration may appear less an ideal and more a blessed necessity. A deepening of our knowledge of the prophets should, in fact, help us to understand how unholy our present abuse of the creation is.

Drinking from the Spring

Now, what of beauty—the modern springboard for seeking the divine in nature? The Bible leaves no doubt that contemplation of nature can provide improved insight into the person of God. This type of exercise is found primarily in the poetic literature and is not directly tied to the descriptions of wilderness sojourns. The Bible reaches beyond scenic grandeur, into diversity, order, and power. Wild nature not only instructs people in God's role as Creator, it also demonstrates God's majesty, love, joy in creation, righteousness, transcendence, and omnipotence. Wild nature can also say something to humans about their relationship to God; we can experience God's deep love for the creation in God's continuing interaction with all that is living. Unfortunately, our contemporary taste for the grand or dramatic and a tendency to limit our observation of the diminutive or day-to-day to scientific study fall short of the psalmist's vision and the Song of Songs insight.

As previously mentioned, literary convention may play a role in the separation of the scriptural wilderness sojourns from discussions of beauty. The two may have begun as different exercises. Wilderness experience in the Bible is marked by direct visions of or revelations from God. Gleaning evidence of God's person through observation of creation is a more indirect means of determining God's character. The latter can be accomplished almost anywhere and does not require an audible answer from Yahweh. Nor, for most of us, does appreciation of nature as God's handiwork require withdrawal. The observation of creation is not a source of the personal sorts of revelation, such promises, covenants, or the application of a new name, on which most of the very historically oriented Judeo-Christian tradition is based. Just to be certain Elijah understood this, God made him wait through all the wonders of nature to hear the little breeze.

The disassociation of natural qualities, such as beauty, from the occurrence of theophanies also springs from the Hebrew concept of the supreme being as transcendent. Theophanies are associated with particular landscape features—especially mountains and desert wilderness. But only the elevation, isolation, or expansiveness of the landscape is important; the mountain can be tree covered or treeless, it can be granite or slate. There is nothing about the mountain itself that makes it holy. Neither the mountain's towering ridges, nor its obsidian cliffs produce the spiritual or attract the divine. Once the Lord of Israel appears, the mountain itself becomes insignificant. The biblical split between the poetic descriptions of nature which concern God's qualities or actions reflected in nature, and the sparse, object-oriented discussions of the sites for theophanies, thus, lies partially in the concept of the Creator above and separate from the creation.

The later prophets do recombine the natural imagery of the psalms with the basic prophetic wilderness experience to produce eschatological images of the desert. The result is the idealized wilderness of Hosea and the personified wilderness of Isaiah. In the Bible, this never fully evolves into a simultaneous combination of aesthetic, historic, and visionary experience. Even in Revelation, the beauty remains in the vision, and a couple of adjectives are expended describing the mountain on which the visionary stands.

The historical passages in the Bible describing theophanies not only avoid descriptive adjectives and aesthetics, they also can be vague geographically. This style of writing deemphasizes specific sites as the source of holiness and draws attention to the actions of Yahweh. There can be no doubt the authors of the Bible were dealing with real places—in many cases well-known to the readers or hearers of the text—yet, with the exception of the temple in Jerusalem and a few other scattered priestly sanctuaries (e.g., Bethel), the Bible does not direct worship towards the locations of theophanies or mighty acts of God. Horeb, for all its importance in Hebrew history, is never mentioned as a point of repeated pilgrimage. Most of the passages providing aesthetic descriptions of nature do not mention particular sites, and exceptions, such as the appearance of Hermon and Gilead in the Song of Songs, tend to be free of direct ties with the divine. The separation of material concerning actual theophanies from environmental poetry helps to prevent nature worship or development of shrines tied primarily to natural features and keeps the biblical reader from confusing the spiritual role of the Creator with that of the creation.

In the monastic literatures, we find an increasing tendency to combine the poetic with the experiential. The desert monks certainly associated natural beauty with their residences, and Celtic monasticism strongly associated the beauty of a location for a hermitage or a church with its spiritual value. Although they were still struggling with the old religion, the Celts adapted pre-Christian forms of expression to their Christian mission. The resulting religious poetry, employing extensive natural imagery, is surprisingly orthodox and relatively free of animism. The hagiographies retain shape-changing and "power over nature" motifs, but these do not seem to deify nature, and, if considered offensive to today's reader, it would be more for their claims of extraordinary spiritual power in the hands of the saints than for any presentation of the natural world as divine.

From Thirst to Rest

Although Old and New Testament patterns of wilderness spiritual experience are very similar, the wilderness experience has changed through time, increasing in the diversity of elements and in the variety of possible outcomes. Some types of practice, such as wilderness asceticism, have slowly evolved in the Hebrew-Christian milieu, while others, such as the association of certain natural sites with "seeing" and "hearing," may go back to pre-Hebrew cultures. The early Christian period, in particular, brought changes in emphasis.

In the foundational model of Genesis, the major characters in the stories were not expecting to see God or an angel in the wilderness. By New Testament times, although there were still "accidental" encounters, going into the desert or biblical wilderness had become part of purposeful efforts to induce spiritual change, either in the individual or in the nation. Both Christ and John the Baptist sought out the wilderness or are led or driven by the Spirit. The theme of a new exodus arose among the Old Testament prophets, but seems to have affected personal spiritual practice to a greater extent after the time of Alexander the Great, perhaps because of a continuing spiritual and moral crisis caused by the invaders' social and legal interference in Jewish affairs. The ideal of wilderness experience that would play a role in the life of the Messiah appeared in Judaism well before the birth of Christ, but it wasn't until the Hellenistic period that purposeful withdrawal to the wilderness became "fashionable."

The Old Testament prophets spent time sitting under trees, and the Bible portrays them as partially withdrawn from the mainstream of the culture. The "schools of the prophets" roamed the countryside, but do not appear to have resided in the deep desert, except when under duress. Hair shirts and odd diets were Old Testament developments, but strict dietary control had more of an early association with Nazerites than with prophets and only became the mark of the holy man after the time of the later prophets. Asceticism thus developed slowly. Beginning in the idea that the holy were "different" and free of worldly concerns and desires, ascetic withdrawal finally became ordered and institutionalized as monasticism in post-Biblical times.

The earliest prophets were associated not so much with distant wilderness, but with natural sites or objects. This links them to the visionary patterns seen in Genesis where Hagar and the patriarchs visit sanctuary or visionary sites and experience a direct encounter with the divine. Preference for specific types of sites spans the entire of Christian history (mountains have had continuing popularity as visionary locations), yet by the time of Christ, a tree, a spring, or a peak was no longer necessary. An "isolated place" or the "wilderness" in general had acquired visionary properties.

The first wilderness journeys were stressful. The concept of wilderness as a place of refuge preceded that of wilderness as a place of rest. Christ's battle with Satan and a series of short breaks from ministry in the isolated places represented a major change from Old Testament patterns, both in the breadth of wilderness experience (from intensely stressful to offering relief from stress) and in the alternation of theophanies and demonic encounters with routine prayer practice. The demonic first appears as a major concern in the New Testament, but, rather than displacing the angelic, the dark side appears in biblical books that also present the more holy messengers, while both good and evil manifest themselves in the traditional visionary location—the wilderness.

The Edenic, the Isaianic

Weaving through the Christian wilderness tradition is the ideal of returning to Eden. This theme is obvious in the prophecies of peace between humankind and the carnivorous beasts and the vision of the tree of life in the New Jerusalem. But a more perfect relationship with wild nature is not the primary motif of the scriptural literature. If residing in the wilderness recovers something critical lost by Adam and Eve, the restored condition is uninhibited communication with God. The God who moves "at the time of the day breeze" speaks to Adam, and Adam hears. The God of the Garden passes in front of Eve, and Eve sees. The quality of the human relationship with creation is always contingent on the quality of the relationship with the divine. Thus, when God speaks and humankind hears and obeys, the response of creation will be in harmony with human action.

15

Mountain Crests, Desert Canyons
(Contemporary Questions)

Great Falls

"How cold do you think it will be on the towpath?" I sat contemplating two pairs of socks, while my hiking companion was trying to telephone his parents.

" I have an extra flannel shirt if you need it," Bob offered as he leaned away from the receiver for a few seconds and then dialed for the third time. "The weather should be warmer this afternoon."

"The snow's not that deep," I remarked. "But I doubt it will be above freezing. How long do you think we will stay out?"

My friend didn't answer me. Looking concerned, he hung up. "No one at home. They should be there now . . . I'll try to call my cousins." His father had been seriously ill, and his mother, who was also not in good health, had been staying home to care for her ailing spouse. Bob had often lauded his mother's dedication. He admired her selflessness in the face of her own limited mobility and, at the same time, was concerned the stress might be too much.

Bob bowed his head slightly, as if thinking about what he should do next, and then dialed another long-distance number.

"My brother should arrive at the airport by twelve thirty," he remarked.

I already knew his brother was flying south to visit his parents and suspected the repeat announcement was for reassurance rather than for information.

Having reached his cousin, my friend suddenly turned to speak in to the phone. "Where is he now?" I heard him ask. Bob's father had suffered an angina attack and, having temporarily lost consciousness, had been taken to the hospital.

* * *

Bob opened the trunk of the car, and I reached for my down-filled parka. We had been late starting owing to the extended series of phone calls necessary to determine his father's condition. The hospital had given a favorable report, so we had decided to go ahead on a short expedition. The situation was awkward for me, since, if I had not suggested the hike, Bob would probably have flown south this weekend himself. Now he was miles away from his family, unable to help.

The hazy sun had warmed the snow. The toes of our boots pushed long trails in the slush as we strolled across the parking area towards the line of trees along the river. Bob almost always walked faster than I did, but today he was subdued, maintaining a slow, even pace. "How did you feel when your father died?" he asked, as we reached the foot path.

In any other setting, at any other time, that abrupt inquiry would have threatened me. Looking out over a frozen lock on the old canal and across a field where the bent, dried stems of last summer's flowering herbs poked up through the snow, I accepted the question. "It was sudden, and at first I was very upset. Then, in the end, I felt very settled about it," I replied, carefully choosing the thoughts to express. "The two of us didn't always get along well, especially when I was in high school." The naked trees standing on either side of the path seemed nonjudgemental as I confessed to failed and fractured relationships. "He was always busy working and didn't have much time for us kids. When I was older I could appreciate him better. Sometimes we even talked about Christianity. . . ."

I watched the small muddy patches in the trail break the pattern of snow blanketed leaves beneath my feet. "Christ healed a lot of it . . . I don't know what I would have done if my father had died and I hadn't tried to forgive him. . . ." The Potomac roared along beside us, muting my commentary, making it more private, keeping the discussion from other travelers on the trail.

Bob said little. He asked a question here and there, listening, then probing gently. I talked about my parents divorce, my brother's death, everything lost and broken.

"Were you ever seriously ill?" I asked.

No, neither of us had ever been in the hospital for more than outpatient treatment.

We discussed how it felt to be confined and incapacitated, how his father hated being helpless, and how sad it was to see someone slowly lose his strength. I could perceive the small branches breaking under my feet, and the well-trampled snow packing and falling away from the soles of my boots. I could not hear the sound of our footsteps over the roar of the river.

"Your father has served Christ all his adult life. He's been faithful. Now that your father is ill and unhappy, do you think God will be faithful to him?" Bob mulled over the question for a few paces. We passed a large white trunked sycamore, its shining branches arching high over the trail. "If God is faithful to anyone, he will be faithful to my father," he answered.

Stepping down cautiously among the icy rocks, we reached Great Falls. Winter rains and slowly melting snow had filled the channels with foaming brown water. Captivated by the sheer force of the torrent and the size of the eddies at the base of each spray covered ledge, I stood motionless at the edge of the gorge and watched the river surge downwards. Bob nudged me. Did I want to walk further along the path?

We traveled on without a goal and without much conversation. I became distracted by the form of the boulders and the vegetation of the floodplain. My companion said little and seemed to be contemplating things far away from the freezing Potomac. The sky clouded over, and a few flakes of snow fell.

We started back well before we were tired and made a return trip to the falls. I lay down on a cold gray outcrop and found a comfortable position from which I could observe the turbid current pour around the boulders and plunge uncontrolled through the narrow rock chutes. Bob stood behind me and, unable to settle into river watching himself, allowed me a few minutes of appreciative pleasure.

"I need to make a phone call," he said finally.

* * *

Having confirmed that his father was resting quietly and that his brother was with his mother at the hospital, my hiking companion invited me out to dinner.

"Things are difficult right now—your father is very ill, your mother is under terrific stress. I think we ought to pray for awhile before going anywhere," I suggested.

"Pray?" said Bob quietly as if he were talking to himself, "I've been praying all afternoon."

He put a fire in the fireplace, and we finished the day by offering all our cares and concerns to an all-knowing, all-loving God. His father was released from the hospital after an overnight stay. Bob flew south the next week to check on his parents. His father returned to the hospital during the visit and died while his son was home to care for him.

* * *

Bob and I had not planned to go hiking because we were expecting to move into personal spiritual confrontation with sickness and death that afternoon. Once the crisis materialized, however, the quiet of the tow path and our freedom to do whatever we wished for the day gave us the chance not only to seriously pray, but also to slowly wrestle with those deep questions concerning pain and the end of life we tend to put off asking. "Getting away" to a quiet locale—even if it was just for a few hours—helped us to grasp the reality of what was happening and place it firmly back in God's hands. For once in our rushed and harried lives, we had an environment and a schedule free of social concerns when we needed it. One wonders how often Christians avoid going into the lonely places when that is exactly where they will best be able to "see."

Coincidence or Calling?

In order to determine what we should or should not be doing in the wilderness or the lonely places, we must dispel some illusions or misconceptions about wilderness experience in a Christian context. These misconceptions have, in fact, limited modern pursuit of wilderness spiritual experience.

The first misconception: Wilderness experience is for prophets or those with a prophetic calling. Wilderness is associated strongly with prophetic ministry, but it is certainly not exclusively prophetic. The biblical literature reports the wilderness adventures of people such as Hagar and Jonathan, who could hardly be considered prophets. Biblical models of wilderness use are very diverse and include the foundational (Genesis), leadership (Davidic), and poetic models, which may sometimes stimulate prophetic action, but do not occur to train prophets *per se.*

The second misconception: Wilderness is always a vocation requiring a lifelong commitment. This is a post-Biblical idea and is integrally tied to the way one views calling and the role of tradition in the church. Orthodox and Roman Catholic theology, which consider the patterns established by the early church to be continuing precedents, accept the wilderness vocation in the form of organized monasticism and withdrawl from the world. Protestantism, in general, does not accept a lifetime of withdrawal for prayer and spiritual exercise and emphasizes continuing integration into the denominational community. Even if one accepts wilderness as a legitimate vocation, however, there is no reason to confine wilderness experience to those who are willing to enter the desert and not return. Many of the biblical wilderness sojourns were short—a few hours or days. Among the major scriptural figures, only John the Baptist seems to have spent a year or more in ascetic withdrawal.

The third misconception: It is necessary to stay in the wilderness for a long period, such as forty days, to develop a wilderness spirituality. As mentioned above and in chapter 8, shorter periods, such as a day or two, may be valuable for rest and prayer. Long periods may indeed be necessary for some purposes, such as removing cultural influences, preparing for ministry, or developing skills in more serious spiritual exercises. The Bible presents a wide spectrum of wilderness experience—from a few hours to forty years. The duration of a stay is determined by individual needs and by God's purposes.

The fourth misconception: Withdrawal to the wilderness requires a special calling. In the Bible, some of the people who had spiritual experiences in the wilderness did not enter the desert expecting to see God. Christ's use of the lonely places for rest appears to have been regular practice and not a matter of being intensely "driven by the Spirit," as he was during his initial forty days in the wilderness. The poetic, leadership, and portions of the contemplative model can all be pursued without some special form of calling. The foundational model includes elements of surprise, and, although Hagar and Jacob were in one sense called, they did not enter the desert or attempt to cross the wadi because they felt driven by God to seek a visionary experience.

The fifth misconception: Wilderness experience is exclusively a form of conversion experience or marks the passage from spiritual immaturity to maturity. Wilderness experience is often formative and is very much associated with transition. These transitions are not always at the beginning of one's spiritual life, however, and may involve the very mature. On one hand, we find the boy David developing strength by fighting the lion and the bear, the Hebrew people turning from a disorganized company of slaves into a nation with a law and a priesthood, and Christ entering the wilderness before assuming ministry. These cases are all related to initiation or transition to mature roles. On the other hand, however, we find the already very faithful Abraham taking Isaac to Moriah and encounter Elijah, who had already defeated the prophets of Baal, sitting under the broom tree—both experiences in the middle of well-developed walks with God.

With the exception of wilderness baptism, biblical wilderness experience often concerns the knowledgeable, the specially called, or the mature. The spectrum of wilderness experience ranges from conversion experiences to repeated wilderness rest as a balanced element in a mature ministry. The association of wilderness with revelation, in itself, suggests wilderness experience is often for the better-prepared. Confusion over the formative role of wilderness may result from the Protestant tendency to see conversion as the key experience and not to expect further major transitions to take place. Conversion is often viewed as dramatic change, while building maturity is often thought of as slow growth. Mature experience can be dramatic or life changing (as with Elijah), and wilderness can be a continuing theme in one's spiritual journey (Moses or Christ).

Many cultures cultivate an isolated wilderness experience as an introduction to adulthood. The "walkabout" of the Australian aborigines precedes assuming an adult role, as do some of the isolated rites of passage of native Americans. These experiences often have visionary elements, such as prophetic dreams. Hebrew and Christian traditions have never required a period of isolated contemplation for the average believer, although wilderness sojourns have, at times, been required of the religious professional. In the Hebrew traditions, wilderness experience is very scattered in terms of life phase. It incorporates children and adolescents, young adults, people with families, and those nearing the end of their active life. The ancient Israelites thought a person of any age could respond to God, and God might anoint or call the very young (David or the prophet Samuel). In the Bible, wilderness experience has never occupied a set place or taken a set role in personal history. Although it is most common prior to the initiation of a ministry, it does not occur consistently at some specific point of the individual spiritual journey. Biblical wilderness experience occurs as a response to individual or national need for revelation, or forms a continuing pattern of sojourns for strengthening or rest (as in the contemplative, poetic, and leadership models).

The sixth misconception: Wilderness spiritual experience requires repeated participation. Again the tradition varies widely from repeated use of wilderness in the context of well-developed asceticism to one-time stays with no purposeful spiritual exercise. Certainly prayer and fasting improve with structured practice, and our relationship with God deepens the more we pursue it. A single well-placed wilderness sojourn may be effective for some purposes, however. The most intense types of wilderness experience, such as meeting Yahweh in a dream, are rare events that do not recur multiple times in one person's life. Each visionary wilderness sojourn is thus distinctive. Severe need may precipitate a wilderness trip to meet a crisis. Elijah had visited the wilderness prior to his flight to Horeb, but his journey to the mount was a distinctive event, not something he did annually. Someone who is preparing for ministry or is in the midst of a major personal battle might spend a relatively long period in the wilderness and never repeat the lengthy sojourn after the problem has been resolved or the transition has been completed.

The seventh misconception: Christian wilderness tradition sees the wilderness environment as negative or antagonistic. The wilderness often is a stressful place, but this is a spiritually beneficial characteristic. Wild nature is always in the hands of God and responds to God's will. The wilderness is not a competitor or enemy, but rather an aid and inspiration to the sojourner. The tradition expects the wild to be friendly to the person executing God's desires and to the individual who is pursuing holiness. The righteous may be tired, thirsty, or wet, but the prophets expected the person who truly loved and followed God to find rest, receive water, and survive the storm. The wild beasts would not harm the anointed of the Lord.

The eighth misconception: Christian wilderness experience is very individual and has nothing to do with personal function in the Christian community as a whole. The isolated ascetic, purifying himself or herself of sin while ignoring the needs of others and of the church, is always a possibility. Several of the wilderness models, however, are specifically tied to developing leadership for the people of God and for acquiring the clarity of vision necessary to discern God's desires. The leadership, prophetic, and ascetic models all have elements which strengthen those responsible for religious, military, or governmental activities. The development of God-dependence and a heart for God are keys to the leadership model. The prophetic model fosters a new understanding of the self relative to God's will for the people. Time and time again, the wilderness sojourner returns with a new vision, a new mission, or a new ministry. It is also worth noting that some of the most isolated desert monks had a tremendous impact on the spread of Christianity, and they always assumed they were part of a central community—the New Kingdom!

The ninth misconception: Wilderness experience interferes with minis-try. In investigating the biblical models of wilderness use, there is no evidence that wilderness sojourns interfere with ministry. Prophets such as Elijah and Jonah were, in some cases, fleeing responsibilities when they entered the wilderness, but the wilderness experience always returned them to their task for God. Often overlooked is the interdependence between the wilderness sojourns and calls to service. The vision in the wilderness may have a tremendous impact on ministry, while the conflicts and pressures of ministry may cause withdrawal to the wilderness. Christ's intensive involvement with the people precipitated the need to seek the lonely places for prayer. John the Baptist's call to cleanse the people of their sins kept him in the wilderness in preparation. Elijah's confrontations with the royal family forced him back under the broom tree to consider his own relationship to God. Moses received his call at the burning bush and returned to Egypt to confront pharaoh. The Spirit drove Christ into the wilderness to prepare for trials waiting in Jerusalem. Francis found the strength to embrace lepers as he stood among the dark firs.

The wilderness experience has never been separate from ministry and calling, but has always been integral to them. This integration is, in fact, the best test of the Christian wilderness experience—if it brings the individual closer to God, closer to the community, and into the service of Christ, the wilderness sojourn is serving its Christian function. A desire to "see" God is the parent of a desire to serve.

Our contemporary preference is always for "productive" time—be it reading scripture, participating in formal study, joining a prayer group, raising money, or printing tracts. Congregations are often driven by activities, such as Sunday school, weeknight teaching, and all the social gatherings for all the different ages and needs represented in the local church. Everyone is expected to be present at least twice each week, yet nobody can find three days for real rest in the Lord, free of social pressures and expectations. Wilderness sojourns on the biblical models are not, in the overall course of human life, very time consuming. Yet Christians charge ahead, so ready to lead and to act, that seeking God in the wilderness seems to be a waste of time and a distraction from more physically productive spiritual occupations.

The tenth misconception: The wilderness experience is optional, or is one way among several to obtain the same spiritual benefit. Our contemporary use of wilderness is usually recreational and therefore, to the industrial western mind, a matter of taste. Wilderness is for the outdoors types or for those who aren't interested in golf or tennis. The biblical wilderness sojourn was never an alternative to staging pageants at the temple or going on an elders' retreat to Jezreel; it served a special purpose in providing deeper contact with the divine.

Today, wilderness in the sense of a designated wilderness area may not be necessary to our spiritual explorations. For certain purposes, however, "wilderness" as an environment relatively free of human inputs and concerns may be critical. In the contemplative and ascetic models, for example, solitude is absolutely necessary. In the implementation of the leadership model, stress and challenge are the keys, whereas in the poetic model the presence of wild nature, little influenced by people, is highly valuable. For many types of reflective exercise, reducing interhuman communication is not just desirable, it is required. Wilderness can still provide isolation, physical stress, removal of human elements, and intimate con-tact with creation just as it did in biblical times.

The Paths to the Mountain

A final misconception, and a very critical one, is that *wilderness spiritual experience was limited to one period in Christian history, or that it stems from one movement or sequence of events.* For many people, biblical wilderness experience means only the Exodus, or Christ, or John the Baptist in the wilderness. Often, when one speaks of "wilderness spirituality," the immediate assumption is that the lifestyle of the desert fathers is under discussion. Some Protestants will, in fact, reject wilderness spirituality out of hand, under the assumption that it implies severe asceticism. What we have found, however, through this review of the biblical and early church materials is a wide variety of wilderness events and encounters with the divine. Also some patterns of wilderness spiritual encounter evolved or changed historically. Although one might accuse the early church of taking wilderness withdrawal too far, we find Francis, at a much later date, developing a pattern of wilderness retreat followed by return to ministry very similar to that found in the Gospels. Francis might—ironically in this regard—be a good "wilderness" role model for conservative Protestants who have strong commitments to evangelization and ministry and prefer their personal spirituality to conform as closely as possible to biblical models.

The chapters on early church wilderness use make it quite clear, that, even during the "monastic period," wilderness spiritual experience followed diverse pathways. The three branches of monasticism that we have investigated—the desert fathers, the Celts and the Franciscans—had their similarities and their differences. This three-way analysis does not, in fact, cover the entire range of the monastic movement, and there are other groups appearing in other eras.

The three schools of monasticism documented here are very similar in their repeated use of wilderness sites for spiritual exercise. From the time of the desert fathers, withdrawal, first into the desert and then later into wilderness of any type, was idealized. St. Antony of Egypt found his inner mountain with its arid surroundings, rock outcrops, and spring. The Celts prayed in forests and in the cold water of mountain lakes. They set up residence on the outer islands and in quiet, oak-canopied glens. Antony's attachment to site appeared again in Columba's love for Doire and Coemgen's respect for the mountains and wild creatures of Glendalough. The medieval hagiographies describe Francis and his followers as repeatedly, if not preferentially, pursuing prayer and contemplation outdoors, and Franciscan selection of wild sites for spiritual exercise was very much in the Celtic pattern. There is a strong biblical basis for visionary experience in wilderness, and we postindustrials should not write this off as quaint, archaic, or misguided diversion from proper biblical interpretation.

The three schools of monasticism differ, however, in their individual relationship to the wild landscape and in their degree of withdrawal. The three literatures do not, in fact, discuss the wild landscape the same way. For the desert fathers, isolation was extremely important; the desert, therefore, became a great "cell" to shelter them from human influence. They loved the desert for its silence and austerity. For the Celts, nature was more of a companion and an indication of God's blessing. Rather than sit in silence, they listened to the songs of the forest birds. They did not rest in eternity on an unchanging rock outcrop, but enjoyed the passing of the seasons. While seeking God within, they sought the work of God without—full of color, music, and diversity. St. Francis was the most consciously concerned about what nature "says" about God. Many natural objects suggested qualities of Christ or reminded Francis of scriptures. Francis's attention to small and gentle animals expanded the sphere of his Christian ministry and was related to his care of those in the lower strata of human society. By his actions, Francis called for both an interest in nature and for mercy for the poor.

Francis's approach to the wilderness also differed from those of the Celts and the desert fathers in temporal terms. Antony and Coemgen were more completely wilderness saints. Francis had no inclination to stay in one place, and he never completely withdrew to the contemplative life. Francis found his first retreat in a cavern among "the black firs" on Mount Subasio outside Assisi. For Francis, however, these initial periods of withdrawal were not for a lifetime—or for seven long years, like Coemgen's stay on the lough—but rather for a few hours of prayer. Francis, like Christ, always returned to the marketplace and to the sin-ridden city. Francis broke away from the cloister and, in doing so, made wilderness less a complete lifestyle and more a place for periodic mystical contact with God. The Celtic saints really lived in nature. Francis, surrounded by growing urbanization, stepped into the modern dichotomy of ministering to the needs of humankind (in the city) and feeding one's own spirit away from others (in the wildernss).

We should not judge the Celts, anymore than we should judge Hagar or Elijah, for not displaying all the elements of "New Testament" wilderness experience. The Irish had no cities and had been spared Roman invasion. Their adaptation of both biblical and patristic traditions generated a beauty of its own and combined comfortably with both their culture and their physical environment. We would do well today to have our spiritual beliefs and vocations so well-integrated with the highest social values and greatest artistic efforts of our society. Each wilderness model has something to tell us about the person of God, about our traditions, and about ourselves.

Going on Your Own

I recognized, as I was writing this book, that discussing "wilderness spirituality" was potentially confusing and implied that there was some special means of faith to be found only on mountain crests or in desert canyons. In the Bible, wilderness experience combines with other types of spiritual experience to complete one's life in Christ and God. The wilderness is but one component—although, perhaps, an underestimated one—in the Christian walk. Wilderness, born of struggle, fulfills itself in community.

As suggested in the introduction, wilderness can have "spiritual, aesthetic, and mystical dimensions." Wilderness can also provide "mental and moral restoration," but more important, it can serve as a site for transformation and deep encounter with God. The wilderness traveler is more than "rehumanized" in discovering his or her relationship to the holy, he or she can engage the divine and assume a new trajectory more completely convergent with God's will. Since wilderness spiritual experience continued into New Testament times and was a critical force in the development of the early Christian church, some forms of wilderness spiritual activity should still be valid today. The question is not whether wilderness is an appropriate place for spiritual exercise, but how biblical models should be adapted for present day missions.

In examining the potential for wilderness spiritual experience, we have to be aware of our own social predispositions. Despite all the discussion of other values, twentieth-century wilderness use has been primarily a mix of recreation and scenic touring. Western culture's current wilderness repertoire is very diverse in terms of sports and places to visit, but it lacks the variety of spirituality found in the Bible and in Christian traditions. Our recreational orientation may, in fact, limit the framework of spiritual perception or forward goals that interfere with more advanced spiritual practice.

Many favored outdoor activities, such as kayaking, rock climbing, and cross-country skiing, require special equipment and some degree of skill and may focus the wilderness traveler on "conquering" the wilderness or developing some form of physical mastery over the environment. These activities fit well into the leadership model of wilderness spiritual experience, but poorly in prophetic, ascetic, or contemplative models, which require extensive time dedicated to meditation and waiting on God. A recreational focus can inhibit spiritual disciplines such as prayer, merely because many forms of recreation emphasize strenuous movement. If an outdoor sport requires intense concentration (i.e., rock climbing, white water canoeing) or is strongly oriented towards achieving a physical goal (even if it's catching a few trout), one's attention will not be turned completely toward the holy. The full spectrum of wilderness spiritual experience incorporates the slow, the stationary and the noncompetitive—be it under a broom tree or in a mountain cave.

In dealing with the recreational, we must also evaluate its effect on our schedule. We can plan to reach the top of a mountain or to walk fifteen miles a day. We cannot plan to see God. We can dictate a recreational time table. We cannot project a spiritual transition and accomplish it in a week.

The recreational, then, must allow time for the nonrecreational, even if it does not succumb completely to the contemplative in the cases of special calling or deep personal spiritual need. There is nothing wrong with integrating spirituality into our twentieth-century preference for wilderness treks and adventures. Some compromises will be required, however, to bring our current patterns of striding across ridges into a more contemplative mode.

Today scenic touring is often removed from direct contact with nature. Train, bus, and automobile tours confine the visitor in the conveyance and the direct attention to the picturesque. Wilderness travel on foot offers more intimate contact with the surroundings, but may still foster our cultural preference for touring as the detached observer. Our self-contained supply and transportation systems encourage this as does our scientific "eye." Touring also has competitive elements, such as seeing "all" of Yellowstone or choosing the prettiest trail for a hike. Aesthetic appreciation, as has already been demonstrated, is an important base for wilderness spirituality. For contemplation of creation to be effective, however, it must be disciplined and God directed. One must have the time and the patience to perceive in depth. The crush of a tour bus or a lightning fast trek to the highest peak in the state are unlikely to generate even a few short psalms.

Another potential interference—and a real threat to Christian spirituality—is western individualism. Part of the twentieth-century wilderness movement has subscribed to the ideal of the rugged individualist climbing to the highest peak alone, or of the hard bitten and cynical cowboy riding the range by himself while looking for rustlers to shoot. A comparison of the cowboy with God-fearing King David finds some similarities—both can survive under rough conditions and both know their herds and their grasslands. King David, however, had a vision far exceeding the immediate rewards of slaying his adversaries, and he had a heart that cried for God. For the climber and the cowboy, the victories are temporary and personal, while, for David, they were eternal and integral to the progress of God's chosen people on the road of holy history. In western society so concerned with "self," a Christian sojourner can easily lose sight of the community, succumb to the temptations of personal achievement, and completely ignore God's timing.

Battling Satan

The Bible has a great deal to say about false prophets and, as mentioned previously, the apostate become wilderness fatalities. The wilderness poses both physical and spiritual threats to the naive and misdirected. The common sense advice usually given to beginning backpackers—don't travel alone, don't travel too far, travel properly equipped, and learn to read the maps—also applies to spiritual novices. If someone asks me how to get started in backpacking, I usually recommend that they begin by taking some day trips in the mountains and not stay out overnight until they are relatively confident hiking without a heavy pack. I also suggest their first overnight trips should be short or in an organized group with experienced leaders. If a beginner wants to participate in something with special hazards, such as winter camping or desert hiking, the need for advance preparation and the guidance of experienced people is doubly important, as is conservatism in planning a first excursion.

By tradition, spiritual exercise also requires preparation and training. In the Bible, younger prophets, such as Elisha, often learned the trade from older ones. David began by facing the bear, not by facing Goliath and Saul (both far more dangerous than any wild predator). Wilderness monastic communities usually require young monks to study and pray with the older and more experienced before going out by themselves into isolated sites. For the Christian, wilderness spiritual exercise without formal Bible study and participation in community worship and activities presents several major hazards, including individualism, introversion, spiritual delusion, and idolatry (usually in the form of nature worship, although prophetic narcissism is also possible). Christ's encounter with Satan at the end of a long solitary fast must be taken seriously. Both internal or external demons can beset the solitary wilderness sojourner.

Before attempting long periods of withdrawal for prayer, fasting, or other serious spiritual disciplines, short sojourns are appropriate. Afternoons spent reading the Bible by a stream or praying on a mountaintop offer a mild form of solitude and can produce personal spiritual discipline. I find, in my present very ordered lifephase, that an occasional two- or three-day contemplative interlude does wonders for my psychological and spiritual state, and that one to two weeks in the wilds is as long as I need to become completely rested. During an important formative period in my younger Christian life, however, I spent two long field seasons working in the mountains for weeks on end by myself. Although I was not completely isolated, spending day after day alone in the forest encouraged prayer (after all, I had no one else to talk to) and made me more God-dependent. The extended solitude was, thus, coincidentally very spiritually appropriate during that period of spiritual transition and was balanced with very supportive Christian fellowship when I returned to the university. Perhaps, as I struggle through mid-life, I will reach a point at which a longer period of withdrawal and contemplation is again necessary. A major difference, in fact, between my younger and my older selves is that the pressures of adult responsibilities make it more difficult to stop and reflect on my chosen path, and I now have to make more of a concerted effort to find time for extended prayer.

When I was younger, wilderness was primarily a refuge from all the non-Christian influences in my collegiate milieu and an environment that encouraged me to develop a personal spiritual life. Now I find a wilderness stay is more of an extension of my Christian life at home. I can reflect and pray more intensely in the wilderness and avoid interference from work and household responsibilities, but the spiritual exercise remains strongly tied to my overall spiritual direction. Then and now, wilderness sometimes brings forth demons, usually in the form of personal fears I have trouble coping with. Since my trust in God has never been complete, being alone can still terrify me.

Despite the potential for problems in wilderness, taking a Bible out for the afternoon for some private prayer and Bible study should not only be safe, it should be spiritually healthy. Most of us can also handle short solo stays and limited group sojourns without difficulty. The longer or more remote the sojourn and the more advanced the spiritual exercise (extended fasts, extended contemplation), the greater the potential for either spiritual or physical difficulties. The Christian who wishes to pursue the spiritual disciplines seriously, in the wilderness or elsewhere, would do well to find a mentor or "guide," or community support. The Bible remains the best map to spiritual trails and deserves careful reading for long trips or short.

A number of techniques may be used to provide safety and support for the beginner, while also providing solitude. One of the simplest methods—and one used frequently by wilderness training schools—is to organize a trip for a group, but to set aside time for individual prayer and "aloneness." This sometimes takes the form of a one- to three-day solo, where individuals leave the group and stay by themselves for a set period of time. A group leader may then check on the soloists to be certain all is well. "Aloneness" can also take the form of a shorter interval (a half hour to a full day) set aside for individual devotions. Two or more people traveling together may separate for part of a day, either by taking different routes or simply by spacing themselves out on the trail or around a campsite. Group leaders, swamped by the demands of group scheduling and need, often overlook the possibility of letting people sit near a stream or on the rim of a canyon by themselves for an hour, even though this is usually quite easy to schedule.

A special warning must be placed concerning the visionary and the prophetic. Although the visionary is a central part of biblical wilderness spiritual experience, it cannot be purposefully initiated and controlled. One enters the wilderness to seek God, not some particular form of otherworldly communication. Someone who desires a vision may in fact be more interested in personal power or social status than in putting Christ in charge. (Paul has a great deal to say about this in 1 and 2 Corinthians.) The central theme in biblical wilderness experience is not that God always provides a theophany, it is that the Creator always provides. If one has a genuine need for spiritual knowledge or wisdom, one should expect the need to be met. The means, however, are in God's hands. On a number of occasions, I have gone out on the mountain to pray about something only to find the necessary answer or direction a week or two later in a conversation with a friend, an event at work, or in a sermon in church. Communication with the divine is always interactive. God has a choice of when and how to speak. Revelation and promises are at God's bidding. We should always, however, expect Christ to care about us and to be present no matter how harsh the environmental or personal circumstances.

The Bible describes the demise of many—from the sons of Korah to the prophets of Baal, to magicians in the New Testament—who did not genuinely speak for God. Going into the wilderness can not by itself precipitate a prophetic call or provide the gift of prophecy: the Holy Spirit inspires prophecy as a divine prerogative. Wilderness may nurture spiritual vision, but does not produce it independent of a mighty act of God. The Christian who tromps off into the desert hoping to obtain a special edge on the prophetic is likely to be disappointed and risks being led by personal pride into false statements or experiences. The Christian who sees a need within the church or is concerned about the relationship of the church to the culture and retreats to the desert for a period of contemplation prior to initiating ministry is taking a safer route and is closer to the biblical prophetic tradition than the person who desires showy theophanies or voices from the clouds. Further, much more than wilderness experience is necessary to generate true prophecy. Christian spiritual gifts, such as prophecy and wisdom, grow from a love of Christ, a love of our neighbors, a deep concern for the church, an understanding of God's Word, an appreciation of the work of the Creator, and obedience to God.

Since we have the Holy Scripture available for reading and study, mature vision arises from an understanding of the contents of the Bible and of the meaning of Christ's ministry, death, and resurrection. Asking God for wisdom or spiritual vision or bringing our problems to Christ in prayer are both healthy and appropriate, however. Unfortunately, our expectations either of the substance of the answer or of the means of the answer often restrict God's options. In the wilderness, we should let God reply through whatever means—be they subtle and natural, such as the gentle little breeze, or dramatic and otherworldly, such as the sapphire floor. In the wilderness, one should be ready to accept the unexpected—a sudden realization about God's character or a clarification of one's relationship to Christ—an unanticipated transformation, or a deep personal insight, such as casting off of old sins or receiving a new calling. We also need to allow providence to operate, both through the creation and through other people, and we need to thank God that it does.

Wandering in the Desert

Throughout the preceding chapters we have discussed possible motives for a wilderness sojourn or contact with wild nature. To convert the biblical models into contemporary categories, we can consolidate the biblical examples into seven spiritual purposes or motives for entering the wilderness or observing the wild. Implementation within a biblical framework justifies these purposes for contemporary Christian practice.

The first, and the most broadly applicable, of these is *developing an appreciation of creation and understanding of God as Creator.* This may vary from simple exercises in natural history, such as enjoying natural beauty or interacting with creation, to holding worship services outdoors or producing psalms and other forms of "creation art." Even a novice can contemplate God as Creator in appropriate settings and investigate the person of God as reflected in the creation as found in Psalms and the Wisdom literature. Bible study can be easily coordinated with encounters with the wild to produce a deeper understanding of the Creator's characteristics and works. An appreciation of creation can be developed without long wilderness stays and is available to Christians in all phase of spiritual maturity.

The second of these purposes is *rest or restoration.* This rest may be spiritual or physical, and there is no reason not to include recreation and play in this category. A wilderness sojourner might need simply to get away for awhile, to exercise or to obtain relief from exercise or physical work. Often in industrialized countries, wilderness provides a break from repetitive labor or from confinement indoors. In Biblical terms, *rest is* primarily a short departure from very stressful ministry or vocation and appears to have emphasized the desert or the lonely places because these provided distance from the pressures of the crowds and day-to-day schedules. The aesthetic, pleasant, or quiet in nature may also be appropriate.

A third purpose is *spiritual exercise and communication with God.* This would include practice of any of the spiritual disciplines, although meditation, prayer, fasting, and seeking solitude are exceptionally well suited to wilderness environments. (See Richard Foster's book, *The Celebration of Discipline: The Path to Spiritual Growth* for a description of the major disciplines and references to classic Christian works on the topic) [1] Disciplines such as prayer can benefit from short intervals of intensive practice in isolation. Short wilderness stays probably do little to encourage simplicity (the shedding of worldly cares and desires), but longer sojourns should nurture an appreciation of the basic elements of life and the richness of God's providence. The discipline of study has been accomplished historically in wilderness, but it is best in the wilds in devotional mode or with attention limited to the Bible or a few inspirational writings. Although difficult to pursue on a weekly or even monthly basis away from the home congregation, two corporate disciplines, worship and celebration, can provide a special experience in wilderness (singing in the cool evening air graced by a desert sunset, for example). Most important, though, among all the choices for activities, is that, in the wilderness, one can freely wait on God and allow time for conversation with the holy.

The fourth purpose of a wilderness sojourn is to *seek spiritual transformation*. This can begin with reflection on one's spiritual life and continue into resolving spiritual or personal problems. In some cases it may include wrestling with a life transition, such as changing jobs, coping with losses, or reaching adulthood. In others, conversion and getting to know Christ may be the issues. For the more mature, wilderness may provide an environment where a divine call can be more clearly heard or new spiritual paths more clearly seen. New directions in ministry may thus arise out of "God's call" in the wilderness.

The fifth purpose of a wilderness sojourn is to *develop leadership skills or prepare for a difficult ministry*. The type of sojourn required is usually lengthy and emphasizes overcoming stress, developing dependence on God, and understanding the divine will. In the biblical examples, wilderness nurtures both civil and religious leadership. Wilderness fosters spiritual growth if the experience is set within an appropriate Christian context. The entire Exodus was an exercise in trusting in God for the future so that the children of Israel could lead other nations to the worship of Yahweh. Developing civil leadership skills is often accomplished in a group, although the biblical pattern for preparing for religious leadership is solitary. This is a wilderness goal very suited to young Christians. It requires more effort and more planning than most other motives for wilderness sojourns.

Related to the development of leaders is the sixth potential purpose for a wilderness sojourn, *resisting sin and temptation or developing a spiritually accurate view of culture or society relative to God's will*. Wilderness can remove sources of temptation temporarily and provide time for reflection on the meaning of one's actions. Wilderness also offers relative freedom from cultural values and norms. In cases of persecution or extreme social difficulty, wilderness may provide refuge from antagonistic social groups or political forces. From the wilderness, the sojourner can contemplate personal failings as well as the work of evil principalities and powers. In the lands without human institutions and communication networks, social ties are temporarily broken, leaving the sojourner free to replace them with more godly bonds. This should be one of our greatest modern

Christian uses for wilderness, but it is actually one of the least emphasized by the contemporary church.

The seventh and last potential purpose for a wilderness sojourn is *development of a corporate response to God and the strengthening of the community*. Groups traveling or residing together in wilderness can develop a sense of community, develop mutual responsibility and concern, learn to respond to God as a body, and raise their level of social and spiritual maturity. Although most corporate spiritual disciplines are best practiced at home or in organized service ventures, the disciplines of submission, service, confession, and guidance have a place in the wilderness also, particularly in work with youth groups and in other types of interpersonal ministry. Most often organized for adolescents, group wilderness experience can be utilized to build community leadership among adults and religious professionals.

Must We Enter the Wilderness?

We must now tackle a difficult question: "Is wilderness or wild nature spiritually necessary?" The question is complicated, perhaps more than any other we have asked, by what we mean by "wilderness." Do we mean only unsettled regions such as remote deserts, mountains, and the two polar icecaps? Do we mean only what the Bible calls wilderness, which is only grasslands and arid, rocky hills? Do we mean designated "American wilderness," which reflects a frontier ideal and modern concepts of nature preservation? Or do we want to use the broad concept presented in this book, which extends from lonely places near town and incorporates all wild nature?

The common pattern among Christians today is to suppose there is *not* any general need for contact with wild nature, but to treat it as an option available for the few believers interested in monastic spirituality or environmental questions. This approach confuses the responsibilities of those with specialized Christian callings with those of the entire church. In the end, nothing is necessary to salvation except the acceptance of Christ's person and work on the cross. (Deathbed conversions are as genuine as any other.) For those of us who have an opportunity to serve Christ in this life, however, some responsibilities are common to all of us, others to only part of the body.

On one hand, there is no reason every Christian should spend forty days in the wilderness with the wild beasts or that every Christian should travel to Horeb since these journeys prepared for prophetic ministry. On the other hand, every Christian needs to comprehend God's role as Creator as well as God's role as Savior (and the two are integrally related). All Christians, therefore, should have some exposure to creation, free from human tinkering and modification, as the work of a loving and caring God. Of the seven motives listed above for a wilderness sojourn or contact with wild nature, only the first, *developing an appreciation of creation and understanding of God as Creator,* is a truly democratic responsibility, at least in regards to the wild. For the other six, wilderness may be the best site or environment, but it is not the only possible location (given free options without political or social repression).

This conclusion does not mean that we should ship all our junior Sunday school classes off to the Grand Canyon or the Amazon River basin. It does mean that Christians should learn to look at all creation, including wild birds, insects, and plants (all of which can be found even in urban areas), as the handiwork of the Sovereign Lord. Contemplation of wild nature instructs us in the diversity and complexity of the Provider's work, and, as the Book of Job points out, teaches us about the Sustainer's prerogatives. Without an appreciation and understanding of God's role as Creator, Christians will mistreat and disregard creation—both wild and tame—and fall short of God's instructions in Genesis 1 and 2 for humankind to care for what the Creator has made. We do not, therefore, have a general need for trips to remote or desert wilderness, but we do have a general need for contact with wild nature.

Wilderness sometimes provides the best locations for the other six purposes listed above, and there are some situations where there is almost no other choice (or the prophets wouldn't have ended up in the wilds as often as they did). This is not necessarily true for day-to-day spiritual exercise, for which the "prayer closet" or the quiet nook at home is the most practical site (although many people have a spot in a garden or at the corner of a lawn that they like to use). God, however, sometimes calls us to retreat from our usual social settings and structures. These may grade from country roads and quiet lake shores (and other lonely places near town) to the inaccessible deserts.

As it was in the time of Moses or John the Baptist, wilderness may still be the best place to free the servant of Yahweh from unrighteous cultural idols. This isn't the wilderness of a skiing vacation, but a "desert" without human support systems. A week spent in isolation, with only the most basic elements of human physical need (food, water, and protective clothing), quickly strips away dependence on Mammon (or sends the sojourner howling back to an air conditioner and a television set). Even evangelical Protestants have recently suggested "remonasticizing" Christianity–because of the Christian preoccupation with wealth, political power, sexuality, and social status. A return to periodic wilderness spiritual practice, might help serious lay Christians to fight worldly influences without requiring them to flee to convents and desert hermitages.

The biblical models of temporary withdrawal not only have worked in the past, they also make psychological sense. Short sojourns break cultural patterns and aid meditation. The longer sojourns (up to forty days) allow time for cultural deprogramming, personal reflection, and moving toward God (see the discussion at the end of chapter 8). Personal deprivation, through fasting and other means, can help to set the material and personal desires in their proper place. Ego trips fade quickly when there is no one nearby except trees, rocks, and God.

A conference center or camp with a bustling cafeteria, a heavy schedule of activities, and a television room will produce a strong interaction with other people and very limited time for contemplation. Better options are a desert campsite, a mountain cabin without a radio or television, or a monastic establishment (or a basic retreat center) with simple rooms and silence. For serious prayer or seeking God, uninterrupted solitude is the best. The typical wilderness environment of roadless forests or open arid lands can nourish contemplative exercise and, in the case of longer sojourns, may be one of the few places where real freedom from social and cultural distractions can be obtained.

Wilderness may also have a special place in contemporary spiritual practice because of an oft forgotten biblical principle—God can work through creation to speak to the wilderness traveler. From Hagar's spring, to the parting of the Reed Sea, to the ravens feeding Elijah, wild nature itself cooperated in God's communication to the people of Yahweh or those with open hearts. God's working through the creation tends to put us in our proper place and can help to provide revelation concerning much deeper spiritual matters, such as our dependence on God or the extent of God's justice. Wilderness can offer greater solitude than many Christian retreat facilities, it removes human interference to a greater extent, and it offers the potential participation of creation in the experience, which would be inhibited by human structures.

Lost in the declarations and identifications of the media and caught in the priorities of materially bounded culture, contemporary Christianity has a desperate need for a self-identity dominated by its relationship to Christ. With a hundred-thousand colorful images flashing by at the speed of light and the sounds of the nations filling the air, spiritual "seeing" and "hearing" encounter massive competition from seemingly infinite earthly sources. The wilderness offers an alternate ancient ground, where the babbling voices are largely silent and cultural clutter rarely intrudes. Perhaps not the entire church, but at least some of the leadership, need to consider the value of wilderness as a platform for speaking to God . . . and for hearing God reply.

In summary, we need to contemplate the overall importance of wilderness spiritual experience to the development of the Christian faith. The wilderness sojourn has never been the predominant mode of Christian worship or of community interaction, yet wilderness spiritual experience receives a great deal of attention in the Bible and in early Christian literature. Spiritual leaders and founders of the faith repeatedly found themselves alone with God. The necessity for wilderness is correlated to the potential of the national social environment to inhibit the execution of God's will or to mute God's voice, which, in the case of contemporary western culture is considerable. Wilderness spiritual experience does not require the natural splendor of Yosemite Valley, nor does it require the drought of the deserts around the Dead Sea. It does, for full development, require solitude, struggle, and contact with creation, and a place expansive enough for God to be God.

16

The Wild and the Kingdom
(Protecting the Wild)

Jesse, the Manatee

It was the height of confusion. I was suddenly facing a south Georgia congressman, two aides, a lobbyist looking for something entertaining to do, and a TV cameraman. The cameraman was unexpected, and the congressional party was late. We had planned how we were going to spend the afternoon, but were now off schedule and potentially over the Coast Guard's passenger limit for the boat waiting at the Cumberland Island National Seashore dock.

"Do you think we ought to drive to the paper company to see the manatees?" I asked one of the park rangers standing by to assist us. " I could ask the congressman if a change in plans is acceptable."

"We'll never see anything from the boat with this many people," the ranger replied. "The park superintendent is coming too."

We had done the boat trip the day before with three aides to a powerful senator, one ranger, Barb (the biological technician), and myself, and had found the manatees resting peacefully in the paper company effluent that had drained into the North River. Jesse, a large male manatee with a radio transmitter attached to a belt around his tail, had been on ebb tide duty at the warm water outfall, so the aides had been able to see one of the animals involved in the research project.

The purpose of the trips was to brief the congressional types on the potential impacts of naval dredging for the nearby nuclear submarine base. The manatee was one of the endangered species using the marshes and the shipping channels in the area. The National Seashore protected some of the creeks used by manatees—but not all of them.

We decided to go by land.

Driving through the paper plant is always interesting and not very aesthetic. I began to regret that we had come this route since the river banks are much more pleasant viewing that the plant's wastewater treatment system. At least the congressman had had some past experience with the paper company and began to talk about the bond issue that paid for the settling ponds.

On arriving at the effluent, the congressman and a plant engineer strode out on the protective pilings in front of the large drainage pipe, while the cameraman wandered around unable to get a clear shot. As expected, the manatees were basking in the warm water coming from the plant, but two men in a small boat were fishing right in their midst, and the water was so turbid that the ten-foot long aquatic creatures appeared to be little more than two brown nostrils, breaking the water's surface every few minutes.

Jesse, the star, was nowhere to be seen—the men in the boat had probably frightened him off. We attempted to pick him up on the radio receiver.

The congressman, interested in both natural history and coastal conservation issues, nobly attempted to view the manatees and to ask meaningful questions. The lobbyist and the aides, who were vigorously attacked by sand gnats, lost interest very quickly and, after a few cynical remarks about all the things they would rather be doing (mostly partaking of cold beverages), they fled back to the vehicles. The cameraman wandered about a bit more; then he too, headed for a van.

The congressman was asking the right sorts of things as he batted at a cloud of biting insects.

"How big are manatees?"

"Up to three meters long and more."

"What do they eat?"

"Plants, in Georgia they seem to use algae and *Spartina*—the salt marsh cord grass."

"Why do you have a radio transmitter on a manatee?"

"To find out how far they travel, where they go, and where they are feeding."

"How many manatees are left?"

"About twelve hundred, although we don't know how many use Georgia marshes."

"Why are they endangered?"

"Habitat destruction in Florida . . . and they are also being hit by boats. Almost all the adult manatees we see have scars on their backs from propellers. A number are killed every year in collisions. Jesse has a distinctive scar pattern. So does the female we put a transmitter on this spring. She has very deep scars on her back from a prop."

I took one final look down at the manatees as the last of our party were leaving the effluent area.

"You guys just aren't any good at public relations," I said under my breath. "This congressman could potentially help you, and all we can see is your nostrils."

A manatee broke the surface and snorted; then it slowly sank.

"And Jesse, where are you when we need you? If I can't get the aides interested in big slimy sea creatures, I could probably get them interested in radio transmitter technology. In Florida, you guys have Jimmy Buffet to hold concerts for you, but, in Georgia, all you've got is me and Barb . . . and we don't sing."

I walked back up the bank frustrated. It was hard to get upset with the manatees when they have been the most innocent of victims. Even though they can weigh over three thousand pounds. they don't bite or attack swimmers. They don't compete with humans for food since they forage on hydrilla, turtle grass, and other aquatic plants. A person could eat one if desperate enough, but they are too rare and have too low a reproductive rate to become a regular part of the human diet. Manatees like to swim, eat, cavort with other manatees, and generally mind their own business.

The manatees' problems with people have nothing to do with one species purposefully attacking the other. We just can't seem to share the rivers and the marshes with the manatees. Manatees get caught in flood-gates and tangled in nylon fishing line. Human development and urbanization have destroyed many of the natural wetlands and stream bottoms on which they depend. Manatees are strong swimmers, but, in a world filled with hundreds of speed boats, sometimes they just can't get out of the way fast enough. We take more and more of Florida, and they get less and less. Now developments in Georgia are moving in on manatee habitat.

"I'm the one who isn't very good at public relations," I thought as I climbed back in a van, "taking these neatly pressed fellows from the District of Columbia to a south Georgia paper mill effluent! But it's a shame we even have to 'sell' the importance of species like these. It's all a question of who gets the sound and the marshes—the navy, the recreational fishermen, the port authority, the paper companies, or the manatees." No one thing by itself was likely to cause a problem for the manatees; in fact, they seemed to be getting along rather well with the paper company, but too many boats and too much development would be the end of them.

What was the congressman's responsibility in all of this? What was mine? The paper company's? The Navy's? Was there some point at which we should just leave the marsh to the manatees and not take any more?

* * *

Having dedicated the first fifteen chapters of this book to the subject of what wilderness and wild nature can provide for Christian spirituality, it is appropriate to conclude with an investigation of what Christianity can or should do for wild nature. If we look back over the literature we have already reviewed, we have very little evidence of human protection of the wild prior to the development of monasticism. Historically, of course, human displacement of wild nature has become greater through time and, during the industrial era, has occurred on a massive scale. If we think back to Psalm 104 (chapter 5), we find that today a number of species of leviathan, whom God made "for amusement," are on the verge of extinction. (And extinction is a problem if we assume leviathan is a whale, a crocodile, or any other large aquatic organism.) The stork is gone from the cedars of Lebanon because the cedars have all been cut and goats have eaten their seedlings. The springs no longer gush through ravines, because too many wells have lowered the water table. Even the lions, who "claim their food from God," find themselves pushed out of even the furthest desert by the hunting activities of human and the overgrazing of the range lands.

The ancient Hebrews could not imagine the world without wilderness, the arid lands, and the great predators. The lion and the bear would always be there, even if driven from the pastures below Bethlehem and forced to take refuge in the inner desert. One of the few passages suggesting a major displacement of wild nature, Ezekiel 34:25 (NIV), promises that the Lord will "rid the land of wild beasts so that they [the people] may live in the desert and sleep in the forests in safety." The passage, however, is probably discussing the fall of Jerusalem and the resultant social disorder due to war (where wild beasts would enter previously cultivated and grazed lands), and not a human occupation of previously unused deserts. It would be interesting to hear what the psalmist would say if he could see the cedars of Lebanon today, devastated by centuries of harvest without replanting. In this case, the Bible becomes a historic document, telling humankind what a valuable resource we have lost through not caring.

If there are few biblical directives to preserve the wild, the early Christian literature we have reviewed presents preservation and protection of the wild from humans as a developing theme, becoming increasingly important through the passage of time. A real preservationist emphasis, at least in terms of protection of the environment from humans, seems to have originated with the Celts. Although stories of the desert fathers mention feeding and healing animals, most of these incidents concerned natural events or conditions, such as a reed in a lion's paw, a cub born blind, or a shortage of water in the desert. Protection of animals or other wilderness features from humans is not an important motif

The literature of the Celts, in contrast, stresses saving large animals from hunters. Saints worried about axes in oak groves and otherwise concerned themselves with forest protection. The later Franciscan stories continue the theme. On more than one occasion, Francis released a hare or a rabbit caught in a trap, and he also freed fishes and waterfowl from the hands of fishermen. The gentle mendicant also talked a boy from Sienna into releasing a flock of turtle doves the boy had trapped and was planning to sell.

Rather than assume the increased Christian interest in preservation was merely due to theological departure from Hebrew values or pagan influences (an assumption made by far too many historians), we have to look at each Christian group in its social and physical environment. The desert fathers, for example, rose to prominence during the decline of Roman power in the Levant and came from countries that had maximized their agricultural productivity and their urban development during the preceding centuries. The drier deserts remained free from much of the thrust of Roman technology and were not, even with the construction of sophisticated aqueducts and water holding systems, available for large-scale agriculture. The Hebrew scriptures treat the wilderness as a constant presence, converted into tillable land only by a mighty act of God. To the first monks, the desert must have seemed unconquerable and unchanging. The lions and ravens would always roam the wadis. Part of the holiness of the hermits relied, in fact, on the premise that the desert belonged to the wild beasts and was difficult to occupy. The early eastern monastic literature, which portrays God protecting monks from lions and crocodiles and lions protecting monks from human marauders, betrays the concerns of the time. The desert offered a minor threat to the hermit, the Saracens and Persians offered a major one. One suspects that the social disorganization following the demise of the Romans was, if anything, beneficial to the lions and detrimental to the monks. The desert monks showed compassion for the wild and assumed the lions would accompany them into the New Kingdom. They did not, however, experience human activity that was a major threat to the local wildlife.

The Celtic monks, in contrast, witnessed the last major intrusions of agriculture into unplowed lowland forests, and the golden era of Celtic monasticism was contemporary with a major period of forest clearing. Legal documents from 800 C.E. indicate judicial protection of individual trees, so it is not surprising to find similar sentiments expressed in the Christian literature (which, unfortunately, is difficult to date). There is no direct evidence the monks thought deer, boar, or wolves were threatened as species, or that they recognized phenomena such as overhunting or habitat destruction. Yet their literature clearly presents hunting and mass starvation as threats to individual animals. The "protection incidents" also seem to concentrate on the elements of the wild landscape that were the most threatened, including the large mammals and oak forests. The Iron Age Celts probably held some animals and trees sacred. The protection extended in the monastic literature, however, was in relation to need rather than divine qualities. Saints like Coemgen (Kevin) could be considered the first Christian wilderness preservationists.

The early biographers of St. Francis suggest that he deviated from the Celts primarily in the focus of the "protection." Francis has little to do with wild boar and deer and only a limited number of encounters with wolves. (Despite the fame of the wolf of Gubbio, the Franciscan legends as a group have very few tales about large animals.) He protected cold-blooded fishes, picked up worms, and fed bees. Francis understood some basic principles of conservation, such as the need to not cut down an entire tree in order that it would continue to produce wood. He had an appreciation for natural diversity, which he displayed when he commanded the gardener to leave the border of the garden uncultivated so that the wild plants would bloom there. The care Francis showed for wild nature hardly exceeded that of Coemgen (who, in a legendary retreat for prayer, stood for days with arms outstretched, so that a blackbird, who had built a nest in his hand, could safely incubate her eggs). Francis did, however, move the emphasis of active nature protection to smaller and more ordinary creatures.

This shift in the principal objects of protection is related to another difference among the monastic literatures. The desert fathers concerned themselves almost entirely with larger and more "noble" beasts—except for an occasional misadventure with a scorpion. The Celts greatly expanded the number of creatures deemed worthy of inclusion in the monastic literature and paid more attention to birds and plants. The Celts, however, continue to emphasize the larger mammals. St. Francis had an appreciation for the diminutive and, with the exception of a few incidents concerning wolves, spent his time with rabbits and doves.

These differences reflect the state of the natural environment as well as theological developments and cultural backgrounds. The first monks saw themselves as warriors for God, and the animals with which they consorted had the same heroic character. In the shadow of the biblical wilderness and the precedents of the Old Testament prophets, a positive (or survivable) interaction with a powerful beast demonstrated the favor of God. The arid environment did not encourage lengthy descriptions of vegetation. The Celts were less limited to the biblical models and less restricted to one type of wilderness habitat. The conventions of Celtic poetry encouraged aesthetic attention to all major components of the natural landscape. The Celts retained a taste for the heroic and an interest in larger animals. They, however, deserve the credit—often given to St. Francis—for shifting Christian attention to the broad spectrum of nature.

Celano reports that "among all the animals he (Francis) preferred the gentle." This may be partially the result of Francis's strongly Christocentric theology. Francis was very taken with the idea of Christ as the lamb of God and encouraged meekness, humility, and submission among his followers. The importance of the more defenseless wild creatures in the Franciscan legends and biographies may also be partially a result of human impact on the landscape. By the time of Francis, the Italian landscape was largely cultivated, grazed, and cut over. Unlike the Celtic monks, Francis and his biographers probably had little opportunity for contact with wild boar. The Franciscan attention to songbirds and insects may be the outgrowth of numerous trips from town to town in the cultivated countryside.

The differences between Irish and Franciscan attitudes and environments may also have influenced the structure of the few Franciscan tales about wolves. In the biographies of Francis, both the wolf of Gubbio and the wolves of Greccio were bothering towns or farmers, and the medieval hagiographers cast them as unrelenting villains. The tale of the wolf of Gubbio varies importantly from similar tales in the older monastic literature in the ferocity of the wolf. The desert fathers do mention animals attacking people, but most of their stories describe a single attack on an individual doing something dangerous—such as the leopard that clawed the monk who threw stones at him—or the behavior of animals that really do attack people—such as crocodiles or a lioness with cubs. There are a few extreme instances of wildlife threatening humans—such as a boa who supposedly devoured farmers—but these are in the minority. In the desert, the most vigorous attacks of the large predators were in defense of the monks. The Celtic literature emphasizes wolf attacks on domestic stock and deer, not on local farmers. (And the Celts preferred, in proving their heroic character, to slay monsters, not wolves.) Both the Celts and the desert fathers seem to have had a better knowledge of the behavior of large predators than did Francis's biographers, and this is reflected in the types of conflict between the wild and the human that they recorded. Again, the more urban and more thoroughly cultivated Franciscan environment may be a source of this difference.

Some branches of Christianity have, over the years, associated the conquering of wilderness or wildness with the overcoming of sin or with the establishment of the New Kingdom of God on earth by displacing the evil and the savage. The Puritans colonizing the New World thought themselves, for example, on an errand into the wilderness. Yet we see in the early Christian literatures very little association between wildness and evil. Instead, the wild responds very positively to the truly holy (not to pass judgment on the Puritans of this time). Stories of saints killing wild animals are in the minority and, even in the Franciscan literature in which the wolf enters the story as an evil character, he is converted rather than being shot, chased, or sent to England to make fur collars. For the desert fathers and for the Celts, a wild animal was not intrinsically evil, but rather had to charge or attack someone unnecessarily or steal a loaf of bread before she could be accused of sin. A lioness with cubs would be considered to be acting according to nature if she attacked an intruder. It was also assumed that, if a saint were truly holy, the lioness would come and set her cubs at his or her feet.

Saints, Scriptures, and Saving the Wild

The historic Christian development of a preservationist ethic concerning wild nature poses some important theological questions. Luther and Calvin would both, of course, inquire, "Is this Scriptural?" Is Kevin a druid in disguise? Is wild nature evil or somehow under satanic control? What does the Bible tell us about wild nature? Is it different from cultivated or domestic nature? Does wild nature only have value if humans utilize it for some practical purpose? Does wild nature have any properties of particular scriptural interest? Is there any biblical indication that we should protect the wild or respect its place in the created order? What was God's intent for our relationship with wild nature? Modern theologians treat the passages in Isaiah about lions lying down with lambs as symbolic and view them as relevant to a coming kingdom, not a principality present on earth today. Do we have any reason to agree with the desert fathers and assume that real lions might be eschatologically important too and should remain present until the end of the age?

We began in chapter 2 with Genesis and the garden of Eden, where God could communicate clearly with humankind. It is appropriate, therefore, to return to the very beginnings of the created order and attempt to perceive God's intentions for the wild portion of creation. In Genesis 1, the movement of God's Word is universal, and the chapter describes all the major elements of what wildlife conservationist and wilderness advocate Aldo Leopold called the "land community."[1] The earth, the heavens, the plants, the fish, the birds, the land animals, and the small creeping things are all present. The creation is a totality of divine handiwork.

According to Claus Westermann, the creation begins with God performing three acts using the divine Word for naming and separation. "And God said: Let there be light! And there was light" (Gen 1:3, WS). God separated the light from darkness, the waters from the heavens, and dry land from the waters. This is followed by the creation of plants, which takes a different form than the previous creation events. Here God said: "Let the earth sprout forth fresh green: plants which produce seed, (and) fruit trees that bear fruit on the earth each of its kind, (fruit) containing its own seed" (Gen. 1:11, *WS*). God's creative power is not employed directly, but what has already been created, the earth, itself puts forth something new.[2]

Westermann suggests the Hebrew account of the origin of plants differs from other ancient Middle Eastern and African creation accounts. In the Sumerian story of Enki, for example, the union of Enki with the goddess of plants, Uti, produces eight ordinary plants and eight healing plants. The point of the Sumerian story is "to point to the meaning these plants have for people, be they edible or healing plants."[3] Other primitive creation stories tend to limit their concerns to plants that provide "the means of life for the group" and dissociate the creation of plants from the creation of the world.[4] The author of this portion of the Genesis narrative is trying to order creation as a whole and incorporates all plants, domestic and wild, in the passage.

Genesis does not classify plants directly by their usefulness to people, but by the type of seed they produce. The passage separates "herbs, which yield seed directly, and trees, which yield fruit in which their seed is contained."[5] The distinction between herbs and trees and among the different kinds of fruits does suggest the importance of plants for humans and for animals and thereby connects them, as the products of the earth, with elements of the creation yet to appear.[6]

After the creation of the planets, God said: "Let the waters teem with living beings, and let birds fly above the earth across the vault of the heavens. And God created the great sea monsters and every living being that moves, with which the waters teem, each of its kind, and every winged bird, each of its kind. And God saw, how good it was" (Gen. 1:20-21, *WS*). The Genesis account again describes species of all sorts, regardless of their direct usefulness to humans. The text emphasizes both the productivity and diversity of creation. The waters "teem" or "swarm" with life. The passages concerning birds really encompass all flying creatures. The text also specifically mentions "great sea monsters." The Hebrew word used might mean a sea monster, serpent, crocodile, or perhaps a large fish.[7] Sea monsters and crocodiles are thought of as dangerous animals and, like the wolf, may be seen as injurious to human welfare. Genesis 1, however, appoints to them a place at the beginning of creation, displaying them as the necessary end of the continuum from small to great.

The narrative proclaims that God then said: "Let the earth bring forth living beings, each of its kind: cattle and reptiles and wild animals, each of its kind. And it was so. And God made the wild animals, each of its kind, and the cattle, each of its kind, and all animals that creep on the ground, each of its kind. And God saw, how good it was" (Gen. 1:24-25, *WS*). The original Hebrew text distin-

guishes three kinds of land animals. In some translations of the Bible the differences are not very clear. The New Revised Standard version, for example, calls
them "cattle and creeping things and beasts of the earth." In a more lucid analysis,
Gerhard von Rad suggests the three classes are: "(1) wild animals (beasts of prey,
but also our "game"), (2) cattle, and (3) all small beasts (reptiles, etc.)."[8] The first
class, then, would include a variety of wild species from ibex to deer, to wolves,
to elephants. The second would include not only cattle *per se,* but other types
of domestic livestock, such as sheep and goats. The third is sometimes translated
reptiles, but doesn't match the scientific definition of a reptile. This group should
encompass amphibians, such as toads and salamanders, and probably a variety of
other creatures, such as arthropods. None of these classes match modern scientific taxonomy. The crocodile, as already noted, probably belongs with "the sea
monsters," despite the fact it is a reptile. The divisions here distinguish wild from
domestic animals and the large from the small. It is noteworthy that domestic
animals are not listed first or further distinguished in any way.

The Hebrew vocabulary used for the wild animals does, as the Revised Standard Version indicates, imply a tie to the earth. Westermann suggests that because,
in the case of animals, the expression used for "let the earth bring forth" does not
necessarily mean direct participation of the earth in the creative act, as it does in
the case of plants. He remarks:

> The earth has a part in the creation of the land animals just as it does in the
> creation of theplants. But the expressions used are different (v. 11: "Let the
> earth put forth fresh green"; v. 12: "And the earth greened forth"). One should
> not press these differences. When [the texts] says "Let the earth bring forth" in
> v. 24, then that cannot mean the direct participation of the earth in the creation
> of animals—there is no sign of this in the action-account—but only that ani-
> mals belong to the earth. The earth with its variety of formations, surfaces and
> structures provides the living conditions for the different species of animals.
> We can say that certain formations bring forth certain fauna. The sentence
> [Genesis 1:24] means something like that.[9]

As Genesis ties the plants to the earth, it also ties the animals. The text presents
them in relationship to their environment.

The Genesis passages provide two further characteristics of the creation.
The first applies without distinction to the land and the sea, to the wild and the
domesticated animals, and to the useful and the "nonuseful" organisms—*they
are all pronounced "good."* The word "good" here does not refer to ethical
virtue, but might be defined as inherently worthy, or valuable in its own right.
The limitations of the English language, in fact, miss the full meaning the He-

brew word, *tob,* which can mean both "good" and "beautiful." Westermann suggests that, in the concluding sentence, "God saw everything He had made, and behold it was very good" : "the listener can thus also hear the echo: 'Behold, it was very beautiful.' The beauty of creation has its foundation in the will of the creator; beauty belongs to God's works. Whoever speaks about the work of the creator also speaks about what is beautiful."[10] Wild nature has inherent worth *because God created it.* God's judgment is not our judgment, and God's valuation is not our valuation. "In human eyes there is much in these works which is not good; much of it incomprehensible, much of it dreadful. . . . A creature can not see the entirety of the whole and thus cannot evaluate it."[11] God holds the sea monsters and the crawling things not only to be good, but also to be beautiful.

For the philosopher, proving inherent worth in nature poses a number of problems. If a being has a "good" of its own (that is, some conditions are good for it, and some are not), it does not necessarily logically follow that it has "inherent worth."[12] In Genesis, the elements of creation assume inherent worth not primarily because they are "deserving of moral concern and consideration" (although this *is* the case), nor because "all moral agents have a *prima facie* duty to promote or preserve the entity's good as an end in itself"[13] (which is also the case), but, *because God made them,* they are a mirror of God's being, and God has judged them good. The moral obligations then follow.

A second characteristic of creation found in the Genesis passages is God's blessing on the animals and on humankind. In Genesis 1:22, God blesses the teeming waters and the creatures of the air, saying: "Be fruitful and increase and fill the waters in the seas, and let the birds increase on the earth" (*WS*). The blessing is repeated in Genesis 1:28 for human beings and, presumably, includes the land animals as well. Biblical scholars agree the blessing "is the power of fertility, i.e., the imperatives of 1:22 are explicative; the blessing which God confers on the creatures he has created is the power to reproduce, multiply and fill the earth. The blessing is not something or other added to this power."[14] Not only wild and domestic animals, but also humans as well, all receive the same blessing, and it continues to operate through the history of earth. It is not human-kind alone that has a divine mandate to populate, but also the other members of the biological community.

In Genesis 2, the imagery changes, but the discussion remains focused on the original creation. Genesis 1 (to Genesis 2:4) is a list of creative acts, with the creation of humankind as the climax and the Sabbath as the conclusion. In Genesis 2, attention switches to the relationship between humankind and God; the setting is the garden of Eden.

The modern reader may have a split image of the garden, seeing it both as a natural paradise, where humans could live without working or growing old, and as a planted garden, surrounded by a wall. (Note that the popular notion that one does not work in paradise is immediately defeated by God's command to till the garden.) Genesis 2:8 says: "Yahweh God planted a garden in Eden, in the east, and put the man that he had formed in it" (*WS*), while Genesis 2:9 continues: "And Yahweh God made all kinds of trees grow out of the ground, pleasant to look at and good to eat . . ." (*WS*). This couplet forms a pair of images. In the first, the vocabulary is agricultural, with God as the gardener, and the attention is on humankind. In the second, the description of the growth of the trees is more "natural," and the emphasis is on the characteristics of the trees themselves. There is no reason to believe the garden was a small walled area; "a park of trees"[15] is a more accurate description. Note the trees are described in terms of two categories of human appreciation—the trees' edible fruit and their beauty. The value of this primeval habitat was far more than a free lunch.

Old Testament scholars have argued about whether Eden was humankind's garden or God's garden. Ezekiel 28:13–14 calls Eden "the garden of God" and says that when people were in the garden, they "were on the holy mountain of God, in the midst of stones of fire." Isaiah 51:3 also refers to Eden as "the garden of the Lord" and contrasts it to the desert. Since God planted the garden and then purposefully placed Eve and Adam in it, the garden was not intended to be God's alone. Before the temptation and fall, both humankind and God walked in the garden and communicated directly (much as Moses attempted to communicate with God on the mountain). The spiritual and the earthly were not separated. The image of a lush natural setting providing amply for humans was probably the intention of the original text.

A final concept found in the beginning of Genesis—and an important one concerning the spiritual role of wild nature—is the relationship of Yahweh to chaos. In the Babylonian accounts of the creation of the universe, the God Marduk fights chaos and, in the process, creates life and order. In the "Enuma Elish" epic of the Babylonians, watery chaos is not only living matter, but also is part of the two first principles, Apsu and Tiamat, "in whom all elements of the future universe were commingled."[16] Thus, in Babylonian myth, the universe is preexisting. In Genesis, God creates all matter and imparts life via the divine breath. Even if one assumes that Genesis 1:2 implies a preexisting chaos, this is in no way part of God. The gods of the Babylonians rise out of the primeval chaos and are, therefore, merely deified natural forces. In the Hebrew accounts, even when Yahweh confronts chaos, "creation does not draw the deity into the flux of the world process,"[17] much less generate God or the godly. The Old Testament presents the universe as a creation of God, which God always transcends. This is in marked contrast to both Babylonian and Canaanite religions, where heavenly bodies, trees, and other natural objects were credited with supernatural power and thereby deified.[18]

Therefore, if we accept creation *ex nihilo*, it becomes very difficult to accept the notions that wild animals and wilderness are functions of primeval spiritual beings or are somehow less spiritually perfected than the domestic. The very existence of wild animals and their interrelationships with plants and the earth are statements of the created order. The sea monster is not an escaped fragment of forces existing before time, but a completed work of God. The Genesis texts imply equality of order and divine control in both the wild and the tame. In Hebrew thought, the wild has no special spiritual power stemming from ancient forces, spiritual beings, or gods. (The Celts, however, gave these properties to some of the monsters they tackled.)

It is true that some poetic and prophetic texts portray Yahweh as fighting chaos in the form of "floods" or "raging waters," but these are not a subset of wild creation. They are instead the primordial chaotic forces that "threaten the order of creation and strive to destroy it."[19] The attack is not led by lions, bears, and crocodiles, but is actually directed towards them. Psalm 93:3 suggests the before-the-beginning battle with chaos when it says: "The floods once rose up, O Lord, the floods lifted their voice and the floods lift their roaring," but it follows with a strong statement of the transcendence of God, when the Psalm declares in verse 4: "Mightier than the thunders of many waters, more glorious than the raging of the sea is the glory of the Lord on high" (*WSR*). At the end of the age, God may release the waters of chaos on sinful humans, but, in the meantime, wild nature is not the arcane vehicle of ancient evil. Human belief that, by removing or destroying the wild we are "fighting for the Lord," is dismally arrogant.

This interpretation of chaos as presently removed from wild nature fits well with the Exodus pattern where natural forces always reflect the will of God. Wild nature may serve as help or provider, or it may execute God's wrath, but it is never a major obstacle or antagonist trying wickedly to thwart Yahweh's creation of a Holy Nation—the people themselves do a more than adequate job of attempting to undermine their mission.

Genesis 1–2 provides, in summary, several concepts concerning wild nature:

1. The creation account in Genesis includes all of Leopold's members in the land community. This community has existed from the beginning, with or without human presence. Wild and domestic are included together, with equal status.

2. The members of this community are linked together in the creation
    account. The plants arose from the earth. The animals are also
    tied to their habitats. The plants are classified according to fruits,
    connecting them to animals and to humans. Both the wild and the
    domestic animals are tied to the vegetation, and both are tied to
    humans.

3. In terms of God's judgment, there is no difference between the
    domestic and the wild, nor between the useful and nonuseful.
    All are good before God. All have inherent worth because they
    were made by God. The Bible declares the worth of "every living
    being that moves" and of "all animals that creep on the ground."
    The wolf, howling in the canyon, is a creature created by God and
    judged good by Him. So are the crocodile and the grizzly bear. So
    are the jellyfish and the snake.

4. If wild nature is good and beautiful because it was made by God, we
    should appreciate it as such. The trees of Eden were not only edible, they
    were beautiful. Our frequent inability to see beauty in the wild is due
    to our inability to see its divine origins or to understand it in a
    context independent of our own needs.

5. The concept of the diversity of living organisms is very important
    in the Genesis account of the creation. In Genesis 1, everything
    exists according to its "kind." Genesis 2 reinforces the concept
    with the variety of trees in the garden. The text makes it clear the
    great variety is the handiwork of God. God's "separating" in the
    cosmos was as important as God's creating. Separation provided
    not only space and time, and therefore history, but also the
    multifaceted ecosystems of the earth. Wild nature is an important, if
    not the major, expression of this aspect of the cosmos.

6. The blessing of fertility was, in the beginning, shared by all. The
    right of humans to multiply at the expense of other species must,
    therefore, be seriously questioned. The blessings and commands, such
    as "let the waters teem with living beings," make productivity an
    important quality of the creation. It is, however, the productivity of
    the whole and not of human beings alone that is important. God's
    command to "fill the waters in the seas," is better understood as a
    filling of different habitats, rather than an overflowing of the
    oceans with fishes, but the implication is there that the biosphere
    would be productive. The garden of Eden is described as a
    fruitful site, with trees identified by their produce.

7. Wild nature is not inherently sinful, nor does it represent a less-formed
    portion of creation. The wild does not secret primeval powers, not does
    it contain some ancient spiritual force leached out of the domestic. Wild

nature responds to the transcendent God. This suggests that "wildness" is an intended part of God's created order, and not a lesser or aberrant state. There is no reason to think that wild nature was ever specifically a reservoir for evil.

Where Are You, Adam?

Having investigated the basic role of wild nature in creation, we now need to determine God's original intent for the relationship of humans to wild nature. The first problem lies with the creation of humankind in God's image (Genesis 1:27). The "image" has suggested to some Christians a God-like role for humans in their relationship to the earth and its creatures. We should note, first, that God did not transfer divine or prerogatives with the image. Basic creative power was, in fact, *not* given to humans. Second, the passage is not concerned primarily with the characteristics of people; it is concerned with the process used in a creative act of God. The passage is not about humans' capabilities and talents, but about God's intent for divine interactions with humankind and for human history. God "created a creature who corresponds to him, to whom he can speak, and who listens to him."[20] Without this, there would be no spiritual history, only natural history. To use creation in God's image as an excuse for environmental destruction is to despiritualize the text and miss the point.

The second problem lies in the troublesome "dominion passage," Genesis 1:26. The Revised Standard Version of the Bible translates the passage: ". . . and let these have dominion over the fish of the sea, and over the birds of the air, and over the cattle, and over all the earth, and over every creeping thing that creeps on the earth." This is followed by a command to "Be fruitful and multiply; and fill the earth and subdue it . . . ."

This portion of Genesis has been interpreted in a variety of ways, from license for humans to do what ever they wished with the natural world, to a much more limited mandate to use land for agriculture. Gerhard von Rad has noted the words for dominion in this passage are "remarkably strong: *rada,* 'tread,' 'trample' (e.g., the wine press); similarly *kabas,* 'stamp.' "[21] James Barr, however, has suggested that nothing more should be read into this choice of vocabulary than "the basic needs of settlement and agriculture," including tilling the ground.[22] Westermann remarks the vocabulary originates with "the court language of the great empires" (Babylon, Egypt), but diverges from the creation narratives of Sumeria and Babylon in which the person is created to "'bear the yoke of the gods' to minister to the gods, to relieve the gods of the burden of everyday work."[23] In Genesis 1:26, humans take a hierarchal position over animals in the world; they are "detached from the life of the gods and directed to the life of this world."[24]

Westermann notes the interpretation of Genesis 1:26 must be consistent with the interpretation of Genesis 1:16: "And God made the two great lights; the greater light to rule over the day, and the lesser light to rule over the night, and the stars too" (*WS*). He states:

> Dominion over animals can not mean killing them for food . . . "to rule" can only have a non-literal meaning—the sun rules over the day, the moon over the night . . . among living beings, humans rule over animals without condition. . . . Dominion over animals certainly does not mean their exploitation by humans. People would forfeit their kingly role among the living . . . were animals to be made an object of their whim. The establishment of a hierarchical order between humans and animals means that the animals are not there just "to vegetate" ; the relationship set up between them is to be understood in a positive sense.[25]

Unlike James Barr, Westermann distinguishes between the "dominion" over animals and the human relationship to plants and others of creation. In Genesis 1:29, God gives humans use of plants, and then, in the following verse, gives the same privilege to animals. Humans receive "every seed-bearing plant over the whole face of the earth and every tree, with fruit-bearing seed in its fruit," while the animals receive use of "every sort of grass and plant for food."[26] At this level, human use of the natural world becomes little different from animal use. The Lord has made provision via plants, which, presumably, will not be hurt by harvest within the levels of God's intent. As many commentators have noted, these passages also imply humankind was not originally intended to kill animals for food.

Returning to Genesis 2, we find further information on God's intent in the placement of Adam and Eve in the garden and in their relationship to the animals. In 2:15, Yahweh puts humankind in the garden to "till and watch over it." The impact of this passage on the English language reader, again, depends to some extent on the translation. The Revised Standard Version says "to till and to keep it," as does the Authorized (King James Version). The verb *abad,* translated "to till," has the connotation not only of work, but also of service and could be translated "to serve." The verb *shamar* might be translated "to keep," "to watch," or "to preserve."[27] Although Eden is intended to be Adam's and Eve's habitation, there is no evidence they removed the vegetation provided by God and replaced it with something else, except for initiating a few basic agricultural activities. Humans, instead, were to watch over the garden, they presumably maintained all God had placed there. It should also be noted that Eden was not a garden of blissful laziness and free fruit. Human labor was invested in its maintenance.

The story of Eden differs from Sumerian and Babylonian creation myths that describe people as created to carry out the heavy burden of work for the gods and to lift the drudgery from their divine overseers. In Genesis, God has no need to assign God's own work to someone else because "the creator is one and there is no world of the gods to which the work of humans is directed."[28] God has instead mandated work, and it becomes part of existence in the space God has designed. The necessity of work is related to the divinely ordained task of "watching" Eden—and to human responsibility before God.

The story of Eden supplies information on Adam's and Eve's relationship to the animals also. In Genesis 2:19–20: "And Yahweh God formed out of earth every kind of animal of the field and every kind of bird of the heavens, and he brought them to the man, to see how he would name them; and just as man would name the living beings so was that to be their name. And the man gave names to all cattle and to the birds of the heavens and to all the animals of the field, yet for man, he found no helper fit for him" (*WS*). Part of the narrative is concerned with Adam's search for worthy company. Adam decides the animals are unsuitable. The animals, however, "retain a positive meaning for man which is described when he names them."[29]

Commentators have differed on what the naming implies for human relationship to animals. At one extreme is the idea that the creation of the animals was an unsuccessful attempt by God to create a companion, and, when man misused the animals, God sent him woman as a punishment. At the other is the idea that the passage is intended only to show the difference between man and beast. Westermann suggests that the animals were created as possible helpers, and God wanted Adam to name the animals. Westermann argues the naming does not give humans power over the animals (since in magic "the one who knows the name of a being can by this knowledge dispose of it").[30] The Creator is rather providing humankind "with autonomy in a certain limited area." When Adam names the animals, he gives them a place in his world and establishes his relationship to them by which he knows and recognizes them individually by kinds. Gerhard von Rad suggests:

> This naming is thus both an act of copying and an act of appropriative ordering, by which man intellectually objectifies the creatures for himself. . . . The emphasis is placed not on the invention of words but on that inner appropriation by recognition and interpretation that takes place in language. Here, interestingly, language is seen not as a means of communication but as intellectual capacity by means of which man brings conceptual order to his sphere of life. Concretely: when man says "ox" he has not simply discovered the word "ox," but rather understood this creature as ox and included it in his imagination and his life as a help to his life. . . . one should note the creaturely proximity of man and

beast to each other. The animal too is taken from the earth and is incorporated by man into his circle of life as the environment nearest him.[31]

Genesis 2:19–20 incorporates not only the cattle (domestic animals), but also includes the birds of the air and the beasts of the field (wild animals). The naming was not intended to convey spiritual control or the power of life and death, but was meant to be the foundational act in developing an understanding of and a relationship to the being of other creatures. We may infer that the wild animals belonged in the garden, rather than being walled outside, and that the divine mandate to "till and to watch" must be executed in their presence.

Genesis 3 goes on to describe the temptation and expulsion from the garden. The serpent, who was "more astute than all the animals of the field which Yahweh had made" (Gen. 3:1, WS), serves as the tempter. Biblical scholars have presented several different interpretations of the serpent:

1. The serpent is really Satan, not a creature at all.
2. The serpent is a symbol, usually of human curiosity.
3. The serpent is a mythological beast. It may have been derived from an Israelite serpent cult or from Canaanite mythology. The serpent could bring prosperity, or it could be a deity of the underworld, and, therefore, a creature of life and death.
4. The serpent really was a clever animal as its ability to talk indicates.[32]

Westermann points out that the Genesis text compares the serpent to all the animals of the field (the other wild animals) that God had made. This would mean the serpent was part of the created order and not demonic or otherworldly. The serpent is a frequent character in African tales about the arrival of death in the world. Its role in these fables has a natural logic: "The serpent can also slough off its skin and renew its life; so it knows what life and death are all about." [33]

The interpretation one accepts about the serpent has interesting implications for the character of "the Fall." If the serpent is not satanic or demonic, nor a mythological beast, then one of the creatures was party to the disruption of peace in Eden. The interpretation of the serpent as part of the created order makes humankind and a sly animal, rather than Satan, the source of sin. The role of the serpent is an interesting contrast to humankind's taking dominion and naming of the animals. Here an animal enters into a conversation with a person, and all feeling for divine order is lost. The fact that part of the temptation is to "be like God" indicates that the authority conferred by "dominion," or by the mandate "to till and to watch," is very limited. Adam and Eve act as if they are dissatisfied with their ordained role. They reach for "a divine and unbridled ability" to master their own lives[34] (and the creation as well).

The two prototype humans have to face their failure. "When then they heard Yahweh God moving about in the garden at the time of the day breeze, the man and his wife hid themselves from the presence of Yahweh God among the trees of the garden" (Gen. 3:8, *WS*). This scene serves as a transition because we see the garden for the last time as it was intended—as a place where both God and humankind could walk and communicate. God moves in the cooler part of the day, and the presence of trees is again specifically mentioned, reflecting the garden's beauty. Eve and Adam, in their last real activity in the garden—other than answering Yahweh—use the trees to hide from God, which was not at all the intended purpose of the flora.

Genesis 3 ends with curse and punishment. A curse is extended to a member of creation, the serpent, and this creature will remain at enmity with humankind. For our purposes, it is not important to decide what effect this curse has had on snakes in particular, but rather to note that a struggle between humankind and another species has erupted. This struggle does not originate solely with God's curse, but is also rooted in Eve's throwing blame on the serpent. By casting her own actions on another creature, she herself is breaking the good relationship intended by God. The life and death sought from the tree in the garden now thrashes between man and beast.

Eve and Adam fare worse than their reptilian accomplice. God declares: "Cursed is the ground because of you; with toil you shall eat from it your whole life long; Thorns and thistles shall it bear you, and you shall eat the plants of the field. In the sweat of your face you shall eat your bread until you return to the ground again, because you were taken out of it. Yes, you are dust and to dust you shall return" (Gen. 3:17–19, *WS*). Assuming that this passage refers to the introduction of work into the world is an incorrect interpretation since Eve and Adam already had to till the garden. Humankind, plants, and the earth were all in a divinely ordained relationship to each other. Von Rad writes: "But a break occurred in this affectionate relationship, an alienation that expresses itself in a silent dogged struggle between man and soil."[35] Now not just an animal, but also organisms far below people—the plants—resist Eve's hoe and Adam's plow. What was easy in Eden becomes difficult outside. The earth and plants still provide for humankind, but not without struggle.

A last passage of importance in Genesis 3 is God's clothing them with animal skins. This is not only the first mention of killing animals for any human purpose, but is also a deviation of the usual Old Testament pattern of the Creator giving "people the ability to make cultural advances for themselves."[36] The passage is a counterpoint to the previous punishment. Since God is putting people

outside the garden, he moves to care for them and, for the first time, acts as preserver rather than just creator.

These passages provide several principles concerning humankind and a potential wild land ethic:

1. Neither the creation of humans in God's image nor the dominion
   passage have anything to do with license to freely exploit wild nature
   in particular or the environment in general. These passages place
   humans in relationship to God and to animals. If anything, the
   creation in God's image suggests additional responsibilities for
   humans, for they, among all God's creatures, are able to hear God.
2. In the Genesis texts, the divine mandate for the use of plants for
   food is similar for both humans and animals.
3. Adam and Eve were supposed not only to till Eden but also to
   watch or preserve it. This implies that the tilling did not require
   displacement of the flora God had planted. Nor does it seem to have
   required displacement of animals. This suggests that we need to
   make room for the remainder of creation, domestic or wild.
4. The naming of animals does not imply the power of life and death
   over them, but rather sets them in relationship to humankind. This
   is a rational relationship, framed in language and based on
   understanding. Wild animals and birds are included. Animals, both
   wild and domestic "retain a positive meaning for man . . ."
5. Humankind worked from the beginning. The balance between
   tilling and watching the environment is important. Tilling alone is
   not enough (and is basically oriented toward producing human
   food). The environment must also be kept or watched, and this
   requires human effort. People tend to stop with the plowing and
   the harvest and sometime have the incorrect impression the rest of
   the living world will care for itself. "Watching" does notnecessarily
   mean interference with natural ecosystems, but it does mean that, in
   light of human habitation and activities, an effort must be made to
   keep the environment healthy. Perhaps one of our greatest
   environmental difficulties is our orientation towards our own
   immediate needs rather than towards the original divine
   mandates. Adam's and Eve's task in Eden was not just to provide
   for themselves, but was also to preserve the garden, the handiwork
   of God. Activities such as agriculture should not be allowed to

destroy the neighboring elements of creation, including the wild and uncultivated.

6. Human work is not for God in the sense that it needs to be done for God's support. It is, instead, earthbound and part of the divine order, intended to benefit both humans and the other living elements of creation.

7. The relationship between humankind and creation was disturbed by the Fall. Relationships with the environment may be difficult. Some species may interfere with human activity. Human needs may require the harvest or removal of many kinds of organisms. Although conditions are not as they were in Eden, God still provides. There is no indication in these passages of a general evil existing in wild nature, independent of human sinfulness. The presence of weeds originated in a curse of God. The weeds themselves are not evil, but humans have lost a relationship with wild nature they had in Eden. Overcoming weeds does not obliterate the Fall, human submission to Christ and the remission of sin does. Even if we assume that wild nature is fallen, it is no more fallen than the rest of creation and is certainly less to blame for the mess than we are.

The Theology of Wild Poetry

Biblical references to wild nature hardly end with the Fall, and God's interest in the work of creation hardly ended with the "closure" of Eden. As Abraham Heschel has stated: "The fundamental thought in the Bible is not creation, but God's continuing care for his creation."[37] Returning to chapter 5, we find the poetic literature that demonstrates how much wild nature tells us about God's person and role as Creator and also tells us about God's intent for wild nature. Psalm 104, for example, reports the perfection of God's timing and continuing provision among both the wild and the tame. And when the psalmist writes: ". . . you bring darkness on, night falls, all the forest animals come out: savage lions roar for their prey, claiming their food from God," the text implies that God ordains both the lions' nocturnal habits and their hunting, since God makes provision for them. Carnivory and nighttime travel are associated with evil in some types of folklore, but Psalm 104 does not present these habits as negative—they are, rather, God's will for the lion. The Psalm also points specifically to the fact of the diversity in creation as originating with God, and lauds God for it. ("Yahweh, what variety you have created, arranging everything so wisely!")

When the Psalmist writes: "Earth is completely full of things you have made: among them the vast expanse of ocean, teeming with countless creatures, creatures large and small, with ships going to and fro and Leviathan whom you made to amuse you," the text tells us the leviathan has a being of her own and belongs in the ocean because God put her there. Human interests are not central. In Psalm 104 we find the inherent worth of wild nature justified by divine origin and continuing divine care. Yahweh is responsible for all life, including the wild ass, the stork, the lion, and leviathan. The earth continues to support living beings because Yahweh continually "breathes" upon it. All the wild creatures that God has created belong on earth, and the Psalmist wishes a blessing on God when he sings: "May Yahweh find joy in what he creates" (Alternate translation: "May the Lord rejoice in his works"). The implication is that God does find joy in all that God has made—including all of wild nature.

The wisdom literature concurs with these themes. The Book of Job, while articulating God's divine rights and power as Creator, describes God's care for wild animals in detail and makes it clear God intends to provide for them. In the Book of Job, Yahweh takes responsibility for the freedom of the wild donkey and declares: " I have given him the desert as a home, the salt plains as his habitat. . . . The mountains are the pastures that he ranges in quest of any type of green blade or leaf." As in Genesis, wildness is in itself part of the divine order, not an undesirable deviation from God's plan. The lesson of the wild ass is, in fact, important enough to be repeated with the example of the wild ox:

> Is the wild ox willing to serve you
>   or spend the night beside your manger?
> If you tie a rope around his neck
>   will he harrow the furrows for you?
> Can you rely on his massive strength
>   and leave him to do your heavy work?
> Can you depend on him to come home
>   carrying your grain to your threshing floor?
>
> (Job 39:9–12, *JB*)

Even the ostrich, initially displayed as unwise because she leaves her eggs exposed on the ground, has her God-given speed, and, "if she bestirs herself to use her height, she can make fools of horse and rider too" (Job 39:18, *JB*). The wild beasts were not intended to be fenced and made human servants. A large portion of creation is not under human control and is not intended to be. If humans observe nature properly, its wonders should humble them before God.

Through the entire divine discourse in Job, Yahweh lets creation speak for divine prerogatives. As von Rad suggests:

... we encounter the idea creation itself has something to say which man can hear. It is to this that Job is referred . . . . this self-witnessing of creation is understood as a flood of urgent questions which refer man back to the mystery of creation and divine guidance. Here what creation says is not understood as praise ascending to God (although this is also mentioned, Job 38: 77), but as a word which turns towards the human mind. While Job is unable to answer any of these questions, the rebel (Job) thus, to a certain extent, puts the whole world back into the hands of God, for whom it exists and who alone supports and sustains it.[38]

How can wild nature be an adversary of the divine program, when wild nature speaks for God? Looking beyond examples from the Hebrew Scriptures, we find that the concept of God's continuing care for creation and its value for human instruction is found in the New Testament also. Perhaps the best example is Christ's exhortation to trust in God, from Matthew 6:25–29:

> Therefore I tell you, do not be anxious about your life, what you shall eat or what you shall drink, nor about your body, what you shall put on. Is not life more than food, and the body more than clothing? Look at the birds of the air: they neither sow nor reap nor gather into barns, and yet your heavenly father feeds them. Are you not of more value than they? And which of you by being anxious can add one cubit to his span of life? And why are you anxious about clothing? Consider the lilies of the field, how they grow; they neither toil nor spin, yet I tell you, even Solomon in all his glory was not arrayed like one of these.

The mention of comparative values here should not be taken as a loss of inherent worth on the part of nature. If the passage were describing humans valuing one thing more than another, it would probably imply some loss of care. Before God, however, birds receive all the attention they need; yet there will be care left over for God's people with whom the Lord has a special relationship. The statement about the lilies here implies not only physical provision, but also the gift of beauty. Human striving cannot match what nature is able to produce by resting peacefully in the hands of God.

These texts concerning God's continuing care allow us to draw several further conclusions for structuring a theological approach to wild nature:

1. The inherent worth of wild nature does not end with the garden of Eden, but continues after the Fall and into the period of "the salvation history" of the people of Israel.
2. God's intent is not that humankind domesticate or gain control over *all nature*. This is not possible. There is nothing in the Bible that indicates that the entire earth began as a wild kingdom and that humans were

placed in it to slowly capture or cultivate every organism and every available space. God ordained some animals and some places to be "wild" and free from direct human control.

3. Psalm 104 reaffirms the importance of diversity or variety in nature. Variety is an important continuing characteristic of creation, not just something found at the beginning.

4. Psalm 104 recognizes the importance of balance in nature and attributes order in nature to God's will. The lion has the right to claim food from God. The lioness is not worthy to eat only because of her function in nature, but is fed because she is a creature of God and God cares about her.

5. God's providence extends to wild nature and is operating continually.

6. The wild animals' habits and habitats belong to them, and this is ordained by God.

7. The concept of God's planting continues after the garden of Eden. Wild animals and plants may be thought of as belonging to God.

8. God has a joy in creation and purposes in the created order that have nothing to do with humankind. Usually, when we manipulate nature, we ask ourselves: "Am I damaging the environment?" Perhaps, we should also ask: "Am I disturbing God?" The Psalmist wants Yahweh to take joy in creation. God appreciates creation far more than we, lost in a veil of sin, will ever be able to. The Creator has an understanding that we, lacking knowledge of the real motives and means of creation, will never have. God, without offering Job deep insight into all the mysteries of his divine will, does strive to share with Job God's personal joy in the cosmos. Job in his repentance accepts.[39]

9. Modern environmental ethics often discusses animal rights or the rights of natural objects. The Bible suggests the concept of rights, which is very Anglo-Saxon, could be replaced by a concept of place— place before God and place in the landscape. God has given the creatures their homes and provisions. We should respect God's choices.

10. Science tempts us to believe that we have divine powers (that we have become as God). Discriminating between the ability to create, in the godly sense, and the ability to manipulate or control is critical to a viable land ethic. If we use or displace part of creation, that is one case. If we completely destroy part of God's handiwork, that is another. The slogan "extinction is forever" states our limitations. We can kill, but not recreate.

The texts cited above provide further evidence that, even if one believes nature fell with Adam, wildness—the ability or will to live independently of human care and control—is not part of the fallen state. The passages in Job suggest "wildness" is part of God's original order in creation, and not something conveyed by human interference. We, in fact, have to separate wildness from fear and antagonism.

During the Age of Discovery, European sailing vessels sometimes landed at islands that had never been visited before by human beings. The sailors going ashore seeking food, water, or entertainment frequently found the various animals inhabiting these islands were totally unafraid of people. Nesting birds would ignore their approach and even marine mammals resting on the beach would allow the sailors to walk right up to them. The outcome of these encounters was usually an easy, fresh dinner for the mariners, and, in some tragic cases, entire colonies of sea birds were wiped out.

Animals who have never previously met humans have to learn their fear of them. An antagonism towards humans is not an inherent quality of many wildlife species. Fear and antagonism are part of the Fall (although probably more human than animal in origin) and are transferred by unpleasant contact or violent interaction. Living in the wild and occupying a habitat free of human activity are not part of this process. Most wild species, in fact, have peaceful temperaments.

Praise the Lord, You Sea Monsters

Recalling the poetry from The Song of Songs in chapter 5, we can further characterize the divine relationship with wild nature: beauty is found in wild nature, love intensifies the recognition of this beauty (including love for God or another person), domestic nature is no more beautiful than the wild, and the essence of beauty is not found in utilitarian value. The psalms further inform us that wild nature praises and glorifies God. This is because of its origin in God's word and has nothing to do with human presence or use. This characteristic of nature is within human perception and should inspire humankind to a similar response. In Psalm 148:7–10, for example, the Psalmist specifically calls on wild nature:

> Praise the Lord from the earth,
>   you sea monsters and all deeps,
> fire, and hail, snow and frost,
>   stormy wind fulfilling his command!
>
> Mountains and all hills,
>   fruit trees and all cedars!
> Beasts and all cattle,
>   creeping things and flying birds!

No creature, no matter how unlikely (the creeping things or snakes) or how awesome (the sea monsters), is exempted from praising God. There is likewise no creature that does not glorify God.

These passages further suggest that we have to reject the notion that wild nature is only valuable if it is useful to humans. If wild nature praises God, that alone justifies its existence.

The Mountains and the Hills Shall Break Forth into Singing

There is considerable evidence in the Scriptures that God does not wish wild nature to be completely displaced by human activity or completely destroyed by human sin. The first example of this in the Bible is the story of Noah and the Flood (Genesis 6:5-8:22). Here God's wrath is directed at humankind, and wild nature is coincidentally threatened by God's judgment. God, who made creation, can also destroy it.

Of importance is God's concern that all the animals, not just those useful to humans, survive. The question, as Westermann points out, is not how Noah got all those animals on the ark. "The story is talking about another sort of reality: humans and the animals stand together in face of catastrophes that threaten life; just as they do in creation. The reality is twofold—the animals perish with the people and the people and the animals are saved together. This is of the utmost significance for the history of humans and animals."[40] If we can remove Noah and his pitch-coated ship from their usual setting as a Sunday school coloring-book exercise, we will recognize the basic issues of sin and salvation, rebellion and obedience, and catastrophe and community are related to the preservation of wild nature:

> And Yahweh saw,
> that the wickedness of humankind was
>   great on the earth,
> and every planning and striving of its heart
>   was always wicked,
> he was sorry that he had made humankind
>   on the earth,
> and he was grieved at heart.
> And Yahweh said:
> I will wipe out humankind whom I have
>   created from the face of
> the earth, from humans to beasts,
>   reptiles and birds of heaven,
> because I am sorry that I have made them.
>   (Gen. 6:5–7, *WS*)

Note that the reason for God's regret appears to be human sinfulness, not the intended destruction. Yet Yahweh experiences pain also over the decision to destroy God's creatures. The sin discussed here must be something more than in the previous chapters of Genesis—it must have been the corruption of a large group, of an entire generation.[41] The passage also displays humankind's integration with the rest of creation. God's initial thought is to destroy all the animals on the face of the earth. "But Noah had found favor in the eyes of Yahweh . . . He was blameless among his contemporaries; Noah walked with God" (Gen. 6: 8–9, WS).

The God who is grieved is also willing to save for the sake of one human being who is righteous—Noah. The story of the ark is a tale of compassion. The narrative does not give an *explanation* as to why God is moved to save the animals as well, but the reason should be obvious. The animals were already tied to humankind: if God was going to save Noah, then Noah needed to take the animals with him. God gave Noah the instructions for the ark, and in so doing, it was God, not Noah, who accomplished the rescue. Noah was not a hero, pulling a little motley herd of beasts out of the drink; God was operating, lifting an entire interrelated section of creation above the flood.

In 7:14 the story echoes the original creation account: "They and wild animals of every kind, and cattle of every kind, and all that crawls on the earth of every kind, and birds of every kind, every winged bird" (WS) entered the ark. This is repeated in 7:21–23, describing the destruction:

> All flesh, all that moved upon the earth, perished, birds and cattle and wild animals, and all that swarmed upon the earth, and all humankind. Everything that had the breath of life in its nostrils, everything that lived on dry land, all died. So he wiped out every existing thing on the face of the earth, man and beast and crawling thing and bird of heaven—all of them were wiped from the earth. Only Noah was left and what was with him in the ark.
>
> (WS)

The Creator also can exercise prerogatives as destroyer.

The motif is then repeated a third time, with the addition of a replication of the Genesis blessing, when the animals finally depart from the ark:

> Go out of the ark, you and your wife and your sons and your sons' wives with you. And all the animals that are with you, all creatures, birds and cattle and all crawling things that spawn upon the earth bring them out with you, that they may breed upon the earth, and increase and multiply upon the earth. And Noah and his sons and his wife and sons' wives went out together. And all the wild animals and all the cattle, and all the birds and everything that crawls on the earth, all of them went out of the ark by families.
>
> (Gen 8:16–19, WS)

This pattern of presentation clearly ties the role of the ark to the first creation events in Genesis. It also ensures the continuation of God's blessing through catastrophe. As Westermann suggests; "The animals that are destined to live, to be fruitful and to multiply are, like the humans, exposed to catastrophes; mass destruction is part of the existence of animals too. The animals that are saved are blessed with a view to this contingency, and no mass destruction or catastrophe in the animal kingdom can abolish this blessing."[42] Again the blessing is shared. God blesses the animals first, and then says to Noah and his sons (Genesis 9:1): "Be fruitful and increase and fill the earth!" The exit from the ark is in good order, and the animals depart in families, thus the text "in the very place where the creatures created by God step into a new life as those saved from catastrophe, has to say again it is life in its entirety, life in community."[43]

The building of the ark could be seen as a test of Noah's obedience, but the text also states that Noah was righteous before the ark construction started. The narrative is interactive. Noah is judged suitable for the preservation of humanity, and, at the same time, humanity was saved because there was some good left in it and some communication remaining with God in the person of Noah. The original creation had not completely failed. God chose the means of salvation, and Noah obeyed because he already walked with God.

From Noah we learn righteousness and justice guided by God lifts creation above the flood, not teakwood technology. When we humans threaten other residents of the creation, it is a reflection of our own heart condition. The moment we sink ourselves completely into the science of wildlife conservation, we are losing the battle. The solution lies neither in human justice, nor in ecojustice, but in God's justice and those who are humble enough to accept it. Over and over again in environmental management, we pursue technology rather than values. Over and over again, our spiritual values—or lack thereof—undercut our conservation technology.

From Noah we also learn that all creation should be saved together; bits and pieces won't do. Noah didn't throw out a life preserver here and a life preserver there. In fact, if he had, both he and the animals would have drowned. He built an ark and loaded everybody under one roof. The animals got on the ark in relationship and marched off in community. One of our major Christian ethical problems today is that we can't believe God wants the animals—all the animals—in the ark with *us*. Our preservation efforts lack the holism necessary to succeed in the face of major catastrophe. Right now, it is just a life preserver here and a life preserver there. Did you ever try to get an elephant to wear a life preserver?

The Wild and the Kingdom

If the story of Noah records a pre-Israelite example of God's judgment followed by salvation, the Book of Isaiah presents a set of prophecies about captivity and restoration both of the people Israel, and if projected into the future, of the entire earth. We have already looked at Isaiah 11 in chapter 7, but it was perhaps the single most important passage in inspirng monastic attitudes towards the wild, and, therefore, is worth repeating (in a different translation):

> The wolf will live with the lamb,
>   and the leopard will lie down with the goat,
> the calf and the lion and the yearling together;
>   and a little child will lead them.
> The cow will feed with the bear,
>   their young will lie down together,
>   and the lion will eat straw like the ox.
> The infant will play near the hole of the cobra,
>   and the young child put his hand into the viper's
>   nest.
> They will neither harm nor destroy
>   on all my holy mountain,
> for the earth will be full of the knowledge of the Lord
>   as the waters cover the sea.   (Isa.11:6-9,*NIV*)[44]

This prophecy unites the cultivated and the wild and speaks of a cosmic peace. Christians have long assumed the imagery refers to Christ's kingdom on earth. The prophet draws his portrait of the kingdom out of a deep creation theology that returns the earth to its original divinely ordained state. Included in this Peaceable Kingdom are large predators and poisonous snakes, the enemies of the Israelite herder or farmer. God's ultimate wish is for all creation, wild and domestic, to coexist in peace. When Isaiah announces "the mountains and the hills will break forth in singing," wild nature not only praises God, but it also participates in an eschatological event, the freeing of Israel from bondage in Babylon. The God of this passage is more than Savior, Yahweh is also Creator, thus, "the whole of creation, the universe, shares in the joy of those set free. "[45] As part of the saving event, the landscape will change. "Instead of the thorn shall come up the cypress, and instead of the brier shall come up the myrtle." The disappearance of the briers here does not imply their worthlessness in creation, but is the conversion of a desolate, damaged land to land with its original flora. It is notable the prophet does not predict conversion of briers to pastures or fields of grain, but to cypress (or pines, in some translations) and myrtles. The implication is a healing of the division spawned by the curse of Genesis 3.

In the prophecies of Hosea, a return to the wilderness is portrayed as a potentially cleansing experience for the people. The Lord will "lure her [the nation, his lover] and lead her out into the wilderness and speak to her heart" (Hosea

2:14, *JB*), and, at that time, "will make a treaty [covenant] on her behalf with the wild animals, with the birds of heaven and the creeping things of the earth" (Hosea 2:18, *JB*). When sin is gone, the antagonism with the rest of creation will disappear. Humankind will appreciate that which is of no immediate use, and even the lowliest of the animals will enter into a covenant partnership. The eschatological theme is continued in the New Testament in Romans 8:19, where "the creation waits in eager expectation for the sons of God to be revealed" [*NIV*]. Although wild nature is not specifically mentioned (Paul writes in a style acceptable to his rational Roman audience), the text affirms the importance of all creation (cultivated and wild) in the *eschaton*.

In both the story of Noah and the writings of the prophets, we find the Righteous Judge acting to remove sin and save creation simultaneously. Directives to protect wild nature are rare in the Bible (there are relevant passages such as Deuteronomy 22:6 that forbids taking a mother bird off her nest, but allows taking the young or eggs). And there are some diversions from the general themes we have just discussed. (Ezekiel 34:25, for example, shows the Lord promising the people of Israel that He will "rid the land of wild beasts so that they may live in the desert and sleep in the forest in safety" [*NIV*]. This passage is one of the few in Scripture that does not seem to give the wild animals their own ordained place or habitat. It may, however, be referring to the results of the fall of Jerusalem and the destruction of the social order due to war. "Yet wild nature is incorporated in some of the most critical saving acts of God. We have to look beyond "thou shalts" and "thou shalt nots." The story of Noah alone makes it quite clear that creation was made to function as a whole. Humans and animals were made to live together on the same planet. We need the remainder of creation, whether we know it or not. And we will be saved together, whether we respect our wild neighbors or not. The Peaceable Kingdom will come—one hopes it will be before the wolves and lions are gone.

We are mistaking the exegetical task when we look for conservation directives in the Scriptures. The Bible is much more basic in its approach to creation. Both the Old and the New Testament tell us about God's care for creation and what our own relationship to creation should be like in terms of appreciation, love, respect, care, execution of justice, and the completion of God's saving acts. The best way to learn to live with wild nature is to understand God as Creator and Savior. St. Antony, in his simplicity, had a wisdom about the basic nature of nature that escapes our modern scientific mindset.

The Wild Justified

The attitudes of Francis and Kevin appear to be biblically justified. There is no reason to think that all wild nature is a reservoir of the evil or chaotic. Their orientation toward living with wild nature and caring for it meets a biblical ideal. If they made any mistake at all in their protection efforts, it was in focusing solely on the individual creature rather than on the created whole. Since their primary motivation was compassion, they oriented themselves toward the threatened or injured dove or wolf. This is only half the battle. They recognized that sin could damage wild nature and that other members of the creation community could suffer harm. Yet they limited their broader spectrum efforts to oak groves, gardens, and an occasional mountain. Kevin's refusal to "suburbanize" Glendalough into a monastic city was one of the few reported cases of a "regional" protection effort (and in this instance Kevin was protecting the landscape from divine action). The ultimate demise of the boar, deer, and wolf in Ireland wasn't the single chase, it was all the human activities combined, including intensive hunting, forest removal, and the expansion of cultivation. The monks understood a key message of the Bible—holiness and righteousness lead to peace with creation. They missed, however, the winning strategy of Noah—if you are going to save part of creation, you must save all of it.

We can in summary, establish some biblically based principles for our relationships with wild nature:

1. God is interested in the welfare of all creatures, wild or domestic.
2. Wild creatures and sites have inherent worth. God made them to be what they are and called them good.
3. Humankind has a responsibility to God to preserve the values that God gave to creation.
4. Values mentioned in the Scripures that humankind should attempt to protect in wild nature are:
    a. its tremendous diversity
    b. its productivity, and God's blessing on the animals to be fruitful and multiply (and fill available habitats)
    c. its interrelatedness, the God-given ties between plants and animals, or between animals and their habitats
    d. its beauty
    e. its role in praising and glorifying God
    f. the special relationship between humankind and animals
    g. the wildness of those creatures God ordained to be wild
5. Several of the above values cannot be preserved or protected unless the integrity of wild ecosystems is protected. This is especially true of interrelatedness, but is also true of diversity, productivity, and beauty.

6. God's providence and wisdom operate in the relationship between creatures and their habits and creatures and their sources of sustenance. We have a responsibility to recognize this and to allow God's providence to continue.

7. We should not disturb God's joy in God's creation.

8. We must recognize the Fall has caused a division between humankind and wild nature. This breech is temporary, and it is God's will that it be healed.

9. We must share creation with other organisms. We should not steal their blessing or interfere in their relationship to God.

10. Our ability to control wild nature through destructive activities should not be mistaken for divine creative power.

11. We must recognize that our relationship with wild nature is sometimes antagonistic and that activities of wild organisms may conflict with our own immediate goals. We may attempt to resolve these through manipulation of nature, but humankind must acknowledge God's ownership of the wild—and God's intentions for it.

12. The presence of the wild is not just economically beneficial to us, it is spiritually beneficial as well.

13. We need the remainder of creation.

14. Our ability to communicate with and understand the wild is contingent on our holiness and our heart condition. Without a deep understanding of God as Creator, we cannot understand God's intent for creation.

We should realize that we may be forced to protect some wilderness territory just to provide for our own spiritual needs. At least some of us need quiet places to walk and contemplate. Some of us need isolated places to sit and pray. These don't necessarily have to be huge expanses of desert, nor do they have to be filled with spectacular canyons and peaks, but they do need to be relatively free of human intrusion. Spiritual values are difficult to accommodate in a world dominated by economics. Wilderness itself is helpful in discerning our unnecessary dependencies and our lifeless idols.

The Stolen Blessing

Since humankind needs managed land for support, there is an obvious basic conflict between the wild and the domestic. We cannot convert land to farmland or timberland without disrupting the natural residents. If James Barr is correct and the "dominion passage" is a mandate for agriculture, then some displacement of wild species is justified. Adam and Eve were, after all, to till the garden of Eden.

As David's conflicts with the lion and the bear indicate, the institution of agriculture often leads to antagonism between humans and wild species in nearby noncultivated or nongrazed areas. To what extent should we leave wild sites untouched?

There is no absolute answer to these questions, and no divine lines in the landscape where the Lord has declared: "Stop here!" The best ethical answers will probably come not by evaluating our economic needs, but by evaluating what will be lost under certain circumstances. We can, for example, attempt to maintain biotic diversity in a region and use this as an indicator of success in finding the delicate balance between tilling and preserving. For the Christian, diversity, productivity, beauty, and maintenance of the relationships among ecosystem elements are probably the best indicators. Displacement of individual organisms will occur and is the normal consequence of agriculture. If entire species and ecosystems are lost, however, and diversity is collapsing, we have stolen creation's blessing. If we take an aesthetic wild landscape, ruin it with mining, and leave it in a pile of acid tailings, we destroy not only the beauty of the site, but also its ability to glorify God. If a site was beautiful, diverse, and productive when we began to harvest its resources, it should be beautiful, diverse, and productive when we finish.

An exceptionally good example of humankind's failure to recognize God's interests is the ongoing extinction of the great whales. Originally the populations of many species of whales were large enough to sustain harvest at some level. Humankind had no intention of farming whales or doing anything to help a depleted population to recover. Despite this, these monsters of the sea have been overhunted to the point of population collapse and extinction. We are in the process of removing one of the great wonders of creation and one of the finest living testimonies to the creative power of God. We are not only reducing diversity, we are also removing a resource that, under better management, we could continue to harvest. "Leviathan," whom, the Psalmist declares, was made for God's amusement, is disappearing from the vast expanse of the oceans. The evil principalities and powers Paul wrote of in Ephesians 6:10 have traveled to the wildest reaches of the earth and, in the process, have moved to destroy as much as they possibly can of God's good work.

We can utilize the Scripture-based principles suggested above in evaluating our treatment of the whales. First, are we maintaining biological diversity? The answer, of course, is no. We are, in fact, threatening not only several species, but marine mammals in general. Even some of the commoner varieties are now in decline and species that are not used for food or commercially harvested, like the manatee, are in serious trouble. "And God created the great sea monsters and every living being that moves, with which the waters teem, each of its kind , . . . And God saw how good it was." We, in turn, do not seem to be able to see the goodness in the teeming waters or in the many kinds of whales. We lack God's insight into and love for creation. To us, if you've seen one whale, you've seen them all. Before God, each whale is different, a distinctive part of a mighty work.

Second, are we maintaining productivity? The answer again is no. God created the waters to "teem" with life. God's blessing for the sea monsters, "Be fruitful and increase and fill the waters in the seas," was given to them before the blessing "to be fruitful and increase and fill the earth" was given to us. When a population collapses due to overhunting and a species that once numbered in the tens of thousands is reduced to a few dozen individuals, the Lord's blessing to be fruitful and multiply is denied.

Third, are we maintaining beauty and wonder in God's creation? Every school child learns about blue whales, the largest living creature on earth. Children delight in drawing pictures of whales frolicking on curly ocean waves. We have a venerable whale and sea monster literature, ranging from the story of Jonah to *Moby Dick*. What is more beautiful than finding a well placed dry ledge at Point Lobos and settling down above the crisp, blue Pacific? Then, in between observations of the sea otters sleeping in the kelp and the sea lions fishing in among the surf spattered rocks, we catch a glimpse of gray whales surfacing and sounding off shore. Where will the beauty and wonder be, when the otters, the sea lions, and the whales are gone?

One of the illusions we seem to accept unquestionably is that, in order for us to use wild nature, we have to lose it. We socially accommodate severe damage as if it were invariably necessary. We have become Adam outside of Eden, struggling so hard with the tillage that we have forgotten to watch and preserve. We have harvested thousands of whales without ever bothering to find out where they lived, what they ate, how fast they were reproducing, and whether their numbers were declining. Yet we can, if we make the effort, slow our boats and not hit whales, find another source of pet food, stop harvesting too many, and monitor the health of whale populations. God gave us the ability to till and to harvest; God also gave us the ability to watch and preserve.

Justice and the Wild

In caring for and protecting wild nature, we have to decide when we should harvest it and when we shouldn't. We also have to decide when we cannot use wild lands productively or when we will disturb their residents irreparably. Some wild species can coexist with humans, and some, at least in the present fallen state of the cosmos, cannot. It is not a question of relinquishing our blessing to them; we already have, by far, the largest portion of the resources and the habitat. Our Christian ethic should provide a living for those organisms which will always be wild, if only because God has blessed them and holds them to be good. Gleaning some wisdom from Noah, we need to organize our efforts to provide continuous and adequate systems for preserving creation. Too often we look only at the whale. Too often the problems affect the entire sea.

We need to avoid an illusion here—that is, if we are able to convert land for agriculture and manage it so that the land is productive and does not erode, that land was intended by God to be put to the plow. The question is not can the land produce, but how may the land best serve God? If a parcel of arable land is the last remaining wild area in a region, it may be better used for preservation of native species, for recreation, or just to remind us what the trees looked like before humans cultivated the last corners of the garden. Putting the same case in a biblical perspective, it is hardly likely God intended Adam to drive the animals out of the Eden by tilling every square foot. Good agriculture management is a great blessing, but it is not the only type of environmental management which fills God's mandates.

One of the commonest arguments made against reserving some space in creation for wild nature is that the same space is needed for human development or, in stronger versions of the argument, to feed poor families with hungry children. This is sometimes true. Opening new lands provides new opportunities and releases population pressure. The same principles operated when part of Europe fled to the New World and established colonies during and after the Age of Exploration. The New World gave the poor a chance to leave crowded cities and helped small farmers acquire land.

The argument that more land is needed to support people basically says, however, that human concerns always come first. We need to analyze how we have gotten into this position of assuming that the needs of wild creatures compete with the needs of human beings. Wasn't it God's intention that there should be room in creation for everyone and that it should be a cooperative venture? Before the Fall, we were to care for creation, not to go to war with it. Even after the Fall, Noah put the animals on the ark. While humankind drowned in sin, the ark preserved the core of God's original effort. Adam's and Eve's problems with thorns in their fields were not intended to result in the slaughter of innocent bystanders, such as snowy egrets (once ruthlessly harvested for plumes) and passenger pigeons (hunted to extinction).

From an ethical point of view, humankind is ever more frequently finding itself choosing between two evils. In the case of the tropical rainforest, for example, cutting and farming stretches of forest does provide some temporary relief for poor families. If the practice were to stop in some regions, people would starve. This type of agriculture, however, is a poor use of the resource. It wastes timber and results in soil erosion. In many cases, cutting and burning rotations for clearing are now too short to allow sites to recover fully before they are farmed again. The real problem here is not that humans must always wrestle with creation and leave a wake of damaged ecosystems, the problem is that those principalities and powers of Ephesians 6. Small farmers in tropical countries often do not own land. Big corporate farms often take the best arable acreage in the fertile lowlands. The small farmer is thus trapped in economic and social systems that encourage poverty. They include exploitation of one nation by another, and of one person by another. When we cheat our neighbor, creation often bears the brunt.

One of the ironies of the concept that wild nature must be sacrificed to relieve human need is that both the destruction of wild nature and the destruction of human dignity through poverty are driven by the same forces. Christians called to help the poor often fail to realize that the future of the starving children they aid may lie in protecting the watershed above the refugee camp. Providing rice today is futile if all the soil in a region is eroded or acidified. Conversely, conservationists who are blind to poverty and socioeconomic issues are ignoring the relationship between poverty, economics, and land abuse. Resource utilization driven by desperation will be ill-planned and aimed at immediate returns. Helping the poor and restoring their humanness may do as much to protect resources as a thousand national parks.

If we have the sinking feeling that Christ's instructions to care for the poor and sick are bringing us in conflict with the Genesis mandate to care for creation, we are attacking the problems on too superficial a level. The "people versus nature conflict" is an illusion. The root issues are really greed and avarice and abuse of political and economic power. Care of the poor, well thought-out and looking towards the future, helps creation. Care of creation, also well thought-out, helps the poor.

The question then becomes, "Who should protect the wild creatures and wilderness?" The third world residents who are encroaching on wild areas have few other economic alternatives. The answer is difficult and requires an extensive social analysis that is beyond the scope of this discussion. The responsibility, however, has to fall on consumers who divert large amounts of the world's resources, on those who control social, political, and economic structures, and on those who can afford to take action. The responsibility, therefore, falls on us.

Coemgen (Kevin) would not have put a poor man out of his hut to feed a wolf, and Francis would not have ignored a leper to rescue a dove. For them, to care for the needy, the injured, and oppressed was one task, encompassing all creation. Our holiness then, lies not just in how often we contemplate creation or seek God in wilderness solitude, but also in how we treat the handiwork of the Creator—human and nonhuman. Our lack of care for our fellow humans betrays a deafness to Christ's call. Our disrespect for nature discloses a deep disinterest in the works of God and in God's will for the cosmos. Perhaps we should call for a "new exodus" and a return to the wilderness to renew our loving covenant with God—and with creation.

May we all desire holiness and pursue it vigorously. May it come not only from conversing with God on a very high mountain, but also from resting with the wild beasts and praying in the desert. May we experience a new exodus and enter the new kingdom with God's chosen people. May we all walk the straight road through the wilderness and see new springs providing water for the cypress, the cedar, the lion, and the jackal. May we all hear the trees as they clap their hands.

The Crystal River

Moving as quietly as possible through the lucid water, I scanned the beds of delicate, twining hydrilla stems: nothing but a few blue crabs, shooting sideways across the murky bottom. I circled back toward the boat. As I stopped to adjust my snorkel, the cold air brushed my face and reminded me it was February. My diving buddy pointed to something surfacing a few yards away. I floated back towards a channel marker and drifted over four large gray lumps nestled in the riverine mud.

One of the rotund objects began to rise slowly, like a child's balloon contemplating escape. As the current pushed me out towards the main channel, I swung around, desperate to mute my own splashing. The plump form met me nose on. He turned his head sideways and gazed into my facemask. The small black centers of his bright yellow eyes stared above chubby, algae tainted cheeks.

"Are you friendly?" the manatee was asking.

He nudged me gently, and then began to roll over.

"He wants to play," I realized, delighted. The manatee, skilled at moving his massive frame without colliding with my less seaworthy body, spun slowly around in the sunlit water.

"Isaiah would rejoice in this," I thought, "although I don't think Revelation places the Crystal River in Florida."

Here among the warm springs, bubbling out from the bed of a clear water stream, the manatee and I were, for a brief span, citizens of the same kingdom.

Then some other snorkelers arrived. Attempting to take pictures, they frightened the manatee and began to chase him. I felt like yelling at the inconsiderate swimmers to stop, but I couldn't talk through the snorkle. The manatee slid down into the hydrilla at high speed and did not return.

"That's what the manatee gets for being curious and friendly—human harassment," I concluded. But pushing back toward the boat, I thanked the Creator of Everything Beautiful for a few minutes of unharried communion with another member of the creation community and hoped that someday wild and human alike would live in a reign of righteousness and peace.

Notes

Chapter 1
Introduction

1. Lynn White, Jr., "The Historical Roots of Our Ecological Crisis," Science 155 (10 March 1967): 1203–7. Religion and Environmental Crisis, ed. Eugene Hargrove (Athens, GA.: University of Georgia Press, 1986).

2. Roderick Nash, Wilderness and the American Mind, 3d ed., (New Haven: Yale University Press, 1982).

3. George Williams, Wilderness and Paradise in Christian Thought (New York: Harper & Brothers, 1962).

4. Ulrich Mauser, Christ in the Wilderness (Naperville, IL.: Alec R. Allenson, 1963).

5. Richard C. Austin, Baptized into Wilderness: A Christian Perspective on John Muir (Atlanta: John Knox Press, 1987).

6. Nash, American Mind, p. 20.

7. John Hendee, George Stankey, and Robert Lucas, "The Need for Wilderness Management: Philosophical Direction," in Wilderness Management, U.S. Department of Agriculture, Forest Service, Miscellaneous Publication, No. 1365 (Washington, D.C.: Government Printing Office, 1978), p. 12.

8. Ibid., p. 13.

9. Aldo Leopold, A Sand County Almanac and Sketches Here and There (London: Oxford University Press, 1949). See specifically the essays "Thinking like a Mountain," pp. 129–33, which provides the cited quote, and "Wilderness," pp. 188–201.

10. See Susan P. Bratton, "Battling Satan in the Wilderness: Antagonism, Spirituality and Wild Nature in the Four Gospels," in Robert C. Lucas, ed., Proceedings—National Wilderness Research Conference: Current Research, Fort Collins, CO, July 23–26, 1985 (Ogden, UT: Intermountain Research Station, U.S. Forest Service, 1986), pp. 406–11.

11. See, for example, Bratton, "Battling Satan," and R. W. Funk, "The Wilderness," J. Bibl. Lit. 78 (1978): 205–14, for more detailed discussions of New Testament vocabulary.

12. Interestingly, all the Biblical commentaries cited here were written by men and translations of the Scriptures utilized were also primarily masculine productions. A number of the translations and secondary sources cited for the monastic period, however, are the work of women, including the well known Celtic scholars, Lady Gregory and Nora Chadwick. The results of this scholarly imbalance may be detectable sporadically

in this volume. The reader interested in gender issues may also notice that the chapters on monasticism mention women relatively infrequently. This is not due to either a lack of monastic texts on the acts and attitudes of females or a lack of female monastic scholars. It is—particularly in the case of the desert fathers (and mothers)—owing to an emphasis in the primary sources of themes other than wilderness and wildlife in relation to women. Remaining chaste under difficult circumstances and overcoming social interference to attain one's spiritual goals are, for example, typical morals in narratives about holy women and probably reflect the actual social problems encountered by female converts in the early monastic era. The Celtic figure Saint Brigit is strongly associated with animals, although many of the stories about her are in agricultural settings and therefore concern domestic creatures. The reader may observe, in chapter 10, that the desert monks describe wildlife very accurately and often record the gender of the animal in question. Since they portray animals as capable of ethical decision making, it is interesting to note that many of the animals in their stories are female and that a she-wolf repents and a lioness protects the chaste. Ironically, when I was first selecting some personal wilderness narratives to use as introductions to some of the chapters, I unintentionally chose examples concerning women. Recognizing my "prejudice," I then composed some introductions concerning men, (including Jeremiah). One early reviewer of the manuscript called the text "sexist" for employing a masculine pronoun in reference to the divine, while another noted the number of the introductory selections that concerned women. One would hope that, if a female writer is stuck with terms such as "desert fathers," she can still present a balanced perspective.

13. "Theophany", Webster's Third New International Dictionary of the English Language, Unabridged (Springfield, MA.: Merriam-Webster, Inc., 1968), p. 2371.

Chapter 2
She Went and Wandered in the Wilderness
(Genesis)

1. Claus Westermann, Genesis 1–11, tr. John J. Scullion (Minneapolis: Augsburg Publishing House, 1984), pp. 190–96.

2. Abram was, after all, the head of the household, and social conventions of the time would have inhibited a man arriving at Abram's camp from speaking directly to Hagar, who was both a woman and of lower social status than Abram and his wife. It is noticeable that, in the Old Testament texts, divine messengers generally follow the desert rules of hospitality, thus the unfortunate incident at Lot's house in Sodom. Although these "men" may appear unexpectedly, they do not fly in and out of closed rooms, like their medieval counterparts. Compare the texts about Hagar to Genesis 18, where the angels greet Abram first, and then speak to Sarai, presumably still in Abram's presence.

3. The interpretations in these two paragraphs come primarily from Claus Westermann, Genesis 12–36 (Minneapolis: Augsburg Publishing House, 1985), pp. 242–48; Walter Brueggemann, Genesis (Atlanta: John Knox Press, 1982), pp. 150–53; and Gerhard von Rad, Genesis (Philadelphia: Westminster Press, 1972), pp. 193–96.

4. Brueggemann, Genesis, p. 244.

5. Compare Brueggemann, Genesis, pp. 266–72, and Westermann, Genesis 12–36,

pp. 514–21.

6. See Genesis 12: 6; Westermann, Genesis 12–36, p. 153–54, and Roland de Vaux, cient Israel, Vol. 2: Religious Institutions (New York: McGraw-Hill, 1965), pp. 278–79.

7. Westermann, Genesis 12–36, p. 154.

8. This analysis is an adaptation of Brueggemann, Genesis, pp. 244–46.

9. See, for example, the Appendix in Richard E. Friedman, Who Wrote the Bible? (San Francisco: Harper and Row, 1987), pp. 246–49. The problem of authorship is also discussed in many introductory textbooks on the Old Testament.

Chapter 3
The Water Became Sweet
(The Exodus)

1. Brevard S. Childs, The Book of Exodus (Philadelphia: Westminster Press, 1974), pp. 47–48. This chapter relies on both Childs's translation of Exodus and on Childs's exegesis of the text.

2. A reviewer criticized my use of the phrase "the person of God" in this text on the grounds that God has three persons. I often use the term person of God in the singular. This is not to undermine Trinitarian theology, but reflects a common practice in theological writing. The term acknowledges that God "is a person" (see, for example, Henry Clarence Thiessen, Lectures in Systematic Theology [Grand Rapids, MI.: William B. Eerdmanns, 1979], pp. 75–88, and especially p. 77). In this book, the "person of God" refers to the personality, nature, and attributes of God. The word "person" is preferable to "personality," in this case, because our human concept of personality is too limiting when considering the biblical understanding of some of the wilderness manifestations of the divine. It should also be pointed out that in the Pentateuch there is very little development of Trinitarian thinking and typological or allegorical interpretation is necessary to extract Christ from these texts, and that biblical terms referring to God appear both in singular and plural forms. In this manuscript it is assumed that God is one and God is also three (although a major theme of the Pentateuch is that God is One). I will use the term "person of God" in reference to Yahweh, Christ, and the Holy Spirit and will assume that understanding the person of any one of the three is also helpful to understanding the other two—that is, the way to the Father is now "through Christ."

3. Also known as the Red Sea. Those who place the events at the Reed Sea assume the crossing was at the northern end of the Red Sea in an area of tidal marsh and very shallow water as opposed to further south where the Red Sea is much deeper. This choice of setting follows Childs.

4. Childs, Exodus, p. 257. See also pp. 254–70.

5. The question of wilderness testing and the meaning of the Exodus become important later in the life of Christ, where these passages are quoted in the temptation in the wilderness.

6. Childs, Exodus, pp. 502–9.

7. Childs, Exodus, pp. 593–97.

8. Childs, Exodus, p. 600.

9. See one of the many discussions by liberation theologians on this topic, or a source such as Michael Walzer, Exodus and Revolution (New York: Basic Books, 1985), pp. 43–70.

10. See Richard J. Clifford, The Cosmic Mountain in Canaan and the Old Testament (Cambridge, MA.: Harvard University Press, 1972).

11. Childs, Exodus, p. 367.

12. Old Testament scholar Walter Brueggemann, on reading an early version of this manuscript, commented that the wilderness really was dangerous and people do get killed out there, suggesting my interpretation underplayed this element. I will leave this question of the degree to which the wilderness as perceived as a threat for the reader to decide for herself or himself after reading the relevant portions of the Pentateuch. Much of the danger for the community, however, lay in the marauding bands of desert dwellers who would only too happily disrupt a people on its way to a theophany. Further, as one might expect from writings intended to portray the person of God, not a single fatality or injury in the Exodus stems from the natural properties of the desert environment operating free of divine intent. The narratives so strongly emphasize the relationship between the people and Yahweh, that wildness is completely subsumed by the divine will.

Chapter 4
The Paw of the Bear, the Paw of the Lion
(David and Jonathan)

1. Hans W. Hertzberg, I & II Samuel (Philadelphia: Westminster, 1964), pp. 111–12. Two other commentaries were used but are not cited specifically here: Ralph W. Klein, 1 Samuel, Word Biblical Commentary Vol. 10 (Waco, TX.: Word Books, 1983), and Peter D. Miscall, 1 Samuel, A Literary Reading (Bloomington: Indiana University Press, 1986).

2. Hertzberg, I & II Samuel, pp. 112–13.

3. Walter Brueggemann, David's Truth in Israel's Imagination & Memory (Philadelphia: Fortress Press, 1985), p. 29.

4. Hertzberg, I & II Samuel, p. 193.

5. Ibid., p. 358.

6. Ibid., pp. 358–59.

7. Barry Rosen and Garland Gooden, The North Carolina Outward Bound School (catalog) (Morganton, NC: Outward Bound, 1986), p. 2.

8. See works such as Sharon Koepke, "The Effects of Outward Bound Participation upon Anxiety and Self Concept" (Master's thesis, Pennsylvania State University, 1973), Marlene Parkhurst, "A Study of the Perceived Influence of a Minnesota Outward Bound Course on the Lives of Selected Women Graduates" (Ph.D. diss., University of Oregon, 1983), Donald Mathias, "An Evaluation of the Outward Bound Solo Experience as an Agent in Enhancing Self Concepts" (Master's thesis, University of Oregon, 1977) and Alan Wright, "Therapeutic Potential of the Outward Bound Process: An Evaluation of a Treatment Program for Juvenile Delinquents" (Ph.D. thesis, Pennsylvania State University, 1982).

9. Rosen and Gooden, Outward Bound, p. 2.

Chapter 5
The Mountain Haunts of the Leopards
(Writings)

1. Hertzberg, I & II Samuel, pp. 140–42.

2. Walter Brueggemann, 1 Kings (Atlanta: John Knox Press, 1982), pp. 17–22. Simon DeVries, 1 Kings, Word Biblical Commentary No. 12 (Waco, TX.: Word Books, 1985), pp. 73–74. "One thousand and five" means "more than one thousand."

3. Abraham Heschel, The Prophets (New York: Harper and Row, 1962), vol. 2, p. 264.

4. The translations are from, respectively: WSR; NIV; and RSV.

5. Gerhard von Rad, Wisdom in Israel (Nashville, TN.: Abingdon, 1972), p. 221.

6. Von Rad, Wisdom, p. 225.

7. Ibid., p. 225.

8. Norman Habel, Job (Atlanta: John Knox Press, 1981), pp. 89–93, and J. Gerald Janzen, Job (Atlanta: Knox Press, 1985), pp. 225–30.

9. Theophile Meek, "Introduction and exegesis: The Song of Songs," in The Interpreter's Bible (Nashville, TN: Abingdon, 1956), vol. 5, pp. 92–97.

Chapter 6
Fed by Ravens
(The Former Prophets, Elijah and Jonah)

1. Heschel, Prophets, vol. 1, p. 10. The entire paragraph is based on pp. 3–26 in Heschel.

2. For a discussion of the setting, see T. R. Hobbs, 2 Kings, no. 13, Word Biblical Commentary (Waco, TX.: World Books, 1985), p. 10.

3. DeVries, 1 Kings, pp. 216–17.

4. Ibid., p. 216.

5. Brueggemann, 1 Kings, pp. 78–81.

6. De Vries, 1 Kings, p. 227.

7. Brueggemann, 1 Kings, p. 89.

8. This exegesis relies heavily on Ibid., pp. 87–91.

9. Hobbs, 2 Kings, p. 21.

10. Bruce Vawter, Job & Jonah: Questioning the Hidden God (New York: Paulist Press, 1983), pp. 110—16.

11. DeVries, 1 Kings, p. 171.

12. Walter Brueggemann, 2 Kings (Atlanta: John Knox Press, 1982), p. 12.

Chapter 7
Jackals in Her Palaces
(Later Prophets)

1. I wrote this description of Cumberland Island the year before Hurricane Hugo hit

a similar area on the coast of South Carolina. Hopefully the projections of the impacts of nuclear submarines will not be as "prophetic" as the projections of the storm surge hitting beachfront housing have been.

2. Amos 4:12. This discussion was inspired by Abraham Heschel, The Prophets, vol. 1, pp. 3–26.

3. James Mays, Hosea (Philadelphia: The Westminster Press, 1969), p. 38. See also Mauser, Christ in the Wilderness, pp. 44–47.

4. Mays, Hosea, pp. 49–52.

5. Keith Carley, Ezekiel among the Prophets, Studies in Biblical Theology, Second Series, no. 31 (Naperville, Ill.: Alec R. Allenson, 1974), pp. 1–2.

6. Ibid., p. 49.

7. Walter Zimmerli, Ezekiel 1 (Philadelphia: Fortress Press, 1979), pp. 415–16.

8. Carley, Ezekiel, pp. 31–32.

9. Claus Westermann, Isaiah 40–66 (Philadelphia: Westminster Press, 1969), pp. 36–39.

Chapter 8
And He Withdrew to an Isolated Place
(Intertestamental Times, Christ, and John the Baptist)

1. Raymond F. Surberg, Introduction to the Intertestamental Period (St. Louis: Concordia, 1975), p. 119.

2. F. W. Beare, The Gospel According to Matthew (San Francisco: Harper and Row, 1981), p. 90.

3. I. Howard Marshall, Commentary on Luke (Grand Rapids, MI: Eerdmanns, 1978), p. 136.

4. From Josephus's "Life," as quoted in William Sanford Lasor, The Dead Sea Scrolls and the New Testament (Grand Rapids, MI: Eerdmans, 1972), p. 146.

5. "The Ascension of Isaiah," translated by R. H. Charles, in The Apocryphal Old Testament, ed. H. F. D. Sparks, (Oxford: Clarendon Press, 1984), p. 787.

6. Ibid., pp. 142–53; W. H. Brownlee, "John the Baptist in the New Light of the Ancient Scrolls," in The Scrolls and the New Testament, ed. Krister Stendahl, (New York: Harper and Brothers, 1957), pp. 33–53.

7. Donald Gowen, Bridge between the Testaments (Pittsburgh: The Pickwick Press, 1976), pp. 211–14.

8. Charles Scobie, "John the Baptist," in The Scrolls and Christianity: Historical and Theological Significance, ed. Matthew Black (London: Society for Promoting Christian Knowledge, 1969), p. 67.

9. Geza Vermes, The Dead Sea Scrolls: Qumran in Perspective (Cleveland, OH: Collins World, 1978), p. 165.

10. Matthew 3, Luke 3, Mark 1, John 1.

11. See the discussion in William Lane, The Gospel of Mark (Grand Rapids, MI: Eerdmanns, 1974), pp. 47–53, and in Mauser, Christ in the Wilderness, pp. 82–89.

12. See Ezekiel 20: 33—36, and the quotes from Isaiah in the preceding chapter.

13. Marshall, Luke, p. 154–56.

14. This discussion of the temptation incorporates ideas in Beare, Matthew, pp. 108–

15, and Marshall, Luke, pp. 168–74.

15. John Hart, in reviewing this manuscript pointed out this important theme, and noted that "the three temptations . . . have to do with Jesus' rejection of a messianic role as political ruler of an independent Israel."

16. Bratton, "Battling Satan."

17. See for example, Mauser's discussion of these issues; Christ in the Wilderness.

18. See Joseph Bonsirven, "Angels," in Palestinian Judaism in the Time of Christ (New York: Holt Rinehart and Winston, 1964), pp. 33–41, for a review of both good and evil spirits in Christ's time.

Chapter 9
The Desert Road
(Acts and Revelation)

1. Ernest Haenchen, The Acts of the Apostles: A Commentary (Philadelphia: The Westminster Press, 1971), p. 276.

2. Ibid.

3. Ibid., p. 291.

4. William Barclay, The Revelation of John (Philadelphia: The Westminster Press, 1976), vol. 2, p. 79.

5. Ibid., pp. 84–86.

6. Robert Conn, Cokesbury Basic Bible Commentary: Revelation (N.p.: Graded Press, 1988), p. 108.

Chapter 10
The Original Desert Solitaire
(Desert Monasticism)

1. Thomas Gannon and George Traub, The Desert and the City (Chicago: Loyola Press, 1969), pp. 28–29.

2. Nash, American Mind, p. 17.

3. Williams, Wilderness and Paradise, pp. 38–46.

4. Nash, American Mind, p. 18.

5. Ibid.

6. St. Athanasius, The Life of Saint Antony, trans. Robert T. Meyer (New York: Newman Press, 1950), pp. 20–31.

7. Ibid., p. 61.

8. Ibid., p. 62.

9. Ibid. The verb used is the third person singular aorist active indicative form of agapao, which is used instead of phileo, or some "weaker" form of "to love."

10. Ibid.

11. Ibid., p. 63.

12. St. Jerome, "The Life of St. Hilarion," tr. Marie Liguori Ewald, in The Fathers of the Church: Early Christian Biographies, ed., Roy J. Defarrai (Washington, D.C.: The Catholic University of America Press, 1952), p. 271.

13. Athanasius, St. Anthony, p. 63.

14. Jerome, "Hilarion," p. 271.

15. Pelagius the Deacon and John the Subdeacon, "The Sayings of the Fathers," in Helen Waddell, The Desert Fathers (Ann Arbor, MI: University of Michigan Press, 1957), p. 129.

16. It is not clear to this author why Roderick Nash finds St. Basil so unusual in his environmental attitudes. It seems likely Nash relied heavily on secondary sources in his research and did not read primary sources. Secondary sources are likely to be concerned with theology or history and would not report information on natural history in detail. The Desert, a City, a history of early monasticism, by Derwas Chitty (Crestwood, NY: St. Vladimir's Press, 1966), not cited by Nash, does mention some of the themes concerning nature, and is one of the best evaluations available in a primarily historical volume.

17. By Rufinus of Aquilea, "History of the Monks of Egypt," in Waddell, Fathers, p. 46.

18. Ibid.

19. St. Jerome, "The Life of St. Paul the First Hermit," in Waddell, Fathers, p. 34.

20. Ibid., p. 35.

21. Ibid., p. 37.

22. Ibid., p. 38.

23. Ibid., p. 46.

24. Jerome, "Hilaron," p. 275.

25. Palladius, The Lausiac History, tr. Robert Meyer, no. 34 of Ancient Christian Writers (London: Longmans, Green, and Co., 1965), p. 61.

26. Benedicta Ward, tr., The Desert Christian: Sayings of the Desert Fathers, The Alphabetical Collection (New York: MacMillian, 1975), p. 4.

27. From the "Vita Georgii Chozibitae," cited in Chitty, Desert, p. 152.

28. Palladius, "Lausiac History," p. 61.

29. Athanasius, St. Anthony, p. 66.

30. Chitty, Desert, p. 127.

31. Palladius, "Lausiac History," pp. 87–88.

32. Ibid., p. 115.

33. John Moschus, "Pratum spirituale," in Waddell, Fathers, p. 168.

34. Ibid., p. 172.

35. From "Vita Johannis Heschastae," cited in Chitty, Desert, p. 112.

36. Helen Waddell, Beasts and Saints (London: Constable and Co., 1949), pp. 17–23.

37. From Sulpinus Severus, "Dialogus," in Ibid., pp. 6–7.

38. Palladius, "Lausiac History," p. 66.

39. From a French translation of a Coptic text, in Waddell, Beasts and Saints, pp. 13–15.

40. Moschus, "Pratum Spirituale," pp. 25–29.

41. Chitty, Desert, p. 173.

42. St. Jerome, "The Life of Malchus, The Captive Monk," tr. Marie Ewald, in Early Christian Biographies, edited by Roy J. Deferrai (Washington, D.C.: The Catholic University Press of America, 1952), pp. 283–97.

43. Ibid., p. 296.

44. Chitty, Desert, p. 106.

45. Athanasius, St. Anthony, p. 28.

46. Jeffrey Burton Russell, Satan: The Early Christian Tradition (Ithaca, NY: Cornell University Press, 1981), p. 166. Note that the idea of the Christians forcing the demons out of the cities seems inconsistent with the apparent attraction between monks and demons. Both concepts are found in the literature, however, and the monks do not seem to have been bothered by the logical difficulties thus presented.

47. Athanasius, St. Anthony, p. 38.

48. Athanasius, St. Anthony, p. 26.

49. Ibid., p. 55.

50. Jerome, "Epistolae," XVI 10, cited in Gannon and Traub, Desert and the City, p. 18.

51. Jerome, "Epistolae," LVIII, 5.

52. Palladius, "Lausiac History," p. 87.

53. Jerome, "Malchus," p. 291.

54. Ward, Desert Christian, pp. 3–4.

55. Russell, Satan, pp. 166–67.

56. J. Donald Hughes, "The Environmental Ethics of the Pythagoreans," Environmental Ethics 2 (1980): 200.

57. Philostratus, The Life of Apollonius of Tyana, trans. F. C. Conybeare (Cambridge, MA: Harvard University Press, 1912), pp. 143–55.

58. Although the monks rejected having women in their cenobia (there were, however, female communities) and considered sexual activity among humans to be worldly, their histories include female animals. The reproductive aspects of creation appear in stories such as Macarius nursing from an antelope or healing a hyena pup. Paul the Hermit could hardly have been said to have had a "Platonic" relationship with a she-wolf—"Isaianic" would be a more appropriate term. Some mention of reproduction may be necessary to the substance of the stories, i.e., a lioness with a cub is a more dangerous animal than just "a lion." In the stories mentioning she-wolves, however, there is no reason for the animal to be female, other than to describe the animal further. The demonic does not seem to be very important in monastic dealing with creation and the reproductive seems to be accepted.

59. Rufinus, "Monks of Egypt," pp. 53–54.

60. Pelagius and John, "Sayings," book II, xvi.

61. Ibid., book II, ix.

62. Ibid., book vii, xxxvi.

63. Chitty, Desert, pp. 82–83, 96.

64. Pelagius and John, "Sayings," book XX, iv.

65. Ibid., pp. 58–146.

66. Rufinus, "Monks of Egypt," p. 57.

Chapter 11
Oaks, Wolves, and Love
(Celtic Monasticism)

1. Gannon and Traub, The Desert and the City, pp. 56–57; and John Ryan, Irish Monasticism: Origins and Early Development (Dublin: Irish Academic Press, 1986),

pp. 59–96. There is uncertainty concerning the exact year of Patrick's arrival. It is also unclear if St. Patrick was one person or two. Very little is known about the roots of Irish monasticism and the first monastic communities. Most of the monastic literature concerns the great flowering of the monasteries from 520 to 660 C.E.

2. Karl S. Bottigheimer, Ireland and the Irish: A Short History (New York: Columbia University Press, 1982), pp. 80–82.

3. Robert T. Meyer, ed. and trans., in the introduction to Athanasius, The Life of Saint Antony (New York: Newman Press, 1950), p. 14.

4. Paul R. Lonigan, Early Irish Church (Woodside, NY: Celtic Heritage Press, 1985), p. 66.

5. Nora Chadwick, The Celts (New York: Penguin Books, 1970), p. 211.

6. For a discussion of pre-Christian mythologies and Irish hagiography, see Charles (Carolus) Plummer, "Heathen Folk-lore and Mythology in the Lives of Celtic Saints," in his

Vitae Sanctorum Hiberniae (London: Oxford University Press, 1968), vol. I, pp.

cxxix–clxxxviii. Plummer produced two major works on Irish saints, the one just cited that includes the Latin saints lives and the other that included lives in the Irish language: Carolus Plummer, Bethada naem nErenn, Lives of Irish Saints, 2 vols. (London: Oxford University Press, 1968). The second volume of the latter work contains English translations.

7. Plummer, Bethada, pp. 176–281 includes a "Life of Maedoc of Ferns (II)" with each segment of prose history repeated in poetry.

8. See, for example, ibid., "The Life of Mochuda," p. 286, and "The Life of Maedoc of Ferns," p. 217. In the former, a man, reports a burial place at Lismore, and says to Mochuda, " . . . thou wilt find a sign that a mound and burial-place has been consecrated by angels. And let it be built and blessed by thyself, for it is there that thy resurrection shall be, and no one shall be doomed to hell if he enters therein." In the latter Maedoc asks four boons from God, and one of these is that "hell should not be closed upon any one who should be buried in any one of his churches to the end of the world."

9. For a review of the environmental thought of Celtic monasticism see, Richard J. Woods, "Environment as Spiritual Horizon: The Legacy of Celtic Monasticism," in Philip N. Joranson and Ken Butigan, eds., Cry of the Environment, Rebuilding the Christian Creation Tradition (Santa Fe, NM: Bear & Co., 1984), pp. 62–84. This article includes a long footnote on wolves. A more detailed source is Sister Mary Donatus MacNickle, Beasts and Birds in the Lives of Early Irish Saints (Philadelphia: University of Pennsylvania Press, 1934). In his review of the Franciscan nature legends, Edward Armstrong (Saint Francis: Nature Mystic [Berkeley: University of California Press, 1973]) includes extensive discussion of Irish material.

10. Plummer, Bethada, pp. 189–90.

11. Ibid., p. 206.

12. Ibid., p. 124.

13. From "Libellus de Vita. . . . S. Godric," by Reginald. Translated in Waddell, Beasts and Saints, pp. 90–91. Note that the Irish evangelized and established monasteries in what are now England and Scotland. Thus Godric and Cuthbert are "English" saints, but they are part of the Irish monastic tradition.

14. Plummer, Bethada, p. 178.

15. Ibid., pp. 190–91.

16. MacNickle, Beasts and Birds, p. 146.

17. Plummer, Bethada, p. 237.

18. Ibid., pp. 32–37.

19. Ibid., p. 79.

20. From an anonymous life of St. Cuthbert by a monk of Lindsfarne, in Bertram Colgrave, trans., Two Lives of Saint Cuthbert (Cambridge: Cambridge University Press, 1940), p. 81.

21. From Bede's life of Cuthbert in Colgrave, Two Lives, pp. 225–27.

22. Colgrave, Two Lives, pp. 85–87.

23. Plummer, Bethada, p. 123.

24. Ibid., p. 215.

25. MacNickle, Beasts and Birds, p. 144.

26. Plummer, Bethada, p. 8.

27. Ibid., p. 125.

28. Ibid.

29. Ibid., p. 31.

30. "Life of St. Declan of Ardmore," trans. P. Power, in his Life of St. Declan of Ardmore and Life of St. Mochuda of Lismore (London: Irish Texts Society, 1914), p. 55.

31. Plummer, Bethada, p. 35.

32. Ibid., p. 207.

33. Ibid., p. 61.

34. Ibid., pp. 148–50.

35. Ibid., p. 202.

36. "The Life of Coemgen (I)," Plummer, Bethada, p. 123, and "The Life of Coemgen (II)," ibid., p. 134.

37. St. Adamnan, The Life of Saint Columba, trans. Wentworth Huyshe (London: George Routledge & Sons, n.d.), p. 136.

38. MacNickle, Beasts and Birds, p. 149.

39. Colgrave, Two Lives, p. 221–23.

40. Plummer, Bethada, p. 5.

41. Ibid., p. 6.

42. This version of the story is from a modern translation from the Irish by Lady Isabella Gregory, A Book of Saints and Wonders put down here by Lady Gregory according to the Old Writings and the Memory of the People of Ireland (Shannon: The Irish University Press, 1971), p. 71. A more literal translation of the ancient texts may be found in Plummer, Bethada, pp. 99–120.

43. Gregory, Book of Saints, pp. 72–73. Lady Gregory may have combined stories about two saints named, Ciaran—Ciaran of Saigher, who kept animals around his cell, and Ciaran of Clonmacnois, who had a fox that carried his book of psalms. Several lives of the latter Ciaran may be found in R. A. Stewart Macalister, The Latin & Irish Lives of Ciaran (New York: The Macmillan Co, 1921).

44. Waddell, Saints, p. 121.

45. MacNickle, Beasts and Birds, pp. 130–32.

46. Colgrave, Two Lives, pp. 222–25.

47. MacNickle, Beasts and Birds, p. 161.

48. Waddell, Beasts and Saints, p. 137. See also Plummer, Bethada, pp. 126–26 and 136–37.

49. MacNickle, Beasts and Birds, p. 162.

50. Waddell, Beasts and Saints, pp. 44–45. See also Adamnan, Saint Columba, pp. 86–87, for a more literal translation of the Latin.

51. MacNickle, Beasts and Birds, p. 162.

52. "Fis Adamnain" (or Adamnan), in C. S. Boswell, An Irish Precursor of Dante (London: David Nutt, 1908), p. 32.

53. Armstrong, Saint Francis, p. 67.

54. Plummer, Bethada, pp. 56–57.

55. Chadwick, Celts, p. 146.

56. Lucan, quoted in ibid., p. 146–47.

57. Ibid. pp. 147–49. See also Stuart Piggott, The Druids (New York: Praeger Publishers, 1968).

58. Chadwick, Celts, p. 147.

59. Nash, American Mind, p. 17.

60. Molua had wood cut for a church; Plummer, Bethada, p. 230.

61. Gregory, Book of Saints, p. 17–18.

62. Ibid., p. 20.

63. Plummer, Bethada, p. 123.

64. Ibid., p. 283.

65. Ibid., p. 177.

66. Ibid., p. 24.

67. Ibid., p. 30.

68. Ibid., pp. 187–88.

69. Ibid., p. 127.

70. Ibid., p. 132.

71. Waddell, Beasts and Saints, pp. 134–36.

72. Plummer, Bethada, p. 105.

73. Power, Declan and Mochuda, p. 215.

74. Colgrave, Two Lives, pp. 214–15.

75. D. D. C. Pochin Mould, Ireland of the Saints (London: B. T. Batsford, 1953), p. 155.

76. Plummer, Bethada, pp. 58–60.

77. Chadwick, Celts, p. 215.

78. Plummer, Bethada, pp. 60, 68–69.

79. Ibid., p. 77.

80. Translated by Douglas Hyde (now in public domain). This translation may be found in several anthologies including Patrick Murray, Treasury of Irish Religious Verse (New York: Crossroad, 1986), p. 15, and Kathleen Hoagland, 1000 Years of Irish Poetry (New York: Devin-Adair Co., 1947), p. 3. Another excellent source for translations of religious and nature poetry, as well as descriptive nature writing, is Kenneth Hurlstone Jackson, A Celtic Miscellany: Translations from the Celtic Literatures (London: Penguin Books, 1988).

81. From translations of Kuno Meyer (now in public domain) found in Murray, Treasury, and other anthologies.

82. Translated by Kuno Meyer; found in Murray, Treasury, and other anthologies.

83. Hoagland, 1000 Years of Irish Poetry, p. 25.

84. Ruth Lehmann, Early Irish Verse (Austin: University of Texas Press, 1982), poem 24.

85. Translated by Kuno Meyer; found in Murray, Treasury, and other anthologies.

86. Lehmann, Early Irish Verse, poem 44.

87. Myles Dillon, Early Irish Literature (Chicago: The University of Chicago Press, 1948), p. 163–64.

88. Jay Vest, "Will-of-the-Land: Wilderness among Primal Indo-Europeans," Environmental Review 9 (Winter 1985): 323–29.

89. Plummer, (Vitae), vol I, pp. cxxxiii–cxlvii.

90. Colgrave, Two Lives, p. 233.

91. Mould, Ireland of the Saints, pp. 153–56.

92. Alexander Penrose Forbes, Lives of S. Ninian and S. Kentigern (Edinburgh: Edmonston and Douglas, 1874), p. 345. Nora Chadwick suggests in The Druids (Cardiff: University of Wales Press, 1966), pp. 44–45 that druidic use of oak groves and caves did occur in response to Roman repression of their schools. Since the Romans never invaded Ireland, the druids' schools would presumably have been able to function openly there. This appears to be the case in the tales of the Ulster cycle where the druids freely travel the countryside. It is possible that druidic use of caves or other locations for teaching did influence the saints, but it remains to be seen if the druids held these protected locals to be sacred.

93. Christopher Bamford and William P. Marsh, Celtic Christianity, Ecology and Holiness (Great Barrington, MA: Lindisfarne Press, 1987) pp. 54–58. Patrick's mountain retreat was clearly an imitation of Christ's forty days in the wilderness, but it might also have been intended to displace worship of old Celtic gods, such as Lugh (the subject of a summer harvest festival), on the heights.

94. Frank Mitchell, The Irish Landscape (London: Collins, 1976), p. 177.

95. Ibid., p. 135.

96. Plummer, Bethada, p. 228.

97. Ryan, Irish Monasticism, p. 333.

98. Ibid., p. 332.

99. Colgrave, Two Lives, p. 97.

100. Count de Montalembert, The Monks of the West: From St. Benedict to St. Bernard (New York: AMS Press, 1966), vol. 2, p. 293.

101. Adamnan, Saint Columba, pp. 153–56.

102. Plummer, Bethada, pp. 32–37.

103. I Kings 17: 20–40.

104. Plummer, Bethada, p. 135.

105. Plummer, Bethada, p. 205–206.

106. Power, Declan and Mochuda, pp. 32–33.

107. Macalister, Ciran, pp. 30–31.

108. Vest, "Will–of–the–Land," pp. 323–24.

109. MacNickle, Beast and Birds, pp. 133–34.

110. Plummer, Bethada, pp. 1, 52, 60.

111. MacNickle, Beasts and Birds, p. 127.

112. Ibid.

113. Plummer, Bethada, p. 33.

114. Mitchell, Irish Landscape, pp. 131, 135, 117. F. H. A. Aalen, Man and the Landscape in Ireland (London: Academic Press, 1978), pp. 66, 69.

115. Aalen, Man and Landscape, p. 92.

116. Mitchell, Irish Landscape, pp. 166, 172.

117. Aalen, Man and Landscape, p. 102. See also E. G. Bowen, Saints, Seaways and Settlements in the Celtic Lands (Cardiff: University of Wales Press, 1969), p. 225.

118. Plummer, Bethada, p. 145.

119. Mitchell, Irish Landscape, p. 172.

120. Kenneth H. Jackson suggests in The Oldest Irish Tradition: A Window on the Iron Age, (Cambridge: Cambridge University Press, 1964) that the Ulster cycle represents pre-Christian culture. The social class structure, armaments, and descriptions of warfare, as well as the discussion of the druids are probably little influenced by later Christian culture, although monks recorded the tales in books such as the "Yellow Book of Lecan" and the "Book of the Dun Cow."

121. Thomas Kinsella, trans., The Tain (Dublin: Dolmen Editions, 1969), pp. 84–92.

122. Ibid., pp. 71–72.

123. Lady Isabella Gregory, trans., "The Book of Invasions," in her Gods and Fighting men: The Story of the Tuatha de Danaan and of the Fianna of Ireland (London: John Murray and Co., 1919), p. 62.

124. "The Fianna," in ibid., p. 168. Note the tales of the Fianna are much later than the Ulster Cycle. The druid in these stories is more of a magician. The tales of Finn probably largely originated during the Christian era.

125. "The Fianna," ibid., pp. 177–78, 430.

126. "The Fianna," ibid., pp. 447–53. Note here the emphasis on heaven rather than on a place of resurrection. This may be due to more Latin influence in the church during the later monastic period.

127. Ibid., pp. 140–58. For other translations of these old tales see Tom P. Cross and Clark H. Slover, Ancient Irish Tales (New York: Barnes and Noble, 1936); P. W. Joyce, Old Celtic Romances (1879; Dublin: The Talbot Press, 1961), and Lady Isabella Gregory, Cuchulain of Muirthemne, The Story of the Red Branch of Ulster (New York: Oxford University Press, 1970).

Chapter 12
A Cave outside the City
(St. Francis)

1. Nash, American Mind, p. 19.

2. Lonigan, Irish Church, p. 69.

3. Anselmo M. Tommasini, Irish Saints in Italy (London: Sands and Co., 1937), pp. 157–89.

4. E. E. Reynolds, The Life of Saint Francis of Assisi (Wheathampstead: Anthony Clarke, 1983), p. 9.

5. Julien Green, God''s Fool: The Life and Times of Francis of Assisi (San Francisco: Harper and Row, 1983), p. 14.

6. C. H. Lawrence, Medieval Monasticism: Forms of Religious Life in Western European in the Middle Ages (London: Longman, 1982), pp. 193–94.

7. Bonaventure, Bonaventure: The Soul''s Journey into God, The Tree of Life, The Life of St. Francis, trans. Ewert Cousins (New York: Paulist Press, 1978), p. 258.

8. Ibid., p. 259.

9. Thomas of Celano, St. Francis of Assisi: First and Second Life of St. Francis, trans. Placid Hermann (Chicago: Franciscan Herald Press, 1963), pp. 54–55.

10. Celano, St. Francis, pp. 55–56, and Bonaventure, Bonaventure, pp. 256–58.

11. E. M. Blaiklock and A. C. Keys, trans., The Little Flowers of St Francis (Ann Arbor, MI: Servant Books, 1985), pp. 63–64.

12. Ibid., p. 59.

13. Bonaventure, Bonaventure, p. 260.

14. Blaiklock and Keys, Little Flowers, p. 60.

15. Ibid., p. 62.

16. Celano, Saint Francis, pp. 72–73.

17. Ibid., p. 72.

18. Ibid., p. 270.

19. Ibid.

20. Regis Armstrong and Ignatius Brady, trans., Francis and Clare: The Complete Works (New York: Paulist Press, 1982), pp. 42–43.

21. Ibid., pp. 38–39.

22. Celano, Saint Francis, p. 77.

23. Blaiklock and Keys, Little Flowers, pp. 42–43.

24. Celano, Saint Francis, p. 69.

25. Blaiklock and Keys, Little Flowers, p. 8.

26. Ibid., pp. 5–6.

27. Ibid., pp. 16–17.

28. Bonaventure, Bonaventure, pp. 189, 191, and 194.

29. Blaiklock and Keys, Little Flowers, p. 23.

30. Ibid., pp. 25–26.

31. Bonaventure, Bonaventure, p. 305.

32. Blaiklock and Keys, Little Flowers, pp. 28–29.

33. Green, God's Fool, p. 69.

34. Johannes Jorgensen, Saint Francis of Assisi (New York: Doubleday, 1955), pp. 240–75.

35. Bonaventure, Bonaventure, p. 303.

36. Carol Field, The Hill Towns of Italy (New York: E. P. Dutton, 1973), p. 66.

37. For range maps of large mammals and birds remaining in Umbria, see Fulco Pratesi and Franco Tassi, Guida alla natura della Toscanna e dell' Umbria (Milan: Libri illustrati Mondadori, 1976).

38. Celano, Saint Francis, p. 270.

39. Loren Wilkinson, ed., Earthkeeping: Christian Stewardship of Natural Resources (Grand Rapids, MI: William B. Eerdmans, 1980), p. 122.

Chapter 13
The Limits of Western Wilderness
(The Reformation)

1. Lewis Spitz, The Renaissance and Reformation, vol. 1, The Renaissance (St. Louis: Concordia, 1971), pp. 10–12.

2. Chitty, Desert, pp. 143–67. Note that the Persians did not have religious reasons for this invasion.

3. Lewis Spitz, The Renaissance and Reformation, vol. 2, The Reformation (St. Louis: Concordia, 1971), 301–27.

4. J. M. Headly, Luther's View of Church History (New Haven: Yale University Press, 1963), p. 211.

5. Martin Luther, "The Judgment of Martin Luther on Monastic Vows," trans. James Atlinson, in Luther's Works, vol. 44, The Christian in Society (Philadelphia: Fortress Press, 1971), p. 291.

6. Martin Luther, Sermons on the Gospel of John, Chapters 14–16, trans. J. Pelikan (St. Louis: Concordia, 1961), p. 62.

7. Luther, "Vows," p. 317.

8. Martin Luther, "To the Christian Nobility of the German Nation," trans. Charles Jacobs, in Martin Luther, Three Treatises (Philadelphia: Fortress Press, 1970), pp. 75–76.

9. John Calvin, The Institutes of the Christian Religion, Books III.XX to IV.XX, ed. John McNeill, trans. Ford Battles (Philadelphia: The Westminster Press, 1960), p. 1271.

10. John Calvin, "Articles agreed upon by the Faculty of Sacred Theology of Paris in reference to matters of faith at present controverted; with The Antidote," trans. Henry Beveridge in John Calvin, Tracts and Treatises on the Reformation of the Church (Grand Rapids, MI.: Wm. B. Eerdmans, 1958), vol. 1, p. 96.

11. Ibid.

12. John Calvin, Commentary on a Harmony of the Evangelists, Matthew, mark and Luke, trans. (Grand Rapids, MI: Eerdmans, 1949), p. 208.

13. Ibid., p. 209.

14. Martin Luther, Lectures on Isaiah 1–39, ed. J. Pelikan (St. Louis: Concordia, 1969), pp. 122–23. Note that John Calvin in his commentary on Isaiah returns to the interpretation of this passage discussing Christ's coming as removing sin from the world.

15. Calvin, Harmony, p. 182.

16. The traditions of the Essenes and other groups are important here. Within Hebrew culture there was a tradition of a wilderness ideal, the desert as a site where the people could be purified. See also "Battling Satan," where my conclusions differ from Calvin's.

17. John Calvin, Commentary on the Book of Psalms, trans. James Anderson, (Grand Rapids, MI: Eerdmans, 1949), vol. 4, pp. 143–71.

18. John Calvin, Institute of the Christian Religion, Books I.I to III.XIX, ed. John McNeil, trans. Ford Battles, (Philadelphia: Westminster Press, 1960), pp. 51–74.

Chapter 15
Mountain Crests, Desert Canyons
(Contemporary Questions)

1. Richard Foster, *The Celebration of Discipline: The Path to Spiritual Growth* (San Francisco: Harper & Row, 1978).

Chapter 16
The Wild and the Kingdom
(Protecting the Wild)

1. From the famous essay "The land ethic," in Leopold, *Sand County Almanac,* pp. 201–26.
2. Westermann, *Genesis 1–11,* pp. 86–88.
3. Ibid., p. 123.
4. Ibid., p. 124.
5. von Rad, *Genesis,* p. 55.
6. Westermann, *Genesis 1–11,* p. 126.
7. Ibid., pp. 136–39.
8. von Rad, *Genesis,* p. 57.
9. Westermann, *Genesis 1–11,* p. 142.
10. Claus Westermann, *Elements of Old Testament Theology,* trans. Douglas Scott (Atlanta: John Knox Press, 1982), p. 93.
11. Ibid., p. 92. See also Westermann, *Genesis,* p. 123, for further discussion.
12. Paul W. Taylor, *Respect for Nature* (Princeton: Princeton University Press, –986), p. 59–80. See also Louis Lombardi, "Inherent Worth, Respect, and Rights," *Environmental Ethics 5* (1983): 257–70.
13. Taylor, *Respect for Nature,* p. 75.
14. Westermann, *Genesis 1–11,* p. 140.
15. von Rad, *Genesis, p. 77.*
16. Alexander Heidel, *The Babylonian Genesis* (Chicago: University of Chicago Press, 1951), p. 97.
17. Walter Eichrodt, *Theology of the Old Testament,* (Philadelphia: Westminster Press, 1967), vol. 2, p. 98.
18. See also Bernhard Anderson, *Creation versus Chaos: The Reinterpretation of Mythical Symbolism in the Bible* (New York: Association Press, 1967).
19. Hans-Joachim Kraus, *Theology of the Psalms,* trans. Keith Crim (Minneapolis: Augsburg, 1986), pp. 128–29.
20. Westermann, *Genesis 1–11,* p. 157.
21. von Rad, *Genesis,* p. 60.
22. James Barr, "Man and Nature: The Ecological Controversy in the Old Testament," in *Ecology and Religion in History,* David and Ellen Springs, eds., (New York: Harper and Row, 1974), pp. 63–64. Note that *rada* and *kabas* may also be transliterated *radah* and *kabash.*
23. Westermann, *Genesis 1–11,* p. 159.
24. Ibid.
25. Ibid.

26. Ibid., p. 77.

27. Wilkinson, *Earthkeeping,* p. 209.

28. Westermann, *Genesis 1–11,* p. 222.

29. Ibid., p. 228.

30. Ibid.

31. von Rad, *Genesis,* p. 83.

32. The list is from Westermann, *Genesis 1–11,* p. 237.

33. Ibid., p. 238.

34. Ibid., p. 248.

35. von Rad, *Genesis,* p. 94.

36. Westermann, *Genesis 1–11, p.* 269.

37. Heschel, *Prophets,* vol. 2, p. 264.

38. von Rad, *Wisdom in Israel,* p. 225.

39. For a more complete discussion of this idea, see ibid., pp. 227–28.

40. Westermann, *Genesis 1–11,* p. 424.

41. Ibid., p. 407.

42. Ibid., p. 451.

43. Ibid., p. 452.

44. Note here that different translations use different animals or terms for the animals in these passages. The "goat" may be a "kid," or the "leopard" may be a "panther." The point, however, is little changed by these versions.

45. Claus Westermann, *Isaiah 40–66: A Commentary,* (Philadelphia: Westminster Press, 1969), p. 292.

Bibliography

Aalen, F H. A. *Man and the Landscape in Ireland.* London: Academic Press, 1978.

Adamnan. *The Life of Saint Columba.* Translated by Wentworth Huyshe. London: George Routledge & Sons, n.d.

Anderson, Bernhard. *Creation Versus Chaos: The Reinterpretation of Mythical Symbolism in the Bible.* New York: Association Press, 1967.

Armstrong, Edward. *Saint Francis: Nature Mystic.* Berkeley: University of California Press, 1973.

Armstrong, Regis, and Ignatius Brady, trans. *Francis and Clare: The Complete Works.* New York: Paulist Press, 1982.

Athanasius. *The Life of Saint Antony.* Translated by Robert T. Meyer. New York: Newman Press, 1950.

Austin, Richard C. *Baptized into Wilderness: A Christian Perspective on John Muir.* Atlanta: John Knox Press, 1987.

Bamford, Christopher, and William P. Marsh. *Celtic Christianity, Ecology and Holiness.* Great Barrington, MA: Lindisfarne Press, 1987.

Barclay, William. *The Revelation of John. Philadelphia*: The Westminster Press, 1976.

Barr, James. "Man and Nature: The Ecological Controversy in the Old Testament." *In Ecology and Religion in History,* edited by David and Ellen Springs. New York: Harper and Row, 1974.

Beare, F. W. *The Gospel According to Matthew.* San Francisco: Harper and Row, 1981.

Blaiklock, E. M., and A. C. Keys, trans. *The Little Flowers of St. Francis.* Ann Arbor, MI: Servant Books, 1985.

Bonaventure. *Bonaventure: The Soul's Journey into God, The Tree of Life, The Life of St. Francis.* Translated by Ewert Cousins. New York: Paulist Press, 1978.

Bonsirven, Joseph. "Angels." In his *Palestinian Judaism in the Time of Christ.* New York: Holt Rinehart and Winston, 1964.

Boswell, C. S. An Irish Precursor of Dante. London: David Nutt, 1908.

Bottigheimer, Karl S. Ireland and the Irish: A Short History. New York: Columbia University Press, 1982.

Bowen, E. G. *Saints, Seaways and Settlements in the Celtic Lands.* Cardiff: University of Wales Press, 1969.

Bratton, Susan P. "Battling Satan in the Wilderness: Antagonism, Spirituality and Wild Nature in the Four Gospels." *Proceedings–National Wilderness Research Conference: Current Research, Fort Collins, CO July 23-26, 1985,* edited by Robert C. Lucas. Ogden, Utah: Intermountain Research Station, U.S. Forest Service, 1986.

Bratton, Susan P. "The Original Desert Solitaire: Early Christian Monasticism and Wilderness." *Environmental Ethics* 10 (1988): 31-53.

Bratton, Susan P. "Oaks, Wolves and Love: Celtic Monks and Northern Forests." *Journal of Forest History* 33 (1989): 4-20.

Brownlee, W. H. "John the Baptist in the New Light of the Ancient Scrolls." In *The Scrolls and the New Testament*, edited by Krister Stendahl. New York: Harper and Brothers, 1957.

Brueggemann, Walter. David's *Truth in Israel's Imagination and Memory.* Philadelphia: Fortress Press, 1985.

Brueggemann, Walter. *Genesis*. Atlanta: John Knox Press, 1982.

Brueggemann, Walter. *1 Kings*. Atlanta: John Knox Press, 1982.

Brueggemann, Walter. *2 Kings*. Atlanta: John Knox Press, 1982.

Calvin, John. *Commentary on the Book of Psalms*. Translated by James Anderson. Grand Rapids, MI: Eerdmans, 1949.

Calvin, John. *Commentary on a Harmony of the Evangelists, Matthew, Mark and Luke*. Translated by William Pringle. Grand Rapids, MI: Eerdmans, 1949.

Calvin, John. *Institutes of the Christian Religion*. Translated by Ford Battles. Philadelphia: Westminster Press, 1960.

Calvin, John. *Tracts and Treatises on the Reformation of the Church*. Translated by Henry Beveridge. Grand Rapids, MI: Eerdmans, 1958.

Carley, Keith. *Ezekiel among the Prophets*. Studies in Biblical Theology, Second Series, No. 31. Naperville, IL: Alec R. Allenson Inc., 1974.

Celano, Thomas of. *St. Francis of Assisi: First and Second Life of St. Francis*. Translated by Placid Hermann. Chicago: Franciscan Herald Press, 1963.

Chadwick, Nora. *The Celts*. New York: Penguin Books, 1970.

Chadwick, Nora. *The Druids*. Cardiff: University of Wales Press, 1966.

Childs, Brevard. *The Book of Exodus*. Philadelphia: Westminster Press, 1974.

Chitty, Derwas. *The Desert a City*. Crestwood, NY: St. Vladimir's Press, 1966.

Clifford, Richard J. *The Cosmic Mountain in Canaan and the Old Testament.* Cambridge, MA: Harvard University Press, 1972.

Colgrave, Bertram, ed. and trans. *Two Lives of Saint Cuthbert*. Cambridge University Press, 1940.

Conn, Robert. *Cokesbury Basic Bible Commentary: Revelation*. N.p.: Graded Press, 1988.

Cross, Tom P., and Clark H. Slover. *Ancient Irish Tales*. New York: Barnes and Nobles, 1936.

Deferrari, Roy J., ed. *The Fathers of the Church: Early Christian Biographies*. Washington, D.C.: The Catholic University of America Press, 1952.

de Montalembert, Count. *The Monks of the West: From St. Benedict to St. Bernard*. New York: AMS Press, 1966.

de Vaux, Roland. Ancient Israel, volume 2: *Religious Institutions*. New York: McGraw-Hill, 1965.

DeVries, Simon. 1 Kings. No. 12 of the Word Biblical Commentary. Waco, TX: Word Books, 1985.

Dillon, Myles. *Early Irish Literature*. Chicago: The University of Chicago Press, 1948.

Eichrodt, Walther. *Theology of the Old Testament*. Philadelphia: Westminster Press, 1967.

Field, Carol. *The Hill Towns of Italy*. New York: E. P. Dutton, 1973.

Forbes, Alexander Penrose, ed. and trans. *Lives of S. Ninian and S. Kintigen*. Edinburgh: Edmonston and Douglas, 1874.

Foster, Richard. *The Celebration of Discipline: The Path to Spiritual Growth*. San Francisco: Harper & Row, 1978.

Feldman, Richard E. *Who Wrote the Bible?* San Francisco: Harper and Row, 1987.

Fulco Pratesi, Fulco, and Franco Tassi. *Guida alla natura della Toscanna e dell' Umbria*. Milan: Libri illustrati Mondadori, 1976.

Funk, R. W. "The wilderness." *Journal of Biblical Literature* 78 (1959): 205-14.

Gannon, Thomas, and George Traub. *The Desert and the City*. Chicago: Loyola Press, 1969.

Gowen, Donald. *Bridge between the Testaments*. Pittsburgh: The Pickwick Press, 1976.

Green, Julien. *God's Fool: The Life and Times of Francis of Assisi*. San Francisco: Harper and Row, 1983.

Gregory, Lady Isabella, ed. and trans. *A Book of Saints and Wonders put down here by Lady Gregory according to the Old Writings and the Memory of the People of Ireland*. Shannon: The Irish University Press, 1971.

Gregory, Lady Isabella, trans. *Cuchulain of Muirthemne: The Story of the Red Branch of Ulster*. New York: Oxford University Press, 1970.

Gregory, Lady Isabella, trans. *Gods and Fighting men: The Story of the Tuatha de Danaan and of the Fianna of Ireland*. London: John Murray and Co., 1919.

Habel, Norman. Job. Atlanta: John Knox Press, 1981.

Haenchen, Ernest. *The Acts of the Apostles: A Commentary*. Philadelphia: The Westminster Press, 1971.

Hargrove, Eugene, ed. *Religion and Environmental Crisis*. Athens: University of Georgia Press, 1986.

Headly, J. M. *Luther's View of Church History*. New Haven: Yale University Press, 1963.

Heidel, Alexander. *The Babylonian Genesis*. Chicago: University of Chicago Press, 1951.

Hendee, John George Stankey, and Robert Lucas, eds. *Wilderness Management*. U.S. Department of Agriculture, U.S. Forest Service, Miscellaneous Publication, no. 1365. Washington, D.C.: Government Printing Office, 1978.

Hertzberg, Hans. *W. I & II Samuel*. Philadelphia: Westminster, 1964. Heschel, Abraham. *The Prophets*. New York: Harper and Row, 1962.

Hoagland, Kathleen, ed. and trans. *One Thousand Years of Irish Poetry*. New York: Devin-Adair Co., 1947.

Hobbs, T. R. *2 Kings*. No. 13 of Word Bible Commentary. Waco, TX: Word Books, 1985.

Hughes, Donald J. "*The Environmental Ethics of the Pythagoreans*." Environmental Ethics 2 (1980): 195-213.

Jackson, Kenneth H. *The Oldest Irish Tradition: A Window on the Iron Age.* Cambridge: Cambridge University Press, 1964.

Jackson, Kenneth H., ed. and trans. *A Celtic Miscellany: Translations from the Celtic Literatures.* London: Penguin Books, 1988.

Janzen, Gerald J. *Job.* Atlanta: Knox Press, 1985.

Jones, Alexander, ed. *The Jerusalem Bible.* Garden City, NY: Doubleday & Co., 1968.

Jorgensen, Johannes. *Saint Francis of Assisi.* New York: Doubleday, 1955.

Josephus. "Life." *In William Sanford Lasor, The Dead Sea Scrolls and the New Testament.* Grand Rapids, MI: Eerdmans, 1972.

Joyce, P. W., ed. and trans. *Old Celtic Romances.* Dublin: The Talbot Press, 1961.

Kinsella, Thomas, trans. *The Tain.* Dublin: Dolmen Editions, 1969.

Klein, Ralph W. *1 Samuel.* No. 10 of the Word Biblical Commentary. Waco, TX: Word Books, 1983.

Koepke, Sharon. "The Effects of Outward Bound Participation upon Anxiety and Self Concept." Master's thesis, The Pennsylvania State University, 1973.

Kraus, Hans-Joachim. *Theology of the Psalms.* Translated by Keim Crim. Minneapolis: Augsburg, 1986.

Lane, William. *The Gospel of Mark.* Grand Rapids, MI: Eerdmans, 1974.

Lawrence, C. H. *Medieval Monasticism: Forms of Religious Life in Western European in the Middle Ages.* London: Longman, 1982.

Lehmann, Ruth, ed. and trans. *Early Irish Verse.* Austin: University of Texas Press, 1982.

Leopold, Aldo. *A Sand County Almanac and Sketches Here and There.* London: Oxford University Press, 1949.

Lombardi, Louis. *"Inherent Worth, Respect and Rights."* Environmental Ethics 5 (1983): 257-70.

Lonigan, Paul R. *The Early Irish Church.* Woodside, NY: Celtic Heritage Press, 1985.

Luther, Martin. *Lectures on Isaiah 1-39.* Translated by J. Pelikan. St. Louis: Concordia, 1969.

Luther, Martin. *Luther's Works.* Vol. 44, The Christian in Society. Translated by James Atlinson. Philadelphia: Fortress Press, 1971.

Luther, Martin. *Sermons on the Gospel of John, Chapters 14-16.* Translated by J. Pelikan. St. Louis: Concordia, 1961.

Luther, Martin. *Three Treatises.* Translated by Charles Jacobs. Philadelphia: Fortress Press, 1970.

Macalister, R. A. Stewart, ed. and trans. *The Latin and Irish Lives of Ciaran.* New York: The Macmillan Co., 1921.

MacNickle, Sister Mary Donatus. *Beasts and Birds in the Lives of Early Irish Saints.* Philadelphia: University of Pennsylvania Press, 1934.

Marshall, Howard I. *Commentary on Luke.* Grand Rapids, MI: Eerdmanns, 1978.

Mathias, Donald. "An evaluation of the Outward Bound solo experience as an agent in enhancing self concepts." Master's thesis, University of Oregon, 1977.

Mauser, Ulrich. *Christ in the Wilderness.* Naperville, IL: Alec R. Allenson, 1963.

May, Herbert G., and Bruce M. Metzger, eds. *The Oxford Annotated Bible, The Holy Bible, Revised Standard Version.* New York: Oxford University Press, 1973.

Mays, James. *Hosea.* Philadelphia: The Westminster Press, 1969.

Meek, Theophile. *"Introduction and Exegesis: The Song of Songs."* The Interpreter's Bible. Nashville: Abingdon, 1956.

Metzger, Bruce M., ed. *The Apocrypha of the Old Testament, Revised Standard Version.* New York: Oxford University Press, 1977.

Miscall, Peter D. *1 Samuel, A Literary Reading.* Bloomington: Indiana University Press, 1986.

Mitchell, Frank. *The Irish Landscape.* London: Collins, 1976.

Mould, D. D. C. Pochin. *Ireland of the Saints.* London: B. T. Batsford, 1953.

Murray, Patrick, ed. *Treasury of Irish Religious Verse.* New York: Crossroads, 1986.

Nash, Roderick, *Wilderness and the American Mind.* 3rd ed. New Haven: Yale University Press, 1982.

New International Version: The Holy Bible. Grand Rapids, MI: Zondervan Bible Publishers, 1978.

Palladius. *"The Lausiac History."* In Ancient Christian Writers, translated by Robert Meyer. London: Longmans, Green and Co., 1965.

Parkhurst, Marlene. *"A Study of the Perceived Influence of a Minnesota Outward Bound Course on the Lives of Selected Women Graduates."* Ph.D. dissertation, University of Oregon, 1983.

Philostratus. *The Life of Apollonius of Tyana.* Translated by F. C. Conybeare. Cambridge, MA: Harvard University Press, 1912.

Piggott, Stuart. *The Druids.* New York: Praeger Publishers, 1968.

Plummer, Carolus, ed. and trans. *Bethada naem nErenn, Lives of Irish Saints.* London: Oxford University Press, 1968.

Plummer, Carolus, ed. *Vitae Sanctorum Hiberniae.* London: Oxford University Press, 1968.

Power, P., ed. and trans. *Life of St. Declan of Ardmore, and Life of St. Mochuda of Lismore.* London: Irish Texts Society, 1914.

Reynolds, E. E. *The Life of Saint Francis of Assisi.* Wheathampstead, England: Anthony Clarke, 1983.

Rosen, Barry, and Garland Gooden. *The North Carolina Outward Bound School.* Morganton, N.C.: Outward Bound, Inc., 1986.

Ryan, John. *Irish Monasticism: Origins and Early Development.* Dublin: Irish Academic Press, 1986.

Russell, Jeffery Burton. *Satan: The Early Christian Tradition.* Ithaca, NY: Cornell University Press, 1981.

Scobie, Charles. *"John the Baptist."* In The Scrolls and Christianity: Historical and Theological Significance, edited by Matthew Black. London: Society for Promoting Christian Knowledge, 1969.

Soggins, J. Alberto. *Judges.* Philadelphia: Westminster Press, 1981.

Sparks, H. F. D., ed. *The Apocryphal Old Testament.* Oxford: Clarendon Press, 1984.

Spitz, Lewis. *The Renaissance and Reformation.* Vol. 1, *The Renaissance.* St. Louis: Concordia, 1971.

Spitz, Lewis. *The Renaissance and Reformation* Vol. 2, *The Reformation.* St. Louis: Concordia, 1971.

Surberg, Raymond F. *Introduction to the Intertestamental Period.* St. Louis: Concordia, 1975.

Taylor, Paul W. *Respect for Nature.* Princeton: Princeton University Press, 1986.

Thiessen, Henry C. *Lectures in Systematic Theology.* Grand Rapids, MI: Eerdmanns, 1979.

Tommasini, Anselmo M. *Irish Saints in Italy.* London: Sands and Co., 1937.

Vawter, Bruce. *Job and Jonah: Questioning the Hidden God.* New York: Paulist Press, 1983.

Vermes, Geza. *The Dead Sea Scrolls: Qumran in Perspective.* Cleveland, OH: Collins World, 1978.

Vest, Jay. *"Will-of-the-Land: Wilderness among Primal Indo-Europeans."* Environmental Review 9 (1985): 323–29.

von Rad, Gerhard. *Genesis.* Philadelphia: Westminster Press, 1972.

von Rad, Gerhard. *Wisdom in Israel.* Nashville: Abingdon, 1972.

Waddell, Helen, ed. and trans. *Beasts and Saints.* London: Constable and Co., 1949.

Waddell, Helen, ed. and trans. *The Desert Fathers.* Ann Arbor: The University of Michigan Press, 1957.

Walzer, Michael. *Exodus and Revolution.* New York: Basic Books, 1985.

Ward, Benedicta, trans. *The Desert Christian: Sayings of the Desert Fathers*: The Alphabetical Collection. New York: MacMillan, 1975.

Webster's Third New International Dictionary of the English Language, Unabridged. Springfield, MA: Merriam-Webster, 1968.

Weiser, Artur. *The Psalms.* Philadelphia: Westminster Press, 1962.

Westermann, Claus. *Elements of Old Testament Theology.* Atlanta: John Knox Press, 1982.

Westermann, Claus. *Genesis 1–11.* Minneapolis: Augsburg Publishing House, 1984.

Westermann, Claus. *Genesis 12–36.* Minneapolis: Augsburg Publishing House, 1985.

Westermann, Claus. *Isaiah 40–66*: A commentary. Philadelphia: Westminster Press, 1969.

White, Lynn. *"The Historical Roots of Our Ecological Crisis."* Science 155 (10 March 1967): 1203-7.

Wilkinson, Loren, ed., *Earthkeeping: Christian Stewardship of Natural Resources.* Grand Rapids, MI: Eerdmans, 1980.

Williams, George. *Wilderness and Paradise in Christian Thought.* New York: Harper & Brothers, 1962.

Woods, Richard J. *"Environment as Spiritual Horizon: The Legacy of Celtic Monasticism."* Cry of the Environment, Rebuilding the Christian Creation Tradition, edited by Philip N. Joranson and Ken Butigan. Santa Fe: Bear & Co., 1984.

Wright, Alan. "Therapeutic Potential of the Outward Bound Process: An Evaluation of a Treatment Program for Juvenile Delinquents." Ph.D. dissertation, The Pennsylvania State University, 1982.

Zimmerli, Walter. *Ezekiel 1.* Philadelphia: Fortress Press, 1979.

Index

Index

Biblical Citations

347